BIODIVERSITY *in a* CHANGING CLIMATE

BIODIVERSITY *in a* CHANGING CLIMATE

Linking Science and Management in Conservation

Edited by

Terry L. Root, Kimberly R. Hall,
Mark P. Herzog, and Christine A. Howell

UNIVERSITY OF CALIFORNIA PRESS

The publisher gratefully acknowledges the generous contribution to this book provided by the Stephen Bechtel Fund.

University of California Press, one of the most distinguished university presses in the United States, enriches lives around the world by advancing scholarship in the humanities, social sciences, and natural sciences. Its activities are supported by the UC Press Foundation and by philanthropic contributions from individuals and institutions. For more information, visit www.ucpress.edu.

University of California Press
Oakland, California

Library of Congress Cataloging-in-Publication Data

    Biodiversity in a changing climate: Linking science and management in conservation / edited by Terry L. Root, Kimberly R. Hall, Mark P. Herzog, Christine A. Howell. -- First edition.
        p.   cm.
    Includes bibliographical references and index.
    ISBN 978-0-520-27885-1 (cloth : alk. paper)—ISBN 0-520-27885-2 (cloth : alk. paper)—ISBN 978-0-520-28671-9 (pbk. : alk. paper)—ISBN 0-520-28671-5 (pbk. : alk. paper)—ISBN 978-0-520-96180-7 (ebook)—ISBN 0-520-96180-3 (ebook)
    1. Biodiversity—Climatic factors—California. 2. Biodiversity conservation—California.   I. Root, Terry Louise, 1954– editor.   II. Hall, Kimberly Reade, 1967- editor.   III. Herzog, Mark Paul, 1969- editor.   IV. Howell, Christine Ann, 1969– editor.
    QH105.C2B55 2015
    577.2'209794--dc23   2014041571

Manufactured in China

24 23 22 21 20 19 18 17 16 15
10 9 8 7 6 5 4 3 2 1

The paper used in this publication meets the minimum requirements of ANSI/NISO Z39.48-1992 (R 2002) (Permanence of Paper).♾

CONTENTS

Part III • Perspectives for Framing Biological Impacts of Rapid Climate Change / 159

EDITORS

KIMBERLY R. HALL
Michigan State University
hallkim@msu.edu

MARK P. HERZOG
US Geological Survey
mherzog@usgs.gov

CHRISTINE A. HOWELL
US Forest Service
cahowell@fs.fed.us

TERRY L. ROOT
Stanford University
troot@stanford.edu

CONTRIBUTING AUTHORS

WILLIAM R. L. ANDEREGG
Princeton University
anderegg@princeton.edu

JESSICA L. BLOIS
University of California, Merced
jblois@ucmerced.edu

BRENDAN COLLORAN
Archimedes Incorporated
San Francisco, CA
brendan@colloran.net

JEFFREY G. DORMAN
University of California, Berkeley
dorman@berkeley.edu

SASHA GENNET
The Nature Conservancy
San Francisco, CA
sgennet@tnc.org

ALDEN B. GRIFFITH
Wellesley College
agriffit@wellesley.edu

ELIZABETH A. HADLY
Stanford University
hadly@stanford.edu

LORI HARGROVE
University of California, Riverside
lhargrove@sdnhm.org

GRETCHEN E. HOFMANN
University of California, Santa Barbara
hofmann@lifesci.ucsb.edu

LAURA KOTEEN
University of California, Berkeley
lkoteen@berkeley.edu

GRETCHEN LEBUHN
San Francisco State University
lebuhn@sfsu.edu

MICHAEL D. MASTRANDREA
Stanford University
mikemas@stanford.edu

CHRISTOPHER J. OSOVITZ
University of South Florida
osovitz@usf.edu

REBECCA M. QUIÑONES
University of California, Davis
rmquinones@ucdavis.edu

MARK REYNOLDS
The Nature Conservancy
San Francisco, CA
mreynolds@tnc.org

JOHN T. ROTENBERRY
University of California, Riverside
john.rotenberry@ucr.edu

JASON P. SEXTON
University of California, Merced
sexton.jp@gmail.com

K. BLAKE SUTTLE
Imperial College London
kbsuttle@gmail.com

MORGAN W. TINGLEY
University of Connecticut
morgan.tingley@uconn.edu

ERIKA S. ZAVALETA
University of California, Santa Cruz
zavaleta@ucsc.edu

MANAGERS

DEBORAH ASELTINE-NEILSON
CA Department of Fish and Wildlife
daseltine@dfg.ca.gov

ANDREA CRAIG
The Nature Conservancy
Los Molinos CA
acraig@tnc.org

MARK FISHER
University of California Natural Reserve
    System
mfisher@ucr.edu

DAN HOWARD
Cordell Bank National Marine Sanctuary
dan.howard@noaa.gov

ROB KLINGER
US Geological Survey
Yosemite Field Station
rcklinger@usgs.gov

TINA MARK
US Forest Service
Tahoe National Forest
tmark@fs.fed.us

ALLAN MUTH
University of California Natural Reserve
    System
allan.muth@ucr.edu

JULIE PERROCHET
US Forest Service
Klamath National Forest
jperrochet@fs.fed.us

ANNE POOPATANAPONG
US Forest Service
Idyllwild Ranger Station
apoopatanapong@fs.fed.us

JENNY RECHEL
US Forest Service
Pacific Southwest Research Station
jrechel@fs.fed.us

MARK STROMBERG
Hastings Natural History Reservation
mark.stromberg@gmail.com

Terry L. Root and Kimberly R. Hall

Since the mid-1970s, our planet has been experiencing noticeable changes in global climate, such as rapid warming, which have strongly affected humans and natural systems. Current climate projections for the future suggest continued acceleration in the pace of these changes, leading to greater risk of harm, especially for those species and human populations that are least able to cope with change. Until collaborative efforts of governments find ways to reduce substantially the emissions of greenhouse gases and stabilize global climate patterns, those of us interested in conserving biodiversity will need to find direct ways to facilitate species' adaptations to ever-changing climate conditions. As scientists and resource managers, it is absolutely essential that we communicate the impacts that we are seeing, as this is one way to help promote climate change mitigation efforts to replace fossil fuel energy with that generated via renewable energy. Further, we must use what we are learning to help develop and implement strategies that prepare people and species for the inevitable continued changes in climate that will occur before mitigation strategies can produce results. Indeed, if we fail to act in these ways, the number of species extinctions will likely be higher than humanity can afford to lose.

The challenge of helping natural systems adapt is huge, but not insurmountable, if all concerned pool their knowledge, expertise, and resources. Specifically, we are referring to scientists conducting decision-relevant research, natural resource managers, and conservation practitioners implementing climate-updated protocols in the field, and, especially, decision-makers who help shape policy. Strengthening the dialogue between researchers and practitioners will inevitably inspire new, proactive approaches to management. This kind of partnership also provides opportunities to identify constraints to the kinds of innovative actions needed to reduce risks to natural (and human) systems, as some actions may require changes in existing policies or new legislative tools before they can be implemented. By strengthening scientist–manager partnerships and aligning our messaging, we also improve our ability to promote needed updates to those in a position to implement policy change. As we strengthen these relationships, we also increase the odds that we will save a greater number of species from extinction, and will protect more people from climate-related risks.

The information and approaches in this book can be applied globally, even though most of the content refers to species and ecosystems occurring in California, USA. The California focus is a direct result of forward thinking by the California Energy Commission, which established the Public Interest Energy Research (PIER) program. PIER has supported a wide range of innovative and important research for understanding and planning for climate change. In this case, rather than investing directly in new research, the PIER program supported emerging scientists—including advanced graduate students and postdoctoral fellows researching the effect of climate change on various species in California—as they gained experience producing and communicating "decision-relevant" climate science. A team of us coordinated efforts to help these young scientists connect with, and learn from, each other, and to exchange information and ideas with field managers. Our primary goals were (1) to facilitate constructive dialogues between scientists and conservation managers and (2) to encourage this cohort of scientists to reach out and communicate (without jargon) how their results can be used to inform natural system management and conservation efforts.

Guidance on how to update conservation planning and management actions using scientific information about climate change observations and projections, linked with knowledge on species and system sensitivities to climate drivers, is emerging (e.g., Stein et al. 2014). Indeed, we see this volume as an important extension of this effort to help move the field of climate change adaptation toward on-the-ground changes. This book not only provides examples of how various species and ecosystems in California are, and will likely be, influenced by the present and forecast changes in climate, but also demonstrates how interactions between scientists and wildlife managers can provide new insights to both groups. It is our hope that by publishing a compelling and diverse array of scientific findings, along with discussions between scientists and natural resource managers, we will inspire similar conversations among many other scientists, practitioners, and decision-makers elsewhere.

Following an introductory chapter, the book is divided into three sections.

Chapter 1 lays out guidelines on how scientists and managers can work together. It explains how scientists, who are doing cutting-edge research, need to make their findings accessible to practitioners, and need to better understand what information is needed to inform decisions. By establishing such partnerships, natural resource managers will have timely access to science that could help them take actions necessary to aid the adaptation of species to the rapidly changing climate.

Part I: Key Changes in Climate and Life

Chapter 2 reviews projections made by several global climate change models that incorporate data from the atmosphere, oceans, and land surfaces. The authors extrapolate information, conclusions, and trends from models that can be potentially useful in the design of management protocols.

Chapter 3 reviews how rapid change in climate can affect specific species and ecosystems. Not only are species ranges and population densities in flux, but scientists are also documenting changes in behavior, phenology, genetics, and morphology.

Part II: Learning from Case Studies and Dialogues between Scientists and Resource Managers

Seven case studies address how particular species living in marine and terrestrial habitats have been, or are likely to be, affected by rapid changes in climate. While doing the research discussed in each chapter, the primary authors, who were senior graduate students or postdoctoral fellows at California universities, not only investigated the effects of climate change on species, but also endeavored to make their work relevant to managers in the field. At the end of each chapter, the primary authors discuss their research with manager(s), demonstrating one simple means of forming a partnership or linkage.

Chapter 4 focuses on regions of upwelling along the California coastline, modeling the ecological impact of krill abundance that can vary greatly in response to atmospheric changes associated with climate change. A scarcity of krill can adversely affect the survival of species that feed on krill, such as salmon and various seabirds (e.g., murres). Consequently, the abundance of krill has direct implications in the management of its predator species.

Chapter 5 examines the gene expression of the purple sea urchin to determine the genetic differences among populations along the Pacific Coast in waters at different temperatures. The authors found that individuals in populations many kilometers apart have more similar genomes than those taken from two populations on either side of Point Conception in California. This unexpected result can most likely be attributed to the large temperature differences off Point Conception. These findings undoubtedly can be applied to the very similar red sea urchin, which is one of California's largest fisheries.

Chapter 6 provides recommendations on how to manage populations of salmon and trout living in the Klamath River Basin. There are real concerns that a rise in global average temperatures will dramatically depress the numbers of both species, or worse, lead to their extinction.

Chapter 7 presents population models of montane bumble bee communities in the Sierra Nevada Mountains. The authors conclude that restoration of the meadows the bumble bees frequent can be an effective management tool to slow the loss of bumble bee species and the plants that depend on them for pollination.

Chapter 8 looks at 28 species of birds occurring in the Santa Rosa Mountains of southern California where average temperatures have increased by as much as 5°C (9°F) since the early 1960s. The authors report upward shifts in elevational ranges, especially for desert species, which were about three times farther than the distance montane species had shifted.

Chapter 9 focuses on the community dynamics of grasslands. The authors report that it is difficult to predict the response of individual species to changes in climate. Managing entire communities may indeed be the best approach for grasslands, and incorporating flexibility into plans will likely facilitate adaptation. Monitoring a diversity of species is probably the best strategy to avoid unwelcome surprises.

Chapter 10 examines the carbon storage and cycling between native perennial and nonnative annual grasses, and the atmosphere. Over the last 250 years, the transfer of 30 to 60 Mg of carbon per hectare from the soil to the atmosphere has occurred in coastal California grasslands, owing to the invasion and near replacement of native perennial grasses by nonnative annual grasses. Presumably, restoration of native perennials could help reverse this loss of carbon into the atmosphere.

Part III: Perspectives for Framing Biological Impacts of Rapid Climate Change

Chapter 11 investigates how facilitating evolutionary mechanisms through management could possibly lower the extinction probabilities of select species. For example, could breeding between different populations or closely related species bolster vitality in changing environments? Or can establishment of corridors facilitate gene flow between separated populations? This review includes many examples from across the southwest.

Chapter 12 focuses on paleobiology and examines how information from prehistoric settings can aid managers by providing additional context for the changes they are seeing. The fossil record indicates that past changes in climate have modified the composition of ecological communities as well as the geographical range, population densities, genetics, morphology, and macroevolution of individual species.

Chapter 13 discusses how preexisting historic data on where species occurred can inform long-term monitoring of species responses to climate change. Such a temporal perspective can allow managers to detect long-term range shifts, colonization–extinction dynamics, and community changes within monitored landscapes. Often these data lay hidden in drawers and notebooks, and while there are important considerations with respect to appropriate use, they can provide unique and extremely valuable insights.

In 2002, one of the editors (TLR) and her late husband, Stephen H. Schneider, edited a book entitled *Wildlife Responses to Climate Change,* in which junior scientists provided case studies. At the time, they had two goals in mind: (1) "establish a credible scientific link between the health of natural systems (wildlife in particular) and human-induced climate change," and (2) "help create a community of young scholars who can demonstrate that . . . connecting wildlife and climate change disciplines can be accomplished with a high level of scientific quality" (Schneider and Root 2002). Now over a decade later, there is abundant, rigorous science demonstrating that natural systems and species are at risk. This current book builds on that now well-established link between the survival of natural systems and human-induced climate change by providing information on a wide range of systems, and views from multiple ecological disciplines. Indeed, we have learned new and different information, and have often been amazed by the rate of responses shown by some species, or surprised by unanticipated shifts or interactions that we observe as we apply different tools to the substantial problem of rapid climate change. This increase in our understanding of species responses has allowed us to add another critical component to the mix: a focus on clear communication and partnerships that enables stronger linkages between scientists and practitioners engaged in natural resource manage-

ment and conservation. We believe that these connections can lead to the creation of new legislative tools that impact management-agency actions, and "on the ground" conservation activities. While we hope that the information presented in this book will help inform management and inspire decision-relevant science, we have an even greater hope—that this book will help inspire stronger relationships, and greater trust, among people with different roles in the overall process of understanding, managing, and reducing climate-related risks. Through better communication, and more investment in working together to identify and apply the science, we will likely improve our odds of achieving the necessary goal of protecting species and systems. Getting this done is and will continue to be hard, and much of this work is very depressing. Consequently, we need to strengthen our support for science, and each other, in order to get the job done. Indeed, this book is offered with great appreciation to those who work daily to help sustain critical ecosystems and natural processes, and decrease the number of species facing extinction in California and elsewhere.

ACKNOWLEDGMENTS

The editors wish to thank Shambhavi Singh for voluntarily reading, editing, summarizing, and formatting the different chapters of this book.

LITERATURE CITED

Schneider, S. H. and T. L. Root (eds). 2002. *Wildlife Responses to Climate Change: North American Case Studies.* Island Press, Washington, DC.
Stein, B. A., P. Glick, N. Edelson, and A. Staudt (eds). 2014. *Climate-Smart Conservation: Putting Adaptation Principles into Practice.* National Wildlife Federation, Washington, DC.

CHAPTER 1

A New Era for Ecologists

INCORPORATING CLIMATE CHANGE INTO
NATURAL RESOURCE MANAGEMENT

Kimberly R. Hall

RAPID CLIMATE CHANGE IS one of the most pressing challenges facing resource managers and conservation practitioners in California and around the globe. Since the 1880s, the linear trend in average global surface temperature suggests an increase of approximately 0.85°C in the Northern Hemisphere, and the last 30 years were likely the warmest period in the last 1400 years (IPCC 2013). It is critical that we accelerate efforts to reduce the accumulation of greenhouse gases in our atmosphere (mitigate the causes of climate change). However, even if drastic reductions are achieved, the emissions that have already been released through the burning of fossil fuels, compounded by the loss of forests and other natural systems that store carbon, commit us to continued changes in climate for many decades to come (Solomon et al. 2009). The rapid pace of changes, combined with the complexity of potential responses of species and natural systems to different climatic factors, suggests that we will often need to transform, rather than just update, our management approaches (Kates et al. 2012, Park et al. 2012).

The extent to which ecologists in the research, conservation, and management fields are able to contribute viable strategies to address these challenges, and promote transformation in our approaches to management, has important implications for biodiversity, natural systems, and the ecological services that support all species, including humans.

The goals of this book are to help motivate efforts to reduce greenhouse gas emissions by describing observed and likely vulnerabilities of species and natural systems to climate change, and to help accelerate the pace of climate change adaptation in the natural resource management sector. The focus of this chapter is on framing how scientists and managers can work together to design and implement updates to our management and conservation practices that increase the odds that species and systems adapt to climate change. While most of the chapters in this book focus on observed impacts in California ecosystems, here we emphasize adaptation, and provide an introduction to the frameworks and tools available in the emerging field of adaptation planning. These frameworks and tools provide the pathway forward for incorporating what we learn from the study of responses of species and systems to climate change into natural resource management and conservation strategies.

Steps for adaptation planning include identifying likely changes in key climatic factors, characterizing the risks that these changes pose to things we care about, prioritizing those risks, evaluating the consequences of various strategies to reduce risk, implementing preferred actions, and tracking and learning from these actions (Moser and Ekstrom 2010, Poiani et al. 2011, Cross et al. 2012, Stein et al. 2014). While science is a critical input to these tasks, it is not by itself sufficient (Gregory et al. 2006), and lack of information should not be used to delay action. Given that the need for "more science" is often described by practitioners as a barrier to adaptation (Heller and Zavaleta 2009, Moser and Ekstrom 2010, Bierbaum et al. 2013, Petersen et al. 2013), a key step that scientists, managers, and policy-makers can partner on is reminding each other that we make decisions under uncertainty all of the time, and there are methods we can use to help make these decisions more rigorous and more transparent. In addition to investing in more science, to make progress on adaptation, we need to think more broadly about the skills and processes that can facilitate society's ability to act on what we observe, and plan for the changes that our climate and ecological models suggest are likely to occur.

Making decisions on how to address climate change risks to species, natural systems, and the people that depend on these systems requires that we integrate science with information on societal values and account for many types of uncertainties (Schneider et al. 2007, National Research Council 2009). Integration of science with values and the collaborative determination of likely costs and benefits of various adaptation actions require that science be presented clearly, with key thresholds identified where possible. As ecologists, we need to explain the logic behind our expectations for the future, and explain our assumptions in ways that help nonscientists understand the relationship between various climate drivers and the sensitivities of species and systems. When we are able to communicate science clearly, we can play an essential role in promoting science-based decisions: We enable a broader group of stakeholders to act as partners in the evaluation of the risks, costs, and benefits associated with different actions (Gregory et al. 2006, Schneider et al. 2007, National Research Council 2009, Moser and Ekstrom 2010). While the myriad of uncertainties associated with climate change impacts (and human responses to these impacts) present a major challenge, tools and frameworks for handling uncertainty continue to grow (National Research Council 2009, Kujala et al. 2013, Hoffman et al. 2014). For those of us trained in the natural sciences, uncertainty and complexity are not new concerns, and learning new ways to handle these elements will likely make us better scientists.

Although learning about tools for handling uncertainty is important, actually engaging in the process of identifying management options that could promote adaptation, and framing out the costs and benefits of those options will likely provide insights that can only be gained through this experience. One lesson that is likely to emerge is that not all "unknowns" that could be addressed with new scientific research are equally relevant (Hoffman et al. 2014). Processes like structured decision-making are particularly suited for helping illustrate this point (e.g., Keeney 2004, Martin et al. 2009), but this honing in on a smaller set of critical uncertainties will likely occur in most situations where policy-makers and managers are working through a management strategy decision, updating management goals, or re-prioritizing investments. Given the urgency of addressing climate change risks, our goal should be to emphasize research into the uncertainties that have the most influence in terms of helping us choose the best options for protecting the things we value. When "decision-critical" science is identified during the process of choosing among alternative management actions, scientists engaged in the process can greatly enhance the decision-relevance of their work by developing new projects that target these uncertainties (Martin et al. 2009, Mastrandrea et al. 2010). This type of partnership has the added benefit to researchers of providing a ready audience and application for their work, which can greatly improve the likelihood of support from funders, and promote interest in and critical feedback on the work from stakeholders. In California, there are many entry points for engagement, such as through workgroups for the California Climate Change Assessment process, or through contributions to the adaptation strategy (information available at the California Climate Change Portal, www.climatechange.ca.gov).

A main barrier to successful partnerships among scientists, managers, stakeholders, and decision-makers is lack of communication. While strong communication skills are often recognized as being essential for creating and maintaining partnerships, few academic scientists have direct opportunities to build the full complement of needed skills. These skills go far beyond writing an article for a broad audience, or being able to give a presentation describing one's work without jargon. We also need to be able to listen and ask questions, so that through dialogue we can identify information gaps, misunderstandings, and understand the bounds of the problem to be solved. One way for students in the sciences to build these skills and build relationships that promote long-term engagement in management decisions is for them to take advantage of opportunities to learn from managers. Ideally, these managers would be addressing questions that could potentially be informed by their area of research, but there is much to be learned even if this is not the case. While textbooks and journal articles may promote particular management actions or conservation strategies, direct engagement with decision-makers who are working toward a particular set of management objectives can provide critical insight into how

scientific information is integrated with information on societal values, and social, technological and financial constraints.

This book represents one example of graduate and postdoctoral researchers taking steps to make these connections with resource managers and conservation practitioners. As noted in the Preface, encouraging students to engage with researchers was the focus of the project funded by the California Energy Commission through the Public Interest Energy Research (PIER) program. The four editors of this book served as mentors in the project, which we called Biological Impacts of Climate Change in California (BICCCA). Our goal was to give graduate and postgraduate students at California universities encouragement and opportunities to meet and engage with on-the-ground resource managers and conservation practitioners, and to work with them to improve their oral and written communication skills. All chapters in this book were written to be accessible and relevant to both applied and academic audiences. Further, many of the chapters are accompanied by "conversations" between students in the BICCCA program and practitioners responsible for managing ecological systems under climate change. Though these conversations only scratch the surface of topics that long-term collaborations to sustain biodiversity will need to address, they demonstrate some of the key opportunities and constraints that arise as scientists and managers work more closely together. Our hope is that as students, established academic scientists, and natural resource professionals read these chapters, they will be inspired to initiate or increase their engagement in similar conversations.

RESPONDING TO CLIMATE CHANGE: A CALL FOR STRONGER COLLABORATIONS

Our desire to help "bridge the gap" between science and practice in the context of updating resource management is driven by the magnitude of climate-related risks to biodiversity. The bridges we construct are likely to be most stable if we recognize that fundamental assumptions underlying how we think about nature will have to change. Most are likely to agree with the idea that "stationarity is dead" (Milly et al. 2008); we can no longer make decisions about resource management with an expectation that patterns of variation in climate-sensitive systems will stay within historic ranges. But how do we think through what might happen over the next decade, or next century? Practitioners often establish "desired future conditions" for a site based on a historic reference condition, or try to achieve a changing mosaic of ecological site conditions that would have occurred under historic ranges of variability. However, as we move away from the assumption of climate stationarity, we need to reconsider what it is we are managing toward. This is one of the many ways the challenge of responding to climate change requires that we change our way of thinking, and stretch our imaginations, as much as it calls us to action.

Until we achieve stabilization of greenhouse gas concentrations, we face a future of continued change. This unfortunate reality suggests we should design resource management plans that are flexible enough to account for continued change over time, which explicitly address multiple forms of uncertainty, and likely increases in the frequency of many types of extreme climate events. One challenge to overcome as we work toward these goals is that traditional scientific and management approaches are not well suited to helping us prepare for directional change, variation across space and time, and extremes. Historically, ecological research has focused on simplified model systems in which we can rigorously test hypotheses, which has also meant that work is typically carried out on small focal areas or study plots, with a focus on relatively narrow ranges of variation in one or a few environmental variables. Similarly, this type of training is likely to have framed the perspectives of practitioners, and shapes the monitoring and other tools we use to inform our work. Depending on our experiences thus far in terms of observed eco-logical responses, this lack of points of reference for thinking about change can be a key barrier to conceptualization of how to move forward.

Our challenges in addressing change and variation highlight another tendency of the resource management/conservation practitioner community, of which three of the four editors of this book are a part. Recent research suggests that practitioners tend to be risk averse with respect to investing in actions perceived as risky, untested, or outside of the norm (West et al. 2009, Hagerman and Satterfield 2013). While we often claim to be engaged in an adaptive management approach, suggesting that our management actions are embedded in a study design that allows the comparison of different options through targeted monitoring, our visions for implementing adaptive management are often not well supported by available resources. The degree to which managers promote different types of adaptation actions may be influenced by both scientific support for the premise underlying the specific action and the degree to which it represents a digression from past activities (West et al. 2009). For practitioners, there is the added challenge of deciding how much to invest in addressing climate-related risks, given that other stressors such as habitat loss, pollution, or impacts of invasive species often seem more pressing (Lawler 2009, Hagerman and Satterfield 2014). We may also have greater confidence that scarce resources will actually do more good if spent on these time-tested approaches (West et al. 2012), rather than on risky new ideas. Further, we may feel overwhelmed by the idea of being responsible for unintended consequences following the application of a novel approach (Hagerman and Satterfield 2013, Hagerman and Satterfield 2014). Thus, in addition to developing modes of research that are a better fit to the challenges we need to address, a second key role for scientists may be developing explicit strategies for dealing with this bias toward the familiar. With respect to the knowledge and certainty thresholds that need to be attained to support implementation of adaptation actions "the burden of truth is much higher for unconventional actions" (Hagerman and Satterfield 2013, p.561), suggesting that working with practitioners, to first frame potential benefits and risks and second evaluate innovative actions when they are chosen, is a critical need.

As we frame risks and benefits, an approach that considers multiple lines of evidence should help us anticipate a broader range of species and system responses. Different fields of study, from investigations of genetics, through population dynamics, to landscape ecology, provide different lines of questioning and different ways of exploring responses. In some cases, looking at a problem through a different lens may suggest fewer uncertainties regarding a plan of action, while in other cases bringing in new tools and frameworks may open up even more questions. As a mentor in the BICCCA program, I saw the diversity of scientific disciplines and perspectives represented by the students and mentors as our greatest strength. Focal ecological systems for students in the program ranged from marine systems (Chapters 4 and 5) through rivers (Chapter 6), alpine systems (Chapters 7 and 8), and grasslands (Chapters 9 and 10). The collection of work here employed a wide range of tools and approaches, addressed many different taxonomic groups, and spans temporal scales from a few years for most work through decades (Chapter 8 and 13) and even millennia (Chapter 12). Given the complexity of the relationships between species life histories and climate drivers, we should expect that we will need multiple approaches to help answer our questions on how to help systems adapt. By drawing from many lines of evidence, we can likely better anticipate a broader spectrum of species and system responses, and more effectively work through processes that allow us to compare risks and prioritize management actions that most effectively reduce these risks. As this idea suggests, there is often a need to bridge across different fields within the biological sciences,

again emphasizing the need for clear communication skills, and a commitment to engaging in processes that promote cross-disciplinary learning.

Operationalizing Adaptation

Most resource managers, especially in California where a wide range of climate-related responses have been observed and projected (e.g., McLaughlin et al. 2002, Kelly and Goulden 2008, Loarie et al. 2008, Stralberg et al. 2011), are likely to agree that climate change presents a key stressor to ecological systems and the services they provide. The chapters in this book add numerous examples that help shape and expand our understanding of vulnerable species and systems. However, even when we recognize the threat of climate change, steps to reduce risks can be hard to identify. The complexities inherent to predicting ecological responses are likely key contributors to climate change being addressed "more than before, but less than needed" (Bierbaum et al. 2013) in the natural resources management / conservation sector. Long before climate change emerged as a conservation game changer, managing and protecting biodiversity given all of the other stressors that drive species declines represented a daunting task. As before, our work is tightly constrained by costs, and at times is in conflict with other stakeholder values. The interviews with resource managers that accompany many chapters give voice to these barriers to implementation of adaptation actions.

Designing and updating management approaches in ways that allow us to reach our project goals as climatic conditions change requires that most, if not all, resource managers move past these barriers and get into the habit of "asking the climate change question." Some may feel they are not prepared to take on this role. However, local knowledge on the ecology, management practices, site history, and social opportunities and constraints represent critical areas of expertise in this process—no "climate adaptation specialist" can step in and reduce the need for this knowledge. Updating management to account for climate change requires that we integrate this site-specific information with information on how changes in climatic drivers are likely to influence species and systems. Ideally, this integration will occur through a long-term partnership between climate specialists, ecological researchers, and practitioners. Integration may take many forms, and how to get started will depend on who is sitting at the table. How we go about integrating climate change is also likely to change over time: While investing time in planning efforts that focus solely on climate impacts and adaptation can play an important initial role in helping stakeholders get up to speed on new ways of thinking, over the long-term adaptation actions are more likely to be implemented if they are part of ongoing planning and management cycles. While integrating climate change into all decisions may seem overwhelming, this idea of repeatedly "asking the climate question" reminds us that we will have many opportunities to promote adaptation, and we can (and must) learn as we go.

While there is no one-size-fits-all answer on how to modify management to address climate change, or even one set of climate change questions that are appropriate for all situations, one way to help train ourselves to make adjustments is to follow a consistent, systematic process as we get started. A shared process, and shared tools, can also help different stakeholders work together by clarifying ideas on impacts and assumptions that underlie any proposed changes in goals or plans for action. Efforts to promote adaptation to key stakeholder groups have been rapidly increasing: Guidance documents, tools, workshops, and websites are being developed at all levels of government, in nongovernmental organizations, and in the private sector (Bierbaum et al. 2013). Within the conservation and resource management literature, there are many sources of recommendations and general principles for how to help natural systems adapt (e.g., AFWA 2009, Joyce et al. 2009, West

et al. 2009, Hobbs et al. 2010, Lawler et al. 2010, Hansen and Hoffman 2011, Poiani et al. 2011, Cross et al. 2012, Stein et al. 2014). Like the specific strategies needed to reduce climate-related risks in a given geography, the guidance, tools, and information that will be most useful in helping stakeholders move forward on adaptation will likely vary from place to place.

To provide context for how information in the rest of this book can be applied, we present a slightly modified version of an approach used by The Nature Conservancy (TNC) to update existing conservation plans (Poiani et al. 2011). This approach assumes that a team of scientists, managers, and policy specialists has already gone through the process of creating, and likely implementing, a conservation plan designed to protect species, systems, and / or ecological processes at a specific location. When presented in a general way, this framework's focus on updating an existing plan, which typically includes multiple conservation objectives, and strategies for addressing a wide range of stressors (i.e., invasive species, pollution, habitat loss or fragmentation) may be a good fit for others interested in sequentially reconsidering how to increase the likelihood of success of actions taken within an ongoing management or conservation effort. In addition, I like the fact that this framework includes the step of considering human responses to climate change as an additional challenge, and that it highlights the idea of looking for opportunities to demonstrate how protecting nature can protect people. When this framework is followed at a more detailed level, familiarity with TNC's conservation planning approach is helpful, which may require more of a time investment than makes sense for practitioners that use other planning approaches. This is one example of why exploring the range of adaptation frameworks and tools available before selecting methods to follow, especially if one is organizing a major effort with multiple partners, is a good idea. For example, for those that are engaged in new efforts, such as an emerging partnership to address climate change impacts in a place where goals and strategies have not already been created, broader frameworks with more flexible approaches may be more helpful (e.g., Hansen and Hoffman 2011, Cross et al. 2012, Stein et al. 2014). However, there are many similarities in steps for different processes, so while the TNC approach (Poiani et al. 2011) was developed for a specific mode of application, it still serves well as a method for connecting information from later chapters to a general process for updating conservation and management.

STEP 1. UNDERSTAND THE POTENTIAL ECOLOGICAL IMPACTS OF CLIMATE CHANGE This step involves gathering information on observed and projected climatic changes, and linking these to potential sensitivities in the focal ecological systems, species, and management actions. As noted above, the step of understanding vulnerabilities is most efficient, and likely most effective, when information such as temperature-change projections is connected to local sensitivities through active discussions, rather than through more passive transfers of information from scientists to managers. When climate scientists, ecologists, managers, and others familiar with the focal species or system work together, they can explore and eventually refine ideas about the specific aspects of some kind of change that is most relevant for considering specific risks. We might have a sense that a species is sensitive to temperature, but there are many ways that temperature can change, leading to different levels of exposure and different risks. For some species, the most meaningful climate data metric might be summer maximum temperatures, as these may correlate with heat-related mortality. Perhaps less intuitively, others may be most at risk from increases in winter low temperatures, which may have in the past limited the spread of an important pathogen or invasive species. Or, maybe the strongest influence of temperature increases is through a less direct driver, such as length of the growing season, increase in stream temperature, drought stress, or timing of snowmelt. In many cases, multiple mecha-

nisms are likely, and they may act in the same direction (i.e., suggesting increasing risk), or may suggest both risks and benefits.

Once useful metrics are identified, ecologists and practitioners can work with climate scientists to identify the best climate data sources for evaluating changes in the metrics over time. Given the many sources and methods for downscaling climate data, and the rapid increases in availability of these data (Barsugli et al. 2013), spending some time thinking critically about what data are most useful before trying to retrieve them should lead to more productive conversations with climate scientists. Other important components of this step include identifying one or more workable time frames for considering impacts (e.g., 20 years, or a short- and long-term reference point), identifying ways to handle the variation across different emissions scenarios and models, and framing some form of conceptual model to help with the identification of likely impacts. While described very briefly here, this step is essentially a climate change vulnerability analysis, and can be undertaken at many different levels of investment. A more detailed description is beyond the scope of this chapter, but many resources and tools are available to help with this process (e.g., Glick et al. 2011, Klausmeyer et al. 2011, Rowland et al. 2011, Thomas et al. 2011, Gardali et al. 2012, Wilsey et al. 2013).

A key point to recognize when working with climate specialists to identify which changes in climate drivers are most important is that these changes can affect many aspects of management. Thus, besides thinking about how changes in factors like the timing of spring, or peak water temperatures, may put species at risk, it's important to also consider how changing conditions can change the effectiveness of some form of management action, or change the influence of neighboring land uses on a focal property. For example, too much or too little moisture can restrict use of tools like prescribed burning, which may reduce a manager's ability to control invasive species. Changes in climate factors may lead to new problems with existing infrastructure, such as reduced stream connectivity due to perching of culverts in road-stream crossings during times of high drought stress/low stream flow. Similarly, paved areas in a neighboring parcel may be more likely to contribute to flooding as storm intensities increase, which can promote overland flow and erosion that leads to degradation of the nearby aquatic systems.

In these conversations, we should also consider the potential for "surprises" (Root and Schneider 2006) of at least three forms: (1) Exceedance of thresholds in some kind of response to climatic changes (e.g., if conditions exceed key thermal or drought tolerances, this may lead to rapid, hard-to-predict declines in fitness); (2) new interactions among species, and/or new or synergistic impacts related to interactions with climate and other stressors (e.g., climate change favoring an invasive species that had not been able to become established in the past); and (3) higher frequency of extreme weather events with catastrophic impacts on focal systems or management infrastructure (floods, ice storms, extreme cold periods in spring). Through interactive discussions, and by testing different ideas over time, scientists familiar with the climate data and managers most familiar with the systems can hone in on climate metrics that best inform consideration of risks. We hope that this book helps encourage these interactions and provides useful information and examples that help move the process along.

STEP 2. FORMULATE SPECIFIC "HYPOTHESES OF CHANGE" After engaging in a broad consideration of possible impacts, the next step in the TNC framework (Poiani et al. 2011) involves focusing in on a set of important ways in which climate change might impact a species, system, or management strategy. The step of creating "hypotheses of change" can help tame the complexity of the adaptation process by asking participants to craft statements that describe specific, management-relevant shifts in focal systems or species. For example, in the multi-

team workshop described by Poiani et al. (2011, p.189), the Moses Coulee project team from Washington state developed the following statement:

> A 2–3°C rise in annual temperature coupled with a 10–30% decrease in summer rainfall and a 5–10% increase in winter precipitation will lead to a greater frequency and intensity of wildfires, which create openings for expansion of invasive cheatgrass, and increased spring productivity of cheatgrass, resulting in the decreased cover of key native shrubs and bunchgrasses.

Generating lists of possible changes and linking them in a conceptual model can help organize a group's thinking, making it easier to identify a subset of vulnerabilities that can potentially be reduced through management actions. The premise here is that each hypothesis of change should be specific, and should illustrate clearly the connection between one or more specific climate drivers and a change in our ability to meet some conservation objective. Clearly stating these relationships helps ensure that we "link actions to climate impacts" as we develop adaptation strategies, one of the "key characteristics to climate-smart conservation" identified in comprehensive guidance from Stein et al. (2014, p.38). Ideally, each of these statements would also be associated with some descriptor of how likely the impact is (high-medium-low probability). When combined with factors such as the magnitude of the impact and its reversibility, the likelihood can help identify which risks are of highest priority for being addressed with adaptation actions.

STEP 3. EXPLORE POTENTIAL SOCIETAL RESPONSES TO CLIMATE CHANGE Many, if not most, of the ecosystem stressors addressed through management are strongly tied to human actions, such as land conversion, pollution, and introduction of nonnative species. Thus, part of updating our approach to include climate stressors in management involves thinking about the responses of people to climate change, as these will likely contribute to changes in stressors, and in future management priorities. The goal of this step is to anticipate, and where possible prevent, responses that could cause more problems for ecosystems. In some cases, this step may lead to work on quantifying the ecological services that current natural systems are providing, or that could be provided through restoration, which help reduce climate-related risks. Examples include investing more of our resources into protecting forests to maintain sources of freshwater, and protecting or restoring floodplains to help contain spring floods (Kousky 2010, Downing et al. 2013, Gartner et al., 2013). By proactively thinking about societal responses, we can hopefully identify and communicate ways that natural areas help promote adaptation for people, and anticipate and help avoid actions taken to protect people that could damage natural systems or impede ecological processes. Sea walls and levees are two specific examples of "gray infrastructure" that communities may identify as needs for protection from sea level rise and river flooding, yet these structures can reduce habitat values, and put species and natural processes at risk. Actions that are identified during this step are likely to focus on improving communication with stakeholders that have a need for services that natural systems could provide, and helping foster partnerships that encourage investments in nature-based rather than gray infrastructure. This is an example of where fostering of long-term relationships across different sectors is especially needed so that people with appropriate expertise can be involved early and often. Gray infrastructure-type responses to the risks of climate change have long planning periods and are very costly to retrofit (Kousky 2010, Downing et al. 2013). This step should also serve to help remind managers to invest time in summarizing their investments in activities such as restoring wetlands that help prevent floods, or in increasing the size of culverts that support road-stream crossings to prepare for flashier flows. When extreme events take place, these data coupled with comparative analyses from economists or hazard management specialists on comparative costs across areas with fewer wetlands or smaller pipes can be some of our strongest tools for inspiring greater investments in actions that benefit both people and nature.

STEP 4. DETERMINE WHICH CLIMATE IMPACTS ARE CRITICAL, AND CAN BE ADDRESSED In the process described by Poiani et al. (2011), the next step involves ranking the risks posed by climate change (as stated in step 2), and integrating these lists with existing management priorities. The goal here is to focus in on a small number of risks that can feasibly be addressed through changes in management, so that these can become the focus of adaptation strategies. This prioritization step can be addressed in many different ways—it may be useful to focus in on one of the most tractable risks first, simply to help provide momentum to the process of investing in adaptation efforts. Over the long term, it will likely be important to develop clear criteria for addressing climate risks, as there is no doubt that the resources needed to address even a limited number of vulnerabilities will exceed the funds available. This is the most obvious step where science must be integrated with societal values, which can be easier if we have some structure to help us frame comparisons, and communicate options to the public. To this end, Schneider et al. (2007, p.785–786) outlined seven criteria for defining "key vulnerabilities" as part of the Fourth Intergovernmental Panel on Climate Change (IPCC) Assessment. The concept of key vulnerabilities can be applied to any type of system that is susceptible to the impacts of climate change (e.g., ecological, human infrastructure, public health). These factors provide a thought-provoking way to filter through long lists of impacts to produced prioritized adaptation actions, and related discussions are likely to range far beyond science and management.

• Magnitude of impact (area, number of species / systems, costs to restore or promote adaptation);

- Timing (rate of change, near or far-off in time);
- Persistence and reversibility (i.e., extinctions vs. population declines; climate change as a trigger for change in land use that would limit future ecological values);
- Likelihood of impacts and vulnerabilities, and confidence in those estimates (probability that the impact will occur, as assessed through statistical tools and/or expert opinion);
- Potential for adaptation (to what extent do species have the genetic variability, dispersal abilities, etc., needed to adapt to changes, and can we implement needed strategies given current regulatory constraints?);
- Distribution of impacts and vulnerabilities (IPCC focuses on equity issues; this could be seen as pattern of impacts relative to the distribution of a population, or could provide a mechanism for considering variation across different stakeholder groups in the impact of climate-related changes in ecosystem services to people);
- Importance of the system at risk (e.g., cultural value, economic value, ecological value).

STEP 5. EVALUATE IF POTENTIAL CLIMATE CHANGE IMPACTS ARE LIKELY TO CHANGE FUN-DAMENTALLY THE PROJECT, OR PRECLUDE PROJECT SUCCESS   This step serves as a reminder that static goals may no longer make sense with respect to managing systems in a nonstationary climate, and some aspects of change may be in direct conflict with sustaining particular species or systems at a site. As in the step above, this step requires integration of science with values and has the potential to lead to discussions that range far beyond the science and on-the-ground management practices. As we delve into our goals, we are likely to uncover that not all share the same perspective on principles that underlie our work, such as the value of "naturalness," the purpose of parks, national forests, or other types of public-trust resources, and the appropriateness of various levels of

intervention (Hobbs et al. 2010). As these ideas suggest, managers should expect the steps in any approach to adaptation to be iterative and nonlinear—sometimes you need to step backward to move forward. With respect to thinking about how goals can be reframed, three terms are commonly used to frame approaches to adaptation: Resistance (act to slow the rate of change and protect valued resources); resilience (act to promote ability to bounce back from disturbance, and stay within some range of conditions); and response/trans-formation to a different state or system type (e.g., Millar et al. 2007, Heller and Zavaleta 2009, West et al. 2009). In their work on Medi-terranean forests of California, Stephens et al. (2010) present a fourth category, realignment, that addresses using management to reset the trajectory of systems to account for reduced abil-ity to adapt to new conditions due to factors such as past management history. Depending on the missions of particular management organizations, choices of how to proceed may be constrained due to conflicts with core princi-ples, or partnerships may be strained due to dif-ferences in perspective that are brought to the forefront by climate change. In other cases, a key strategy that emerges from these conversa-tions may be to revisit project goals, or even an organization's mission.

STEP 6.   DEVELOP ADAPTATION STRATEGIES AND EVALUATE THEIR FEASIBILITY AND COST Adaptation strategies can range from minor "tweaks" in practices (e.g., changing the timing of some actions, like mowing or prescribed burns) to more costly changes, or entirely new practices. Again, in this short review we hope to provide context for how the impact informa-tion, and management goals and constraints, connects with adaptation planning. However, many resources to help with the framing of adaptation strategies are available, ranging from general ideas (Heller and Zavaleta 2009, Lawler 2009, West et al. 2009, Hansen and Hoffman 2011, Stein et al. 2014) to system-specific examples and case studies for California (e.g., Millar et al. 2007, Stephens et al. 2010, Downing et al. 2013). This growing literature on climate change adaptation has identified the need to take a very proactive approach, especially in the context of protecting species and ecological systems that are already at risk (West et al. 2009, Lawler et al. 2010, Poiani et al. 2011). Being proactive entails identifying climate change as a significant threat, indicating consequences it would likely have, and identifying specifically how to address those threats through a set of actions designed to reduce probable future risks.

The step of identifying adaptation strategies reminds us to assess the costs and benefits of changes in management, a process that should link back to the hypotheses of change that we defined above. A key checkpoint to consider is, do the actions you have identified as adaptation directly address a priority impact? The most successful adaptation actions will clearly show this logical flow, and our ability to explain the rationale behind our choices will help build public support for actions that may be more costly than "business as usual." As different forms of guidance have emerged, one can sense tension with respect to what kind of management actions "count" as adaptation. For example, increasing connectivity is often one of the first concepts to emerge in climate adaptation conversations, and it may be tempting to then say that any effort to promote connectivity also promotes adaptation. The idea of acting "with intention" (e.g., Stein et al. 2014) provides an important cross-check on our list of strategies. If, rather than linking each adaptation action to a specific reduced risk, we call any action that possibly increases connectivity or reduces the impact of some stressor (like invasive species) a climate adaptation strategy, we will be less effective, and weaken support and understanding for our efforts. These activities are still vital, and often do promote adaptation, but we should be prepared to describe how, and for what species or system, future viability is likely to be enhanced. Clearly, they may also be vital for other reasons beyond climate change adaptation, and being clear in our logic can help us make the case that they are still important, even if there are few climate-related benefits.

If the task of overhauling management for an area to address climate change seems daunting at first, a useful first step may be to evaluate and modify current actions that are clearly counterproductive or are at high risk of having benefits negated as a result of climate change. For example, it may be straightforward to compare the longevity of some investments, as in the case of restoration of coastal wetlands that vary in terms of exposure and resilience to increasing sea levels (Stralberg et al. 2011). Similarly, another fruitful first strategy may be to update ongoing monitoring work to ensure that data collection efforts are useful for evaluating changes attributable to climate change. In some cases, this is as simple as including local climatic measures in the monitoring scheme or extending the temporal extent of monitoring such that changes in timing of events can be captured. This form of coordination can provide the data needed to develop additional strategies in the future. Similarly, an early step to help identify site-specific changes that could inform on-the-ground strategies could involve a systematic review of information collected in the past, as described by Tingley (Chapter 13).

STEP 7. DEVELOP MEASURES OF SUCCESS, IMPLEMENT, ADAPT, AND LEARN To restore, maintain, and protect biodiversity as the climate changes, it is more important than ever to have clear ways of defining and measuring the success of our actions. Although full coverage of monitoring and data sharing among resource managers is beyond the scope of this chapter, these activities represent an essential follow-up to "asking the climate question" of our management goals and decisions. Strengthening our inference and defining measures of success that help us communicate with a wide range of stakeholders are urgent needs. As such, they represent challenging but likely fruitful areas for collaborations among scientists, managers, and policy-makers.

To build upon the chapters here, which are part of a rapidly growing body of information on species and system responses, we need to step up collaborative, multi-stakeholder efforts that support implementation of key actions, and changes in policy. Specifically, to build the momentum that moves science-based ideas into actions on the ground, we need to involve a broad range of stakeholders, and be clear on how various adaptation strategies may benefit, or harm, the things that each group values. The decisions that need to be made now and in the future require that as scientists we understand that science needs to be framed in such a way that a wide range of stakeholders can evaluate our results. When scientific information is presented well, stakeholders can use it to inform value judgements on how scarce resources are used. We also need to invest in communication and become more comfortable with stating our assumptions and values, as these are critical building blocks for improving understanding. To do the best we can in supporting functioning natural systems, we will need to prepare for surprises in how species and systems respond, and think about rules for "triage," as not all species will be viable (at a defensible cost) in future climates (Root and Schneider 2006, Lawler 2009). At the risk of oversimplification, progress on the collaboration side can be measured by answering three questions: (1) Are we talking to each other? (2) Is there shared understanding on the key issues and options? (3) Can we identify shared goals, and ways to achieve them? Collaborations can break down in any of these areas, and tools for promoting collaboration seem especially needed when trying to bring people together around issues with this level of complexity and uncertainty.

Taken together, all of these components of responding to climate change suggest that we have entered a new era with respect to the skill sets that will be most useful to scientists that enter ecological fields, either as academic or agency researchers, practitioners within non-governmental organizations, or as resource managers. While no one person, or even organization, is likely to include the full range of expertise needed to meet this challenge, we can benefit both nature and ourselves if we broaden our perspectives and strengthen our connections.

LITERATURE CITED

AFWA. 2009. *Voluntary Guidance for States to Incorporate Climate Change into State Wildlife Action Plans and Other Management Plans.* A Collaboration of the AFWA Climate Change and Teaming with Wildlife Committees, Association of Fish and Wildlife Agencies. Washington, DC. http://www.fishwildlife.org/files/AFWA-Voluntary-Guidance-Incorporating-Climate-Change_SWAP.pdf.

Barsugli, J. J., G. Guentchev, R. M. Horton, A. Wood, L. O. Mearns, X.-Z. Liang, J. A. Winkler, K. Dixon, K. Hayhoe, R. B. Rood et al. 2013. The practitioner's dilemma: How to assess the credibility of downscaled climate projections. *Eos, Transactions American Geophysical Union* 94:424–425.

Bierbaum, R., J. B. Smith, A. Lee, M. Blair, L. Carter, F. S. Chapin III, P. Fleming, S. Ruffo, M. Stults, S. McNeeley et al. 2013. A comprehensive review of climate adaptation in the United States: More than before, but less than needed. *Mitigation and Adaptation Strategies for Global Change* 18:361–406.

Cross, M. S., E. S. Zavaleta, D. Bachelet, M. L. Brooks, C. A. F. Enquist, E. Fleishman, L. Graumlich, C. R. Groves, L. Hannah, L. Hansen et al. 2012. The adaptation for conservation targets (ACT) framework: A tool for incorporating climate change into natural resource management. *Environmental Management* 50:341–351.

Downing, J., L. Blumberg, and E. Hallstein. 2013. *Reducing Climate Risks with Natural Infrastructure.* The Nature Conservancy, San Francisco, CA. 29 pp. http://www.nature.org/ourinitiatives/regions/northamerica/unitedstates/california/ca-green-vs-gray-report-2.pdf.

Gardali, T., N. E. Seavy, R. T. DiGaudio, and L. A. Comrack. 2012. A climate change vulnerability assessment of California's at-risk birds. *PLOS ONE* 7:e29507.

Gartner, T., J. Mulligan, R. Schmidt, and J. Gunn. 2013. *Natural Infrastructure: Investing in Forested Landscapes for Source Water Protection in the*

United States. World Resources Institute, Washington, DC. http://www.wri.org /publication/natural-infrastructure.

Glick, P., B.A. Stein, and N.A. Edelson (eds). 2011. *Scanning the Conservation Horizon: A Guide to Climate Change Vulnerability Assessment.* National Wildlife Federation, Washington, DC.

Gregory, R., L. Failing, D. Ohlson, and T.L. McDaniels. 2006. Some pitfalls of an overemphasis on science in environmental risk management decisions. *Journal of Risk Research* 9:717–735.

Hagerman, S.M. and T. Satterfield. 2013. Entangled judgments: Expert preferences for adapting biodiversity conservation to climate change. *Journal of Environmental Management* 129:555–563.

Hagerman, S.M. and T. Satterfield. 2014. Agreed but not preferred: Expert views on taboo options for biodiversity conservation, given climate change. *Ecological Applications* 24:548–559.

Hansen, L.J. and J.R. Hoffman. 2011. *Climate Savvy: Adapting Conservation and Resource Management to a Changing World.* Island Press, Washington, DC. 245 pp.

Heller, N.E. and E.S. Zavaleta. 2009. Biodiversity management in the face of climate change: A review of 22 years of recommendations. *Biological Conservation* 142:14–32.

Hobbs, R.J., D.N. Cole, L. Yung, E.S. Zavaleta, G.H. Aplet, F.S. Chapin, P.B. Landres, D.J. Parsons, N.L. Stephenson, P.S. White et al. 2010. Guiding concepts for park and wilderness stewardship in an era of global environmental change. *Frontiers in Ecology and the Environment* 8:483–490.

Hoffman, J., E. Rowland, C.H. Hoffman, J. West, S. Herrod-Julius, and M. Hayes. 2014. Managing under uncertainty, Chapter 12. In B.A. Stein, P. Glick, N. Edelson, and A. Staudt (eds), *Climate-Smart Conservation: Putting Adaptation Principles into Practice.* National Wildlife Federation, Washington, DC.

IPCC. 2013. Summary for policymakers. In T.F. Stocker, D. Qin, G.-K. Plattner, M. Tignor, S.K. Allen, J. Boschung, A. Nauels, Y. Xia, V. Bex, abd P.M. Midgley (eds), *Climate Change 2013: The Physical Science Basis. Contribution of Working Group I to the Fifth Assessment Report of the Intergovernmental Panel on Climate Change.* Cambridge University Press, Cambridge, UK and New York, NY. 1535 pp.

Joyce, L.A., G.M. Blate, S.G. McNulty, C.I. Millar, S. Moser, R.P. Neilson, and D.L. Peterson. 2009. Managing for multiple resources under climate change: National forests. *Environmental Management* 44:1022–1032.

Kates, R.W., W.R. Travis, and T.J. Wilbanks. 2012. Transformational adaptation when incremental adaptations to climate change are insufficient. *Proceedings of the National Academy of Sciences of the United States of America* 109:7156–7161.

Keeney, R.L. 2004. Making better decision makers. *Decision Analysis* 1:193–204.

Kelly, A.E. and M.L. Goulden. 2008. Rapid shifts in plant distribution with recent climate change. *Proceedings of the National Academy of Sciences of the United States of America* 105:11823–11826.

Klausmeyer, K.R., M.R. Shaw, J.B. MacKenzie, and D.R. Cameron. 2011. Landscape-scale indicators of biodiversity's vulnerability to climate change. *Ecosphere* 2:art88.

Kousky, C. 2010. Using natural capital to reduce disaster risk. *Journal of Natural Resources Policy Research* 2:343–356.

Kujala, H., M.A. Burgman, and A. Moilanen. 2013. Treatment of uncertainty in conservation under climate change. *Conservation Letters* 6:73–85.

Lawler, J.J. 2009. Climate change adaptation strategies for resource management and conservation planning. *Annals of the New York Academy of Sciences* 1162:79–98. Year in Ecology and Conservation Biology 2009.

Lawler, J.J., T.H. Tear, C. Pyke, M.R. Shaw, P. Gonzalez, P. Kareiva, L. Hansen, L. Hannah, K. Klausmeyer, A. Aldous et al. 2010. Resource management in a changing and uncertain climate. *Frontiers in Ecology and the Environment* 8:35–43.

Loarie, S.R., B.E. Carter, K. Hayhoe, S. McMahon, R. Moe, C.A. Knight, and D.D. Ackerly. 2008. Climate change and the future of California's endemic flora. *PLOS ONE* 3:e2502.

Martin, J., M.C. Runge, J.D. Nichols, B.C. Lubow, and W.L. Kendall. 2009. Structured decision making as a conceptual framework to identify thresholds for conservation and management. *Ecological Applications* 19:1079–1090.

Mastrandrea, M., N. Heller, T. Root, and S. Schneider. 2010. Bridging the gap: Linking climate-impacts research with adaptation planning and management. *Climatic Change* 100:87–101.

McLaughlin, J.F., J.J. Hellmann, C.L. Boggs, and P.R. Ehrlich. 2002. Climate change hastens population extinctions. *Proceedings of the National Academy of Sciences of the United States of America* 99:6070–6074.

Millar, C.I., N.L. Stephenson, and S.L. Stephens. 2007. Climate change and forests of the future:

Managing in the face of uncertainty. *Ecological Applications* 17:2145–2151.

Milly, P. C. D., J. Betancourt, M. Falkenmark, R. M. Hirsch, Z. W. Kundzewicz, D. P. Lettenmaier, and R. J. Stouffer. 2008. Stationarity is dead: Whither water management? *Science* 319:573–574.

Moser, S. C. and J. A. Ekstrom. 2010. A framework to diagnose barriers to climate change adaptation. *Proceedings of the National Academy of Sciences of the United States of America* 107:22026–22031.

National Research Council. 2009. *Informing Decisions in a Changing Climate. Panel on Strategies and Methods for Climate-Related Decision Support, Committee on the Human Dimensions of Global Change. Division of Behavioral and Social Sciences and Education.* The National Academies Press, Washington, DC. 188 pp.

Park, S. E., N. A. Marshall, E. Jakku, A. M. Dowd, S. M. Howden, E. Mendham, and A. Fleming. 2012. Informing adaptation responses to climate change through theories of transformation. *Global Environmental Change-Human and Policy Dimensions* 22:115–126.

Petersen, B. C., K. R. Hall, K. J. Kahl, and P. J. Doran. 2013. In their own words: Perceptions of climate change adaptation from the Great Lakes region's resource management community. *Environmental Practice* 15:377–392.

Poiani, K. A., R. L. Goldman, J. Hobson, J. M. Hoekstra, and K. S. Nelson. 2011. Redesigning biodiversity conservation projects for climate change: Examples from the field. *Biodiversity Conservation* 20:185–201.

Root, T. L. and S. H. Schneider. 2006. Conservation and climate change: The challenges ahead. *Conservation Biology* 20:706–708.

Rowland, E. L., J. E. Davison, and L. J. Graumlich. 2011. Approaches to evaluating climate change impacts on species: A guide to initiating the adaptation planning process. *Environmental Management* 47:322–337.

Schneider, S. H., S. Semenov, A. Patwardhan, I. Burton, C. H. D. Magadza, M. Oppenheimer, A. B. Pittock, A. Rahman, J. B. Smith, A. Suarez et al. 2007. Assessing key vulnerabilities and the risk from climate change. In M. L. Parry, O. F. Canziani, J. P. Palutikof, P. J. van der Linden, and C. E. Hanson (eds), *Climate Change 2007:*

Impacts, Adaptation and Vulnerability. Contribution of Working Group II to the Fourth Assessment Report of the IPCC. Cambridge University Press, Cambridge, UK. 779–810.

Solomon, S., G. K. Plattner, R. Knutti, and P. Friedlingstein. 2009. Irreversible climate change due to carbon dioxide emissions. *Proceedings of the National Academy of Sciences of the United States of America* 106:1704–1709.

Stein, B. A., P. Glick, N. Edelson, and A. Staudt (eds). 2014. *Climate-Smart Conservation: Putting Adaptation Principles into Practice.* National Wildlife Federation, Washington, DC.

Stephens, S. L., C. I. Millar, and B. M. Collins. 2010. Operational approaches to managing forests of the future in Mediterranean regions within a context of changing climates. *Environmental Research Letters* 5(2):024003. doi: 10.1088/1748-9326/1085/1082/024003.

Stralberg, D., M. Brennan, J. C. Callaway, J. K. Wood, L. M. Schile, D. Jongsomjit, M. Kelly, V. T. Parker, and S. Crooks. 2011. Evaluating tidal marsh sustainability in the face of sea-level rise: A hybrid modeling approach applied to San Francisco Bay. *PLOS ONE* 6:e27388.

Thomas, C. D., J. K. Hill, B. J. Anderson, S. Bailey, C. M. Beale, R. B. Bradbury, C. R. Bulman, H. Q. P. Crick, F. Eigenbrod, H. M. Griffiths et al. 2011. A framework for assessing threats and benefits to species responding to climate change. *Methods in Ecology and Evolution* 2:125–142.

West, J. M., S. H. Julius, P. Kareiva, C. Enquist, J. J. Lawler, B. Petersen, A. E. Johnson, and M. R. Shaw. 2009. US natural resources and climate change: Concepts and approaches for management adaptation. *Environmental Management* 44:1001–1021.

West, J. M., S. H. Julius, and C. P. Weaver. 2012. Assessing confidence in management adaptation approaches for climate-sensitive ecosystems. *Environmental Research Letters* 7(1):014016. doi:10.1088/1748-9326/7/1/014016.

Wilsey, C. B., J. J. Lawler, E. P. Maurer, D. McKenzie, P. A. Townsend, R. Gwozdz, J. A. Freund, K. Hagmann, and K. M. Hutten. 2013. Tools for assessing climate impacts on fish and wildlife. *Journal of Fish and Wildlife Management* 4:220–241.

PART I

Key Changes in Climate and Life

CHAPTER 2

Climate Change from the Globe to California

Michael D. Mastrandrea and William R. L. Anderegg

Abstract. Projections of future anthropogenic climate change derive from global models of the atmosphere, ocean, and land surface known as General Circulation Models (GCMs). Such models provide output at the scale of 100–200 km boxes, but techniques known as "downscaling" can generate projections at higher spatial resolutions. Downscaling can be done via two very different approaches, statistical and dynamic, and like global projections it must rely on trajectories of future greenhouse gas emissions known as emission scenarios. Projections of temperature change in California are largely consistent across models, but changes in precipitation and extreme events are more difficult to model and are also very relevant to the state's flora and fauna.

INTRODUCTION

California's climate regions range from the coastal, moist redwood forests in the north to the high mountainous regions of the Sierras to the arid deserts of southern California. Over time, the state's communities and economy have developed strategies to manage climate stresses and to prosper within the state's diverse climatic zones. Likewise, California's native flora and fauna have thrived in this variety of climatic regions, making the state one of the world's most important biodiversity hotspots of species found nowhere else (Myers et al. 2000).

However, the rapidly changing climate is now threatening to exceed the limits of species' natural strategies for managing climate conditions. While the effects of changing climate on ecosystems have already been noted in California, future changes may overwhelm ecosystems' natural resilience and cause widespread change. In northern California, the Bay checkerspot butterfly (*Euphydryas editha bayensis*) became the first documented species driven locally extinct by changing climate (McLaughlin et al. 2002). Looking ahead, future temperature changes could threaten up to 66% of California's unique plants with greater than 80% range reduction (Loarie et al. 2008). Temperature changes are expected to drive large shifts in bird ranges, leading to entirely new and no-analogue bird communities (Stralberg et al.

2009). These consequences highlight the vulnerability of California's natural systems to climate variability and change.

Climate modeling and projections of future climate change at the global and regional scale can be used to inform policy decisions to mitigate future climate change by reducing emissions of greenhouse gases, and are a key component of anticipating and managing future risks of climate change through adaptation. In this chapter, we provide an overview of climate modeling and how regional climate projections are produced by downscaling the results of General Circulation Models (GCMs). We discuss climate change scenarios for California, highlighting potential interaction with the state's ecosystems, and touch briefly on how these might be used in state and regional planning efforts in the future.

PROJECTING GLOBAL CLIMATE

Climate projections depend in large part on two factors: (1) How much and how quickly greenhouse gases are emitted into the atmosphere and (2) how the climate, oceans, and terrestrial systems respond to rising atmospheric concentrations of these gases.

Emission Scenarios

In 2000 the Intergovernmental Panel on Climate Change (IPCC) *Special Report on Emission Scenarios* (SRES) developed the most commonly used set of future emission scenarios based on different assumptions about global development paths (Nakicenovic et al. 2000). Scenarios differ in their trajectories for population and economic growth, technological development, and patterns in trade and sharing of technologies, among other things. Each scenario represents a possible "baseline" trajectory of emissions without explicit policy intervention, although some scenarios are more likely to simulate expected "business as usual" trends with continued high emissions and others are closer to a pathway that could be achieved with a stringent emissions reduction policy.

These scenarios are now over 10 years old, and a new set of emission scenarios and climate change projections are now available for use in climate impacts and policy analysis, based on the Representative Concentration Pathways (RCPs) (Moss et al. 2010, Collins et al. 2013). One primary difference is that the RCPs include trajectories that assume policy interventions, yielding a wider range of possible futures. Figure 2.1 compares SRES and RCP scenarios in terms of radiative forcing, a measure of the strength of the warming influence on the climate system from increased atmospheric greenhouse gas concentrations under each emission scenario. A growing number of analyses employ the RCPs, but much of the published literature on climate impacts still relies on the SRES scenarios, including climate and impact projections for California. Thus, we focus in this chapter on the SRES scenarios.

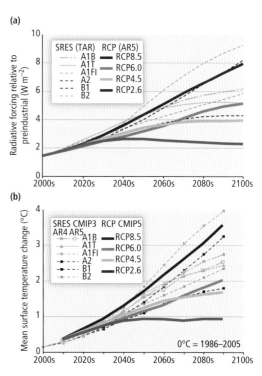

FIGURE 2.1: Range of SRES and RCP emission scenarios. Comparison of projected radiative forcing and temperature trajectories under the SRES and RCP scenarios. Source: Burkett et al. (2014).

These scenarios are run in global GCMs to produce projections of global and regional climate change. GCMs are developed by different research groups around the world and represent many different ways to simulate complex interactions between the atmosphere, ocean, ice sheets, and land surface that determine the sensitivity of the earth system to increasing greenhouse gas concentrations.

Climate Sensitivity

Climate sensitivity is a measure of how much temperatures will rise with a given increase in atmospheric greenhouse gas concentrations. Climate sensitivity is not known with certainty, as it depends on how various earth system processes respond to warming and on various "feedbacks" that either amplify or dampen warming. Scientific understanding of exactly how these processes and feedbacks will interact is still developing. For example, as temperatures rise, the atmosphere can hold more water vapor. More atmospheric water vapor traps heat and increases global temperatures further—a positive feedback. However, the clouds created by this water vapor could either enhance warming by absorbing and radiating outgoing infrared radiation from the earth's surface (another positive feedback) or dampen warming by reflecting more incoming shortwave radiation from the sun back to space before it reaches the earth's surface (a negative feedback).

The "climate sensitivity" represents the response of the climate system to changes in CO_2, and this sensitivity is often expressed as the long-term temperature increase associated with a doubling of atmospheric CO_2 concentrations. The IPCC reports a likely range for climate sensitivity of 1.5–4.5°C (2.7–8.1°F) meaning at least a 66% probability that the climate sensitivity is within this range (Collins et al. 2013). The IPCC also concludes that the climate sensitivity is extremely unlikely to be less than 1°C (meaning less than 5% probability) and very unlikely to be more than 6°C (meaning less than 10% probability). Different cli-

mate models treat the relevant processes and feedbacks differently, which leads to a range of values for climate sensitivity exhibited across models.

Therefore, different models project different levels of global temperature increase even for the same emission scenario, which can serve as an estimate of the uncertainty in the climate system's response to the same greenhouse gas forcing. The projections for the end of the century that follow illustrate this spread across models.

The Projections

Over the next few decades, the projected changes in global mean temperature are fairly similar across emission scenarios due to the inertia of the climate system (Meehl et al. 2007, Collins et al. 2013). By the second half of the century, however, different emission scenarios yield vastly different temperature responses. These characteristics are illustrated in Figure 2.2 for the SRES scenarios. For the highest-emission scenario (A1FI), models project global average warming of 2.4–6.4°C (4.3–11.5°F) by the end of the century (Meehl et al.

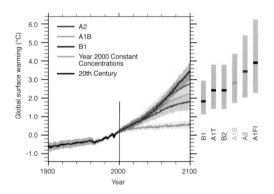

FIGURE 2.2: Projected global average temperature increase through the 21st century under several emission scenarios. The left-hand panel shows the evolution of temperature increase over time (as well as projected temperature increase assuming constant year-2000 atmospheric greenhouse gas concentrations). The right-hand panel shows the ranges of temperature increase at the end of the century (across different climate models) for six emission scenarios. Source: IPCC (2007).

2007). For the lowest-emission scenario (B1), models project warming of 1.1–2.9°C (2–5.2°F) by the end of the century. The difference between these ranges indicates the influence of different emission trajectories on projected climate change. The ranges themselves represent uncertainties associated with how models handle the response of the climate system captured in the climate sensitivity, and how the uptake of carbon dioxide by the ocean and by land ecosystems will be altered by changing temperature and atmospheric greenhouse gas concentrations.

A significant fraction of current greenhouse gas emissions are taken out of the atmosphere by oceanic processes and living plants. The strength of these ocean and land carbon "sinks" is expected to decrease over time, leaving a greater fraction of emissions in the atmosphere to drive further warming. Effects of climate change such as severe drought may also contribute to the switching of land ecosystems from a carbon sink to a carbon source. For example, a recent study found declines between 2000 and 2009 in the amount of global terrestrial ecosystem productivity (carbon uptake), driven by regional drought (Zhao and Running 2010).

Downscaling

Most global climate models are currently limited to representing the earth's surface with "grid cells" of roughly 100–200 km on a side. Climatically important phenomena, such as clouds, occur on much smaller scales, and some areas, including California, have complex topography that cannot be adequately represented at this coarse scale. Projections for precipitation, for instance, are hindered by this lack of spatial detail, that, for example, cannot fully distinguish the moist western side of the Sierras from the much drier eastern side. Indeed, many simulations of the current climate have identified differences in some regions between modeled and observed conditions. Scientists use a variety of tools to address these differences by "downscaling" results

from global models to a regional scale. It is important to note that while generating projections at a finer spatial scale, downscaling does not necessarily lead to greater confidence or accuracy in those projections.

Downscaling techniques generally fall into two categories—dynamic and statistical—and various forms of each method involve different trade-offs in feasibility and model complexity and accuracy. Dynamical downscaling techniques employ a regional climate model running at a finer resolution than global models. These Regional Circulation Models (RCMs) use dynamic and time-varying inputs from atmospheric GCMs as their boundary conditions at the edge of the region of interest. They can be constructed as a one-way nesting within the GCM (inputs travel only from the GCM into the RCM) or with two-way nesting (feedback and inputs move back and forth between the two models). All downscaling techniques must be validated against independent data (not used in constructing the model) and dynamic techniques are usually validated with present-day reanalysis data (Wigley 2005). In essence, the RCM is given a set of boundary conditions in the past, then run up to present day. The model's results are compared to an independent reanalysis of modern-day climate observations. How well the RCM reproduces observed climate trends within its region is of crucial importance for model accuracy. Chief among its disadvantages, dynamic downscaling is computationally intensive, which then limits the feasibility for running a large number of long-term projections under different emission scenarios or using different GCMs as boundary conditions.

Much of the downscaling conducted to date in California has used statistical techniques to downscale GCM projections (Wood et al. 2004, Cayan et al. 2008a, Cayan et al. 2012). The general goal is to develop various transfer functions (statistical relationships) that can be used to link coarse-scale (e.g., 200-km grid boxes) data from GCMs to finer-scale (e.g., 12-km grid boxes) future projections. Statistical downscaling as applied in California has used two approaches. The first, Bias Correction and Spatial Downscaling (BCSD), links observed climate patterns with the patterns represented in global climate model simulations of the same historical period. Distributions of temperature and precipitation for each calendar month are assembled for observed data (e.g., covering 1950–1999). Distributions of the same time period are also assembled from GCM simulations for each grid box, and compared to the observed distributions. Differences in the GCM mean and variance for the historical period at the individual sites are corrected and the same corrections are then applied to future GCM projections. The model is then run for the future and projections can be made on a variable- and site-specific basis. For example, a GCM projection for total precipitation in January of 2050 in a certain grid box is compared to the distribution of total January precipitation values simulated by the GCM for that grid box during the observed period (e.g., 1950–1999). The projected total precipitation will correspond to a certain percentile in that distribution (e.g., the median). In this example, the total precipitation for January in each downscaled area within that particular grid box would be set to the median value of the observed distribution of January precipitation for that downscaled area (Wigley 2005, Cayan et al. 2008a).

The BCSD approach preserves the statistical distribution of temperature and precipitation, but does not yield time series that match the daily progression of weather simulated by the GCM (Maurer et al. 2010, Cayan et al. 2012). The second approach, Bias-Corrected Constructed Analogues (BCCA), preserves the daily evolution of weather that is simulated by the GCM. It combines the initial large-scale bias correction step of BCSD, applied at a daily scale, with a different approach to spatial downscaling. In BCCA, each day of the GCM simulation is compared with a library of observed coarse-scale and corresponding finer-scale daily patterns of the variable of interest (e.g., temperature, precipitation). The 30 coarse-scale observed patterns that are most similar to the simulated day are selected, and a linear combination of these patterns is used to produce a coarse-resolution analogue. The same linear combination is then applied to the corresponding 30 finer-scale patterns to produce the downscaled projection.

Statistical downscaling relies on the fundamental assumption that past spatial and temporal relationships between the climate and area modeled will remain constant in the future. Such relationships may change as the climate changes (Meehl et al. 2000, Alley et al. 2003). Statistical downscaling is much less computationally intensive and more tractable than dynamic downscaling, and is therefore more feasible for long-term projections under multiple emission scenarios. Both statistical and dynamic downscaling methods have been found to be capable of reproducing accurate characterizations of both the mean and variance of known climatic events (Wigley 2005).

CALIFORNIA'S CHANGING CLIMATE

In recent decades, California and the western United States have experienced clear signs of a changing climate. Average temperatures for the state of California have risen about 0.9°C (1.7°F) from 1895 to 2011, sea level has risen more than half a foot, spring snow levels in lower- and mid-elevation mountain areas have dropped and snowpack melting has shifted earlier, flowers are blooming one to two weeks earlier, and warmer temperatures combined with long dry seasons over the last few decades have contributed to more severe wildfires (Cayan et al. 2008a, Cayan et al. 2012, Moser et al. 2012, Garfin et al. 2014). California's climate is expected to change considerably over this century. The severity of these changes depends on the rate at which greenhouse gases accumulate in the atmosphere based on policy and development choices and how the climate responds to the rising concentrations of these gases. Projected changes include further increases in average temperatures, changes in precipitation patterns, rising sea levels, and changes in the frequency and / or severity of extreme events such as heat waves, droughts, and fires.

Temperature

By mid-century, the average annual temperature of California is projected to rise ~1–3°C (~1.8–5.4°F) above the 1961–1990 average, regardless of the emission scenario evaluated (Cayan et al. 2012, Moser et al. 2012). By the end of the century, temperatures in California are projected to rise ~2.5–4.8°C (4.6–8.6°F) under the medium-high A2 scenario. Holding emissions to the lower B1 pathway would still lead to warming of ~1.5–3.3°C (2.8–6°F). The divergence of projections for higher- and lower-emission scenarios by the end of the century demonstrates the long-term benefits of mitigation policy.

The rise in average annual temperature has very different implications for seasonal temperatures. In the past five decades, spring and winter temperatures have increased more than the annual average, while summer temperatures have increased more slowly. In contrast, studies predict that this pattern will reverse in the future, with summer temperatures rising most rapidly (Cayan et al. 2008a, Cayan et al. 2012). By the end of the century, summer temperatures are projected to rise 1.5–6°C (2.7–10.8°F), while winter temperatures are projected to rise 1–4°C (1.8–7.2°F) across the higher- and lower-emission scenarios. Inland temperatures are also projected to rise faster than coastal temperatures, due to the stabilizing influence of the ocean.

Rising summer temperatures are particularly of concern in terms of impacts on many ecosystems and species (see Part II of this volume). Coupled climate-vegetation models suggest that alpine and subalpine forests may be strongly affected by rising summer temperatures (Lenihan et al. 2008). Higher summer temperatures also work in conjunction with drought to increase the risk of climate-related disturbances to ecosystems such as wildfires

and drought-induced forest die-off (Anderegg et al. 2013).

Precipitation

For California (and more generally), different climate models produce projections of precipitation change that vary far more widely than projections of temperature. Precipitation patterns are influenced by regional topography, proximity to geographical features such as mountains or bodies of water, regional temperature differences, and larger-scale atmospheric circulation patterns. Precipitation often varies widely at scales below the grid-box scale of GCMs. While downscaling methods can also be applied to precipitation, uncertainty regarding projections of precipitation remains higher than for temperature.

Projected changes in total annual precipitation in California over this century vary, but the most prevalent pattern suggested is drier conditions (Cayan et al. 2009, Cayan et al. 2012). No model projections suggest a change in the Mediterranean seasonal pattern of precipitation California currently experiences, with most precipitation falling between November and April.

Warming temperatures are projected to decrease the amount of precipitation falling as snow and increase the amount falling as rain. This pattern is expected to continue the already observed trend of decreased spring snow accumulation in the Sierra Nevada (Kapnick and Hall 2009, Cayan et al. 2012), and lead to earlier spring melting of snowpack. By the 2090s, the amount of water stored as snow on April 1 across the Sierra Nevada is projected to be reduced on average to 25% of its historical (1961–1990) level under the medium-high A2 scenario, and be reduced on average to 51% of historical levels under the lower B1 scenario (Cayan et al. 2012).

Sea Level Rise

Warming temperatures contribute to global sea level rise through two main processes. First, a hotter atmosphere causes the ocean to warm, leading to thermal expansion of ocean water. Second, warmer temperatures melt mountain glaciers and the large ice sheets in Greenland and Antarctica, adding water to the ocean that has been stored on land as ice. In California, records suggest an observed rate of sea level rise of 17–20 cm (6.7–7.9 in) per century, which is similar to the global estimate (Cayan et al. 2012).

The magnitude of future sea level rise depends on the level of future warming, uncertainties in the response of the system to warming, and uncertainties in the rates of ice sheet melting. While sea level rise due to thermal expansion and some components of melting ice can be reliably projected (though not without some uncertainty), the rates of melting of the large ice sheets in Greenland and Antarctica are more uncertain—specifically, in terms of quantifying the rate of discharge of ice from these ice sheets into the surrounding oceans, which has accelerated in recent years. The IPCC Fifth Assessment Report projects the global sea level to rise 45–82 cm (1.5–2.7 ft) by the last two decades of the century for the highest RCP scenario discussed above (RCP8.5; somewhat comparable to A2; see Figure 2.1), 32–63 cm (1–2.1 ft) for RCP4.5 (somewhat comparable to B1) and 26–55 cm (0.9–1.8 ft) for the lowest RCP scenario (RCP2.6; a low-emission mitigation scenario), compared with 1986–2005 (Church et al. 2013). Another analysis based on the observed relationship between temperature increase and the rate of sea level rise over the 20th century suggests a larger range (across the A2 and B1 scenarios) of ~30–45 cm (1–1.5 ft) by mid-century and 90–140 cm (3–4.6 ft) by the end of the century above 2000 levels (Cayan et al. 2012).

Extreme Events

While changes in average temperature, precipitation, and sea level will very likely occur gradually, the frequency and intensity of extreme events such as heat waves, droughts, and floods can change substantially with even small average changes in temperature or precipitation. Such extreme events can greatly impact natural systems through population crashes or alterations to the physical environment such as flood erosion (Parmesan et al. 2000). Rising average temperature will lead to more frequent and longer periods of extreme heat, and the potential for temperatures above the range of historical experience. For example, in Sacramento, the frequency of extreme temperatures currently estimated to occur once every 50 years (a severe heat wave by current standards) is projected to increase at least tenfold by the end of the century under the higher A2 scenario and fivefold under the lower B1 scenario (Cayan et al. 2012). Additionally, the length of individual events is expected to increase.

Extremes in coastal sea levels are already increasing in many, but not all, parts of California. For example, Crescent City has experienced a slight decrease in the occurrence of extreme sea levels due to coastal uplift along parts of the northern California coast (Cayan et al. 2008b). The occurrence of extreme sea levels has increased 20-fold since 1915 in San Francisco, however, and 30-fold in La Jolla since 1933. With climate change, increasing frequency and duration of sea levels that exceed current extreme thresholds threaten coastal infrastructure and flood defenses that were designed to protect against the historical range of sea levels and storm intensities (Cayan et al. 2008b, Cayan et al. 2012, Moser et al. 2012).

CONCLUSION

California's diversity in climate regions makes the state vulnerable to changing conditions in many ways. Global models and downscaled regional projections provide a framework for selecting policies to dampen the impacts of future changes in climate. Many such models have been run for California and hold the potential to project future statewide impacts on ecosystems and species (e.g., Loarie et al. 2008, Stralberg et al. 2009) and inform on-the-ground management at individual sites (e.g., Mastran-drea et al. 2010).

Projected trends in temperature and precipitation, as well as sea level rise and changes in the frequency and / or intensity of some types of extreme events, will stress the state's social and natural systems even under lower-emission scenarios. Concurrent and synergistic interactions between climatic changes (i.e., longer heat waves coupled with more intense summer droughts, fires, etc.) and extreme events present perhaps the largest potential to threaten natural systems in the near future.

LITERATURE CITED

Alley, R. B., J. Marotzke, W. D. Nordhaus, J. T. Overpeck, D. M. Peteet, R. A. Pielke Jr, R. T. Pierrehumbert, P. B. Rhines, T. F. Stocker, L. D. Talley et al. 2003. Abrupt climate change. *Science* 299:2005–2010.

Anderegg, W. R. L., J. Kane, and L. D. L. Anderegg. 2013. Consequences of widespread tree mortality triggered by drought and temperature stress. *Nature Climate Change* 3:30–36.

Burkett, V. R., A. G. Suarez, M. Bindi, C. Conde, R. Mukerji, M. J. Prather, A. L. St. Clair, and G. W. Yohe. 2014. Point of departure. In C. B. Field, V. R. Barros, D. J. Dokken, K. J. Mach, M. D. Mastrandrea, T. E. Bilir, M. Chatterjee, K. L. Ebi, Y. O. Estrada, R. C. Genova et al. (eds), *Climate Change 2014: Impacts, Adaptation, and Vulnerability. Part A: Global and Sectoral Aspects. Contribution of Working Group II to the Fifth Assessment Report of the Intergovernmental Panel on Climate Change.* Cambridge University Press, Cambridge, UK.

Cayan, D. R., E. P. Maurer, M. D. Dettinger, M. Tyree, and K. Hayhoe. 2008a. Climate change scenarios for the California region. *Climatic Change* 87:21–43.

Cayan, D. R., P. D. Bromirski, K. Hayhoe, M. Tyree, M. D. Dettinger, and R. E. Flick. 2008b. Climate change projections of sea level extremes along the California coast. *Climatic Change* 87:57–73.

Cayan, D. R., M. Tyree, M. D. Dettinger, H. Hidalgo, T. Das, E. Maurer, P. D. Bromirski, N. Graham, and R. E. Flick. 2009. *Climate Change Scenarios and Sea Level Rise Estimates for California: 2008 Climate Change Scenarios Assessment.* PIER Report CEC-500-2009-014-F, California Energy Commission, Sacramento, CA.

Cayan, D. R., M. Tyree, D. Pierce, and T. Das. 2012. *Climate Change and Sea Level Rise Scenarios for California Vulnerability and Adaptation Assessment.* California Energy Commission Report CEC-500-2012-008, California Energy Commission, Sacramento, CA.

Church, J. A., P. U. Clark, A. Cazenave, J. M. Gregory, S. Jevrejeva, A. Levermann, M. A. Merrifield, G. A. Milne, R. S. Nerem, P. D. Nunn et al. 2013. Sea level change. In T. F. Stocker, D. Qin, G.-K. Plattner, M. Tignor, S. K. Allen, J. Boschung, A. Nauels, Y. Xia, V. Bex, and P. M. Midgley (eds), *Climate Change 2013: The Physical Science Basis. Contribution of Working Group I to the Fifth Assessment Report of the IPCC.* Cambridge University Press, Cambridge, UK.

Collins, M., R. Knutti, J. Arblaster, J.-L. Dufresne, T. Fichefet, P. Friedlingstein, X. Gao, W. J. Gutowski, T. Johns, G. Krinner et al. 2013. Long-term climate change: Projections, commitments and irreversibility. In T. F. Stocker, D. Qin, G.-K. Plattner, M. Tignor, S. K. Allen, J. Boschung, A. Nauels, Y. Xia, V. Bex, and P. M. Midgley (eds), *Climate Change 2013: The Physical Science Basis. Contribution of Working Group I to the Fifth Assessment Report of the IPCC.* Cambridge University Press, Cambridge, UK.

Garfin, G., G. Franco, H. Blanco, A. Comrie, P. Gonzalez, T. Piechota, R. Smyth, and R. Waskom. 2014. Southwest. Chapter 20. In J. M. Melillo, Terese (T. C.) Richmond, and G. W. Yohe (eds), *Climate Change Impacts in the United States: The Third National Climate Assessment.* U. S. Global Change Research Program, Washington, DC. 462–486. doi:10.7930 / J08G8HMN.

IPCC. 2007. Summary for policymakers. In S. Solomon, D. Qin, M. Manning, Z. Chen, M. Marquis, K. B. Averyt, M. Tignor, and H. L. Miller (eds), *Climate Change 2007: The Physical Science Basis. Contribution of Working Group I to the Fourth Assessment Report of the IPCC.* Cambridge University Press, Cambridge, UK.

Kapnick, S. and A. Hall. 2009. *Observed Changes in the Sierra Nevada Snow Pack: Potential Causes and Concerns.* PIER Technical Report CEC-500-2009-016-F, California Climate Change Center, Santa Barbara, CA.

Lenihan, J. M., D. Bachelet, R. P. Neilson, and R. Drapek. 2008. Response of vegetation distribution, ecosystem productivity, and fire to climate change scenarios for California. *Climatic Change* 87:S215–S230.

Loarie, S. R., B. E. Carter, K. Hayhoe, S. McMahon, R. Moe, C. A. Knight, and D. D. Ackerly. 2008. Climate change and the future of California's endemic flora. *PLOS ONE* 3(6):e2502. doi:10.1371/journal.pone.0002502.

Mastrandrea, M. D., N. E. Heller, T. L. Root, and S. H. Schneider. 2010. Bridging the gap: Linking climate-impacts research with adaptation planning and management. *Climatic Change* 100:87–101.

Maurer, E. P., H. G. Hidalgo, T. Das, M. D. Dettinger, and D. R. Cayan. 2010. The utility of daily large-scale climate data in the assessment of climate change impacts on daily streamflow in California. *Hydrology and Earth System Sciences* 14:1125–1138.

McLaughlin, J. F., J. Hellman, C. Boggs, and P. R. Ehrlich. 2002. Climate change hastens population extinctions. *Proceedings of the National Academy of Sciences of the United States of America* 99:670–674.

Meehl, G. A., F. Zwiers, J. Evans, T. Knutson, L. Mearns, and P. Whetton. 2000. Trends in extreme weather and climate events: Issues related to modeling extremes in projections of future climate change. *Bulletin of American Meteorological Society* 81:427–436.

Meehl, G. A., T. F. Stocker, W. D. Collins, P. Friedlingstein, A. T. Gaye, J. M. Gregory, A. Kitoh, R. Knutti, J. M. Murphy, A. Noda et al. 2007. Global climate projections. In S. Solomon, D. Qin, M. Manning, Z. Chen, M. Marquis, K. B. Averyt, M. Tignor, and H. L. Miller (eds), *Climate Change 2007: The Physical Science Basis. Contribution of Working Group I to the Fourth Assessment Report of the IPCC*. Cambridge University Press, Cambridge, UK.

Moser, S., J. Ekstrom, and G. Franco. 2012. *Our Changing Climate 2012: Vulnerability & Adaptation to the Increasing Risks from Climate Change in California*. California Energy Commission Report CEC-500-2012-007, California Energy Commission, Sacramento, CA.

Moss, R. H., J. A. Edmonds, K. A. Hibbard, M. R. Manning, S. K. Rose, D. P. van Vuuren, T. R. Carter, S. Emori, M. Kainuma, T. Kram et al. 2010. The next generation of scenarios for climate change research and assessment. *Nature* 463:747–756.

Myers, N., R. A. Mittermeie, C. G. Mittermeier, G. A. B. de Fonseca, and J. Kent. 2000. Biodiversity hotspots for conservation priorities. *Nature* 403:853–858.

Nakicenovic, N., J. Alcamo, G. Davis, B. de Vries, J. Fenhann, S. Gaffin, K. Gregory, A. Grübler, T. Y. Jung, T. Kram et al. 2000. *IPCC Special Report on Emission Scenarios*. Cambridge University Press, Cambridge, UK.

Parmesan, C., T. L. Root, and M. R. Willig. 2000. Impacts of extreme weather and climate on terrestrial biota. *Bulletin of the American Meteorological Society* 81:443–450.

Stralberg, D., D. Jongsomjit, C. A. Howell, M. A. Snyder, J. D. Alexander, J. A. Wiens, and T. L. Root. 2009. Re-shuffling of species with climate disruption: A no-analog future for California birds? *PLOS ONE* 4:e6825.

Wigley, T. M. L. 2005. Input Needs for Downscaling Climate of Climate data. California Energy Commission Discussion Paper.

Wood, A. W., L. R. Leung, V. Sridhar, and D. P. Lettenmaier. 2004. Hydrologic implications of dynamical and statistical approaches to downscaling climate model outputs. *Climate Change* 62:189–216.

Zhao, M. and S. Running. 2010. Drought-induced reduction in global terrestrial net primary production from 2000 through 2009. *Science* 329:940–943.

CHAPTER 3

Climatic Influences on Ecosystems

William R. L. Anderegg and Terry L. Root

Abstract. Climate drivers act on ecosystems and species in multiple, interconnected and synergistic ways. In this chapter, we describe the primary climatic influences on species and ecosystems around the globe. Within the context of these drivers, the observed trends in species' response to recent climate change are examined, focusing on western North America. We highlight issues and systems particularly pertinent to California, but provide a global framework and context for a mechanistic understanding of many of the major effects of climate on ecosystems and species. Extinction of the most vulnerable species is already occurring. Global policies of mitigation and adaption will likely be necessary to stave off future species extinctions and significant changes in ecosystems.

INTRODUCTION

All across the planet, from the shallow waters of tropical coral reefs to boreal forests and arctic tundra, species are already changing in response to anthropogenic climate change.

Birds and butterflies are moving poleward and higher in elevation with warmer temperatures (Root et al. 2003). Amphibians in tropical cloud forests have gone extinct with warming, drying, and having no habitat into which to expand (Pounds et al. 2006). Alpine mammal populations in North America have been disappearing at the lower edges of their ranges due to temperature stress (Beever et al. 2003). The observed impacts of climate change on ecosystems presage more dramatic impacts in the future, depending on the magnitude, rate, and pathway of future warming (Root et al. 2005, Parmesan 2006, IPCC 2014). Future anthropogenic rapid changes in the climate are expected to have profound consequences for the earth's biota. Our ability to predict changes, facilitate adaptation, and lessen future ecological disasters and extinctions will be aided by understanding of how climate affects species, populations, communities, and ecosystems.

In this chapter, we describe the primary climatic influences on ecosystems around the globe and, within the context of these drivers, the observed trends in species' response to recent climate change, focusing on western North America. We highlight issues and systems particularly pertinent to California, but provide a global framework and context for a mechanistic understanding of many of the major effects of climate on ecosystems and species.

In many cases, non-climate human-caused influences, such as habitat loss and fragmentation, have already compromised the potential resilience and resistance of ecosystems to future change, rendering the systems more vulnerable to shifting climate (Scholes et al. 2014). Various facets of development in California over the past 150 years have threatened large numbers of species (Myers et al. 2000). Habitat loss and fragmentation due to urban growth, agricultural expansion, forestry activities, and water diversion and damming, coupled with competition and other pressures from nonnative species, have placed up to 35% of the state's flora (1700 species) and 15% of the state's vertebrate fauna (111 species) at increased risk of extinction (Stein et al. 2000, Greer 2004, OEHHA 2004, Bunn et al. 2007). These stressors are likely to act in synergistic ways with pressures from climate change, for instance, by restricting the ability of species to move in response to temperature changes (Loarie et al. 2008, Burrows et al. 2014).

CLIMATE INFLUENCES IN ECOSYSTEMS—PRIMARY DRIVERS

Climate has always been a major driver in shaping and changing ecosystems throughout the earth's history (Schneider and Londer 1984). Both the general climatic conditions and patterns of extreme events in an area influence species and ecosystems in many ways. They affect individual fitness, population dynamics, species ranges and densities, community assembly, and ecosystem structure and function, among many others. Variance in climate over regional spatial scales can drive locally adapted physiologies, morphologies, and behav-

ioral patterns. We provide an overview of the broad climatic driving forces that act on ecosystems including temperature, precipitation, atmospheric carbon dioxide, and other secondary drivers (i.e., climate-sensitive disturbance factors or processes) such as fire frequency. Additionally, we provide a brief overview of how species can and have been observed to respond to changing climate. While many of the climate drivers presented in this chapter can affect species, we primarily provide examples of the breadth of possible species' responses to temperature changes because the vast majority of research has focused on temperature and because predictions of future temperature changes have relatively less uncertainty than precipitation, for instance.

Temperature

Temperature plays a critical role in determining the geographical distribution of ecosystems, as well as distribution and abundance of many species. In terrestrial systems, temperature range constraints have been well documented in many plant, insect, bird, and mammal species (Graham 1986, Grace 1987, Root 1988, Lenoir et al. 2008). While suitable resources often exist outside of a species' range, individuals closer to the range boundaries are generally closer to the edge of their physiological tolerances and therefore face higher environmental stresses (Brown et al. 1996, but see Root 1988 and Chapter 9).

Species generally respond to changes in temperature in any combination of four possible manners: (1) Changing the timing of life-history events (e.g., migration, blooming) called phenology; (2) changing their distributions or ranges to more climatically favorable areas; (3) changing aspects of behavior, morphology, reproduction, or genetics; or (4) changing abundance with the potential of undergoing extirpations (local extinctions) or extinctions (Parmesan 2006, Barnosky et al. 2011). Furthermore, because each species has its own individual tolerances and responses to changing climatic conditions, species are affected differentially and exhibit different responses. This can lead to large-scale disruption of ecologic communities and trophic interactions (Stralberg et al. 2009).

Phenology Shifts

Changes in the timing of species' life-history events, or phenology, have been the most comprehensive and widespread of all sets of observations with climate change. Many species of plants have been flowering earlier and growing seasons have lengthened over recent decades. Globally, species that responded with phenology changes to climate change saw an average advance of 5.1 days per decade over the previous 30 or more years (Root et al. 2003). Phenological advances have been largest in early parts of spring (Badeck et al. 2004). A follow-up study of Aldo Leopold's 1930s and 1940s observations on a Wisconsin farm revealed that, while only one species shifted later, 35% of the 55 species (plants and birds) studied showed earlier phenology events in the 1990s, associated with spring temperature increases of 2.8°C (Bradley et al. 1999). Growing season has increased in length in the northern parts of the United States since 1966 (White et al. 1999, Scholes et al. 2014).

Birds provide striking examples of changes in life-history events. Laying date in tree swallows (*Tachycineta bicolor)* has been correlated to May temperature and has gotten earlier by about nine days from 1959 to 1991 over its entire breeding range in the contiguous United States (Dunn and Winkler 1999). In California, of the migratory birds found to have their first arrival date changing significantly ($p < 0.1$), 100% (8 of 8) are arriving earlier over time (MacMynowski et al. 2007). Phenological responses to recent climate change are apparent with many other taxa as well. In 70% of the California butterfly species studied, first-flight date has advanced by an average of 24 days over 31 years, and winter temperature and precipitation explained 85% of the variation in flight date (Forister and Shapiro 2003).

Trophic Interactions and Asynchrony

Differences in life-history strategies and physiological tolerances can lead to species exhibiting vastly different responses when exposed to similar amounts of warming and responses can even differ within species depending on complex interactions of temperature with other drivers such as soil moisture (Wolkovich et al. 2012). In some species, factors other than temperature, such as photoperiod, may trigger phenological events, which can lead to a severe mismatch in trophic interactions among species if a focal species responds to temperature or a temperature-related factor (e.g., snowmelt timing). This asynchrony of life-cycle events can result in, for instance, predators without prey, or herbivorous and egg-laying insects without host plants. While some species have proven able to track such differential shifts and avoid the damages of asynchrony (Charmantier et al. 2008), the majority of cases show that interacting species were out of synchrony due to climate change, leading to decreased fitness in many cases (Visser and Both 2005). For instance, population crashes and local extinctions have been shown to be a direct result of butterfly-host plant asynchrony in especially warm or dry years (Ehrlich et al. 1980, Thomas et al. 1996, McLaughlin et al. 2002).

Distribution Shifts

Species have exhibited changes in distributions, also known as species' ranges, primarily poleward and upward in elevation in response to recent anthropogenic warming. In the Northern Hemisphere, the upper elevational boundaries of species ranges moved an average of 11 m per decade higher and the northern boundaries 17 km per decade north over the past 20–140 years (Chen et al. 2011). The differential spatial movement of various species in a community will likely stress many biotic interactions in the future and could contribute to species occurring together that are not currently together, thereby disrupting current ecological communities and forming "no-analogue" communities (Rosenzweig et al.

2007, Williams and Jackson 2007, Stralberg et al. 2009). Invasion of nonnative species may also be favored by warming temperatures (Pauchard et al. 2009).

Substantial changes in bird, insect, and plant ranges with climate change have been documented in western North America. Bird species in the Sierra Nevada Mountains have moved upward in elevation tracking their temperature envelope (Tingley et al. 2009, Tingley et al. 2012). Large proportions of extirpations along southern range boundaries of Edith's checkerspot butterfly (*Euphydrias editha*) in North America have shifted the mean location of living populations 92 km northward (Parmesan 2006). Dominant vegetation and bird species in Southern California shifted in abundance uphill with warming (Kelly and Goulden 2008, Chapter 8). In Yosemite National Park, alpine mammals have largely moved upward in elevation in response to the 2–3°C increases in temperature (Moritz et al. 2008). Across the western United States, many alpine pika (*Ochotono princeps*) populations at lower elevations of their range have been extirpated between 1930 and present day (Beever et al. 2003).

Marine systems exhibit strong changes in community composition and species abundances with recent warming. In Monterey Bay, California, Sagarin et al. (1999) document a significant increase in southern-range plankton species and decrease of northern-range plankton species between 1931 and 1996, which were accompanied by a 2°C sea surface temperature rise. Similar shifts in fish communities in kelp-forest habitat off the California coast have been found (Holbrook et al. 1997). Using models, Dorman (Chapter 4) found that in years when the coastal upwelling is delayed, the abundance of krill plummets, which in turn results in decreased predator survival (salmon).

Across species, the local velocity of climate change—how fast an organism would need to travel to stay in the same temperature conditions—appears to be the most important pre-

dictor of range shifts, although dispersal ability and available habitat will also likely prove important (Pinsky et al. 2013). Synergistic impacts of land-use that impairs species movement and climate that necessitates movement will likely threaten many less-mobile species. The ability of species to move to more suitable climates, the presence of suitable habitat, and the extent of species current range all have strong conservation implications for the ability of a species to adapt and persist (Loarie et al. 2010). Because extant habitat fragmentation and disturbance may compromise the ability of species to disperse with climate changes, managed relocation may be necessary in some cases to allow species to colonize new locations and avoid extinctions, though this could carry substantial risks as well (McLachlan et al. 2007).

Behavior and Genetics

Species have exhibited several other different types of adaptations, including changes in behavior and morphology, but there are many fewer studies on these. Desert lizard species in Mexico have reduced their foraging activities and remain in the shadows longer than before the temperatures increased. As a consequence, reduced access to food has decreased their reproductive output (Sinervo et al. 2010). The reproduction of sea turtles is being affected because sex determination occurs in the egg and it is temperature dependent (Janzen 1994). The genetics of the pitcher plant mosquito has shown striking changes, which can be linked directly with warming temperatures (Bradshaw and Holzapfel 2006). The average mass and wing length of songbirds in central California has increased over the last 40 years, perhaps due to changing climate variability (Goodman et al. 2011).

Abundance Changes and Extinctions

When species are confronted with a habitat where the conditions are inhospitable, meaning outside their physiological environmental tolerances, and phenological dispersal, behavioral, and genetic mechanisms cannot help the species to sufficiently adapt to the environment, changes in species abundance and, at the extreme, species extirpations or extinctions can result. Species near the poleward edges of continents or near mountaintops will have no habitats into which they can disperse as their environment warms (e.g., Beever et al. 2003). Roughly 20–30% of known species will likely be at increasingly high risk of extinction if global mean temperatures exceed 2–3°C (3.6–5.4°F) above preindustrial temperatures [1.2–2.2°C (2.2–4°F) above current] (Thomas et al. 2004, IPCC 2007a). If the global average temperature goes above 4°C (7.2°F), this would likely commit 40–50% of known species to extinction (IPCC 2007a). Species extinction jeopardizes a large number of ecosystem services such as crop pollination and pest control, and the loss of any species is irreversible.

Prominent examples of climate-induced extirpations and extinctions have been documented in amphibians and reptiles. Wetland desiccation and drought tied to climate change led to local extirpations and rapid population declines in amphibians in Yellowstone National Park (McMenamin et al. 2008). Resurveys of 200 mountaintop sites in Mexico in 2009 reveal that 12% of lizard populations were not found and presumed to have gone extinct due to high ambient air temperatures since 1975 (Sinervo et al. 2010). These probable extinctions have been accurately simulated by modeling the fundamental trade-off these ectotherms (animals with body temperatures determined by ambient temperatures) face in sheltering in shade to avoid hot ambient temperatures versus foraging time. As discussed above, restricted foraging time translates into reduced population growth rates and survival. These models suggest that global extinction of lizards reached 4% by 2009 and will likely reach 16% by 2050 and 30% of lizard populations by 2080 (Sinervo et al. 2010).

Effects on California

Mediterranean-type ecosystems, which cover most of California, are likely to be strongly influenced by temperature increases (Sala et al. 2000). With only a 2°C (3.6°F) increase, models suggest a substantial reduction of alpine and subalpine habitat, a shift from conifer/evergreen to mixed-evergreen forest, expansion of desert and grassland at the expense of shrubland, and substantial decrease of many endemic species' range (Hayhoe et al. 2004, Loarie et al. 2008).

Precipitation

Along with temperature, precipitation shapes many ecosystems. Precipitation often determines the type, structure, density, and diversity of vegetation and ecosystem boundaries (Holdridge 1947, Whittaker 1975). For example, in the tropics rainfall is one of the best predictors of species diversity and biomass in a site (Gentry 1982). Precipitation also plays a vital role in replenishing groundwater, soil moisture, and sustaining lakes, rivers, and other freshwater systems, although both the quantity and timing of precipitation matter greatly, especially for freshwater systems (Trenberth et al. 2007). A critical source of water for many plants around the world is fog (Bruijnzeel et al. 2011). Along with fog, overcast skies are known to significantly reduce drought stress in plants (Fischer et al. 2009). While precipitation is an important driver in ecosystems, changes in precipitation with climate change are harder to predict and have greater spatial heterogeneity than changes in temperature. These two elements make it difficult to generate and test hypotheses regarding climate change alterations in precipitation patterns.

In terrestrial ecosystems, precipitation often limits plant species' ranges and abundances. The response to changes in precipitation is largely system-dependent. Drought plays an important role in forest dynamics in many temperate forests, including the Argentinian Andes, Rocky Mountains, North American woodlands, and the Eastern Mediterranean (Rosenzweig et al. 2007). Severe droughts can trigger widespread tree mortality (Anderegg et al. 2013) and increase vulnerability to pest, fire, and other disturbances (e.g., Breshears et al. 2005). Conversely, increases in precipitation, especially in arid systems, can increase vulnerability to plant diseases, fire risk through increased biomass yielding higher fuel loads, and alien species invasion (Dukes and Mooney 1999, Westerling et al. 2006).

Effects on California

Future changes in precipitation in California are unclear. Models provide mixed projections of future trends with a majority indicating slight decreases in precipitation in southern parts of the state and slight increases in northern parts (IPCC 2013). The most dramatic effects of precipitation changes will likely arise as temperature and rainfall interact, producing extreme events (Box 3.1) such as drought and increased fire risk, or high-volume storm events that scour streams. Even with little change in precipitation volume, changes in precipitation form (rain vs. snow), timing, and variability can still greatly influence ecosystems. For instance, increased precipitation variability in central coastal California has been linked to the local extinction of the Bay checkerspot butterfly (*Euphydryas editha bayensis*) (McLaughlin et al. 2002). With an increasing number and severity of droughts, cloud patterns along the coast will likely change (Still et al. 1999), putting dozens of drought-sensitive plants, many of which are rare, within the fog-dependent ecosystems in jeopardy (Fischer et al. 2009).

Atmospheric Carbon Dioxide

Increased atmospheric carbon dioxide (CO_2) concentrations can directly affect global vegetation distribution, structure, and productivity (Rosenzweig et al. 2007). Carbon, taken up from the atmosphere as CO_2 in photosynthesis through the Calvin Cycle, forms the backbone of organic compounds, especially sugars, synthesized by plants that fuel survival and growth. Opening pores on leaves known as stomata to take up CO_2, however, has an inherent trade-off

---

BOX 3.1 · Extreme events

Extreme weather events, such as heat waves, droughts, floods, wildfires, and severe storms, can be very damaging to ecosystems and are likely to be some of the most profound impacts of anthropogenic climate change (IPCC 2011). Some extreme events, such as heat waves, have already changed in frequency and severity in recent decades and are expected to change much more in years to come, even with relatively small changes in the mean global temperature (Meehl et al. 2007, Trenberth et al. 2007, IPCC 2011).

Changes in extreme events are likely to cause large changes in terrestrial ecosystems. Extremes in temperature or precipitation can act on individuals by influencing fitness, species populations by changes in abundance, phenology or reproduction, and ecosystems through major structural change such as hurricanes, floods, or fire. Past extreme events in temperature and drought have been linked to population crashes or local extinction in several butterfly species (Parmesan et al. 2000). Mass die-offs of multiple tree species, including piñon pine (*Pinus edulis*) and trembling aspen (*Populus tremuloides*), across the southwestern United States in recent years have been tied to severe droughts in the region (Breshears et al. 2005, Anderegg et al. 2012b). These widespread forest mortality events were triggered by extreme "climate change-type drought," in which severe drought is exacerbated by higher summertime temperatures. This indicates that even if drought intensity or severity does not increase, these systems will be vulnerable due to the temperature increases alone (Adams et al. 2009, Williams et al. 2013). Widespread forest mortality events triggered by extreme climate events can alter ecosystem structure, function, and severely impact biodiversity (Allen et al. 2010, Anderegg et al. 2012a, Anderegg et al. 2013).

---

of losing water from the plant. Increased atmospheric CO_2 should, in theory, enhance tree water-use efficiency—how much carbon a tree gets for a certain amount of water loss—and thereby increase growth, net primary productivity, and carbon storage, all of which is termed CO_2 fertilization. In reality, however, studies have found increases in water-use efficiency in forests globally (Keenan et al. 2013) but more cryptic and less consistent increases in growth or carbon storage (e.g., Shaw et al. 2002, Morgan et al. 2004, Peñuelas et al. 2011). Thus, the magnitude and sign of CO_2 enrichment will likely vary vastly among ecosystems. The largest CO_2-fertilization influence has been seen in grassland systems and semiarid ecosystems and is likely due to the benefit of increased water savings from lower stomatal water loss (Morgan et al. 2004, Donohue et al. 2013). Other constraints such as nutrient limitation, water or temperature stress, and changing community assembly could limit carbon fertilization in terrestrial systems (Oren et al. 2001). Observed decreases in terrestrial net primary productivity between 2000 and 2009 due to regional droughts emphasize how other climate stresses may limit or even overwhelm the expected CO_2-fertilization effect (Zhao and Running 2010).

In addition to the effect of increased atmospheric CO_2 on terrestrial plants, high atmospheric CO_2 concentrations have and will significantly alter ocean chemistry through ocean acidification. Since the industrial revolution, dissolved CO_2 from human emissions has already significantly affected ocean chemistry, leading to a decrease of mean surface-ocean pH by 0.1 pH units. Lower oceanic pH decreases the concentration of dissolved calcium carbonate minerals, primarily aragonite and calcite, which holds dire consequences for marine ecosystems (IPCC 2014). As carbonate concentrations drop, coral reef-building organisms will have more difficulty calcifying and creating the physical structures that define coral reefs. Several modeling studies suggest that at atmospheric CO_2 concentrations of 480 parts per million (ppm), erosion could outstrip calcium carbonate buildup, and coral reefs will slowly dissolve (Orr et al. 2005). In addition to lowered calcification rates from decreasing pH, direct effects of acidity and temperature

severely threaten coral reefs and other marine ecosystems (Doney et al. 2009).

Sea-level rise

Global sea levels rise with warmer temperatures due to the thermal expansion of the ocean and melting of continental ice in places like Greenland and the Antarctic. Higher water levels combined with increased storm severity can lead to many more damaging floods and extreme storm impacts in coastal regions. Increasing the number of major storms can more permanently damage coastal ecosystems including coral reefs by reducing recovery time and resilience (Forbes et al. 2004). In addition, rising water tables, levels, and salinity can reduce wetland, estuary, and coastal habitat. This further reduces habitat for many species in systems already pressured by human development (Field et al. 2007).

Effects on California

A 140-year log in San Francisco suggests that major winter storms have become more frequent since 1950 (Bromirski et al. 2003). Primary effects of rising sea level include bluff erosion, loss of beaches and wetlands, increased incidence of severe floods, and salinity movement inward in deltas, such as the Sacramento–San Joaquin delta area, which can severely alter vegetation and species habitat (Lund et al. 2007). High sea-level rise is expected to lead to large losses in tidal marsh in the San Francisco Bay, though all sea-level rise scenarios suggest a decrease of high marsh habitat and an increase in low marsh and mudflat habitats (Stralberg et al. 2011).

Snowpack and runoff

Changes in precipitation patterns and winter–spring temperatures may lead to decreased snowpack, runoff, or changes in runoff timing in many parts of the world. Above average warming on mountain peaks will likely lead to earlier and shorter runoff periods, rapid water release and possible downstream floods, and water shortages during summer growing seasons (Mote et al. 2005, Fischlin et al. 2007). This change in hydrological-cycle timing can affect downstream vegetation, as well as agriculture and human water supplies, and cause increases in flooding. In addition to reducing springtime and summer water availability, earlier snowmelt and runoff in alpine systems exposes plants and animals to frost, increases susceptibility to fire, alter phenology, and can disrupt animal movements, potentially raising wildlife mortality (Inouye et al. 2000, Keller et al 2005, Westerling et al. 2006).

Effects on California

Decreases in mountain snowpack have already been observed in many of the mountains in Northern California (Mote et al. 2005). Mountain snowpack and runoff timing are predicted to change by many climate models and these predicted changes will likely add more stress to the state's social and natural systems that depend on downstream water. Hayhoe et al. (2004, p.12426) suggest that, "Declining Sierra Nevada snowpack, earlier runoff, and reduced spring and summer streamflows will likely affect surface water supplies and shift reliance to groundwater resources, already overdrafted in many agricultural areas in California." Winter floods are more probable with increased runoff that have the potential to damage riparian and wetland systems. Furthermore, earlier snowmelt in the Sierras has been found to lead to increased vulnerability to wildfires in mountain forests (Westerling et al. 2006).

Fire

While it is often assumed that temperature and precipitation determine the boundaries of the world's ecosystems, fire is also a strong driving force in several biomes (Bond et al. 2005). Natural fire cycles have kept some fire-sensitive ecosystems far from the physiological con-

straints of only temperature and precipitation. For instance, many fire-prone grassland, shrubland, and savannah systems transition to forest ecosystems in fire-exclusion experiments (Bond and van Wilgen 1996). Conversely, anthropogenic fires induced in island ecosystems have successfully transformed forested areas into grasslands (D'Antonio and Vitousek 1992, Ogden et al. 1998).

Climate change has already and will likely continue to intensify fire regimes in many different types of ecosystems, driving large changes in vegetation and ecosystems (Gillett et al. 2004, Westerling et al. 2006). Fire alters community structure by favoring species that survive fire or spread with fire (Bond and Keeley 2005). Intensified fire regimes across boreal forests, especially in North America, may drive changes in vegetation structure and composition, as witnessed in the shift from *Picea*-dominant to *Pinus*-dominant forests in Eastern Canada (Lavoie and Sirois 1998).

Effects on California

Frequency and intensity of wildfires with climate change is projected to increase in many areas of California (Westerling et al. 2006). Alpine forests and Mediterranean-type vegetation, two major types of Californian vegetation, are commonly identified as vulnerable systems to wildfires (Fischlin et al. 2007). Increasing summer wildfires could lead to changes in dominant vegetation types or changed community structure. Land management, such as grazing and fire suppression, will also interact strongly with wildfire probability and, in places like Sierra Nevada mixed conifer forest with a natural cycle of small and non-crown fire regime, increase the likelihood of massive crown fires (Westerling et al. 2006).

CONCLUSION

Climate drivers act on ecosystems and species in multiple, interconnected, and synergistic ways. Changes in temperature, precipitation, carbon dioxide concentrations, extreme events, sea-level rise, snowpack, and fire regimes all can ripple through ecosystems and cause profound changes. These changes will likely interact synergistically with other human-caused environmental changes such as habitat fragmentation. California's boreal forest and Mediterranean habitats are vulnerable to a series of climate-induced changes and much of the state's unique flora is at risk of large range contractions.

Species across the globe have already responded to levels of recent anthropogenic warming. They provide a coherent signal, independent climate attribution test, and indication of future trends and possible asynchronies that could drive large numbers of species to extinction. Already, extinction has hit the most vulnerable sets of species, range-restricted amphibians, and global policies of mitigation and adaption will likely be necessary to stave off future species extinctions and significant changes in ecosystem.

LITERATURE CITED

Adams, H. D., M. Guardiola-Claramonte, G. A. Barron-Gafford, J. C. Villegas, D. D. Breshears, C. B. Zou, P. A. Troch, and T. E. Huxman. 2009. Temperature sensitivity of drought-induced tree mortality portends increased regional die-off under global-change-type drought. *Proceedings of the National Academy of Sciences of the United States of America* 106:7063–7066.

Allen, C. D., A. K. Macalady, H. Chenchouni, D. Bachelet, N. McDowell, M. Vennetier, P. Gonzales, T. Hogg, A. Rigling, and D. D. Breshears. 2010. Climate-induced forest mortality: A global overview of emerging risks. *Forest Ecology and Management* 259:660–684.

Anderegg, W. R., L. D. Anderegg, C. Sherman, and D. S. Karp. 2012a. Effects of widespread drought induced aspen mortality on understory plants. *Conservation Biology* 26:1082–1090.

Anderegg, W. R., J. A. Berry, D. D. Smith, J. S. Sperry, L. D. Anderegg, and C. B. Field. 2012b. The roles of hydraulic and carbon stress in a widespread climate-induced forest die-off. *Proceedings of the National Academy of Sciences the United States of America* 109:233–237.

Anderegg, W. R., J. M. Kane, and L. D. Anderegg. 2013. Consequences of widespread tree mortality triggered by drought and temperature stress. *Nature Climate Change* 3:30–36.

Badeck, F. W., A. Bondeau, K. Bottcher, D. Doktor, W. Lucht, J. Schaber, and S. Sitch. 2004. Responses of spring phenology to climate change. *New Phytologist* 162:295–309.

Barnosky, A. D., N. Matzke, S. Tomiya, G. O. U. Wogan, B. Swartz, T. B. Quental, C. Marshall, J. L. McGuire, E. L. Lindsey, K. C. Maguire et al. 2011. Has the Earth's sixth mass extinction already arrived? *Nature* 471:51–57.

Beever, E. A., P. F. Brussard, and J. Berger. 2003. Patterns of apparent extirpation among isolated populations of pikas (*Ochotona princeps*) in the Great Basin. *Journal of Mammalogy* 84:37–54.

Bond, W. J. and J. E. Keeley. 2005. Fire as a global 'herbivore': The ecology and evolution of flammable ecosystems. *Trends in Ecology and Evolution* 20:387–394.

Bond, W. J. and B. W. Van Wilgen. 1996. *Fire and Plants. Population and Community Biology Series* 14. Chapman & Hall, London, UK.

Bond, W. J., F. I. Woodward, and G. F. Midgley. 2005. The global distribution of ecosystems in a world without fire. *New Phytologist* 165:525–537.

Bradley, N. L., A. C. Leopold, J. Ross, and H. Wellington. 1999. Phenological changes reflect climate change in Wisconsin. *Proceedings of the National Academy of Sciences of the United States of America* 96:9701–9704.

Bradshaw, W. E. and C. M. Holzapfel. 2006. Climate change: Evolutionary response to rapid climate change. *Science* 312:1477–1478.

Breshears, D. D., N. S. Cobb, P. M. Rich, K. P. Price, C. D. Allen, R. G. Balice, W. H. Romme, J. H. Kastens, M. L. Floyd, J. Belnap et al. 2005. Regional vegetation die-off in response to global-change-type drought. *Proceedings of the National Academy of Sciences of the United States of America* 102:15144–15148.

Bromirski, P. D., R. E. Flick, and D. R. Cayan. 2003. Storminess variability along the California coast: 1958–2000. *Journal of Climate* 16:982–993.

Brown, J. H., G. C. Stevens, and D. M. Kaufman. 1996. The geographic range: Size, shape, boundaries and internal structure. *Annual Review of Ecology and Systematics* 27:597–623.

Bruijnzeel, L. A., M. Mulligan, and F. N. Scatena. 2011. Hydrometeorology of tropical montane cloud forests. *Hydrological Processes* 25(3): 465–498.

Bunn, D., A. Mummert, M. Hoshovsky, K. Gilardi, and S. Shanks. 2007. *California Wildlife: Conservation Challenges (California's Wildlife Action Plan)*. UC Davis Wildlife Health Center for California Department of Fish and Game, Sacramento, CA.

Burrows, M.T., D.S. Schoeman, A.J. Richardson, J.G. Molinos, A. Hoffmann, L.B. Buckley, and E.S. Poloczanska. 2014. Geographical limits to species-range shifts are suggested by climate velocity. *Nature* 507:492–495.

Charmantier, A., R.H. McCleery, L.R. Cole, C. Perrins, L.E. Kruuk, and B.C. Sheldon. 2008. Adaptive phenotypic plasticity in response to climate change in a wild bird population. *Science* 320:800–803.

Chen, I.C., J.K. Hill, R. Ohlemüller, D.B. Roy, and C.D. Thomas. 2011. Rapid range shifts of species associated with high levels of climate warming. *Science* 333:1024–1026.

D'Antonio, C.M. and P.M. Vitousek. 1992. Biological invasions by exotic grasses, the grass/fire cycle, and global change. *Annual Review of Ecology and Systematics* 23:63–87.

Doney, S.C., V.J. Fabry, R.A. Feely, and J.A. Kleypas. 2009. Ocean acidification: The other CO_2 problem. *Annual Review of Marine Science* 1:169–192.

Donohue, R.J., M.L. Roderick, T.R. McVicar, and G.D. Farquhar. 2013. Impact of CO_2 fertilization on maximum foliage cover across the globe's warm, arid environments. *Geophysical Research Letters* 40:3031–3035.

Dukes, J.S. and H.A. Mooney. 1999. Does global change increase the success of biological invaders? *Trends in Ecology and Evolution* 14:135–139.

Dunn, P.O. and D.W. Winkler. 1999. Climate change has affected the breeding date of tree swallows throughout North America. *Proceedings of the Royal Society of London Series B* 266:2487–2490.

Ehrlich, P.R., D.D. Murphy, M.C. Singer, C.B. Sherwood, R.R. White, and I.L. Brown. 1980. Extinction, reduction, stability and increase: The responses of checkerspot butterfly populations to the California drought. *Oecologia* 46:101–105.

Field, C.B., L.D. Mortsch, M. Brklacich, D.L. Forbes, P. Kovacs, J.A. Patz, S.W. Running, and M.J. Scott. 2007. North America. In M.L. Parry, O.F. Canziani, J.P. Palutikof, P.J. van der Linden, and C.E. Hanson (eds), *Climate Change 2007: Impacts, Adaptation and Vulnerability. Contribution of Working Group II to the Fourth Assessment Report of the IPCC.* Cambridge University Press, Cambridge, UK. 617–652.

Fischer, D.T., C.J. Still, and A.P. Williams. 2009. Significance of summer fog and overcast for drought stress and ecological functioning of coastal California endemic plant species. *Journal of Biogeography* 36:783–799.

Fischlin, A., G.F. Midgley, P. Jeff, L. Rik, G. Brij, T. Carol, R. Mark, D. Pauline, T. Juan, and V. Andrei. 2007. Ecosystems, their properties, goods, and services. In M.L. Parry, O.F. Canziani, J.P. Palutikof, P.J. van der Linden, and C.E. Hanson (eds), *Climate Change 2007: Impacts, Adaptation and Vulnerability. Contribution of Working Group II to the Fourth Assessment Report of the IPCC.* Cambridge University Press, Cambridge, UK. 211–272.

Forbes, D.L., G.S. Parkes, G.K. Manson, and L.A. Ketch. 2004. Storms and shoreline retreat in the southern Gulf of St. Lawrence. *Marine Geology* 210:169–204.

Forister, M.L. and A.M. Shapiro. 2003. Climatic trends and advancing spring flight of butterflies in lowland California. *Global Change Biology* 9:1130–1135.

Gentry, A.H. 1982. Patterns of neotropical plant species diversity. *Evolutionary Biology* 15:1–85.

Gillett, N,P, A.J. Weaver, F.W. Zwiers, and M.D. Flannigan. 2004. Detecting the effect of climate change on Canadian forest fires. *Geophysical Research Letters* 31:18–24.

Goodman, R.E., G. LeBuhn, N.E. Seavy, T. Gardali, and J.D. Bluso-Demers. 2011. Avian body size changes and climate change: Warming or increasing variability? *Global Change Biology* 18:63–73. doi: 10.1111/j.1365-2486.2011 .02538.x.

Grace, J. 1987. Climatic tolerance and the distribution of plants. *New Phytologist*, 106:113–130.

Graham, R.W. 1986. Responses of mammalian communities to environmental changes during the late Quaternary. In J. Diamond and T.J. Case (eds), *Community Ecology.* Harper and Row, New York. 300–313.

Greer, K.A. 2004. Habitat conservation planning in San Diego County, California: Lessons learned after five years of implementation. *Environmental Practice* 6:230–239.

Hayhoe, K., D. Cayan, C.B. Field, P.C. Frumhoff, E.P. Maurer, N.L. Miller, S.C. Moser, S.H. Schneider, K.N. Cahill, E.E. Cleland et al. 2004. Emissions pathways, climate change, and impacts on California. *Proceedings of the National Academy of Sciences of the United States of America* 101:12422–12427.

Holbrook, S.J., R.J. Schmitt, and J.S. Stephens Jr. 1997. Changes in an assemblage of temperate reef fishes associated with a climatic shift. *Ecological Applications* 7:1299–12310.

Holdridge, L. R. 1947. Determination of world plant formations from simple climatic data. *Science* 105:367–368.

Inouye, D. W., B. Barr, K. B. Armitage, and B. D. Inouye. 2000. Climate change is affecting altitudinal migrants and hibernating species. *Proceedings of the National Academy of Sciences of the United States of America* 97:1630–1633.

IPCC. 2007a. Summary for policymakers. In S. Solomon, D. Qin, M. Manning, Z. Chen, M. Marquis, K. B. Averyt, M. Tignor, and H. L. Miller (eds), *Climate Change 2007: The Physical Science Basis. Contribution of Working Group I to the Fourth Assessment Report of the Intergovernmental Panel on Climate Change.* Cambridge University Press, Cambridge, UK.

IPCC. 2011. Summary for policymakers. In C. B. Field (ed.), *Managing the Risks of Extreme Events and Disasters to Advance Climate Change Adaptation: Special Report of the Intergovernmental Panel on Climate Change.* Cambridge University Press, Cambridge, UK.

IPCC. 2013. Annex I: Atlas of global and regional climate projections. In T. Stocker (ed.), *Climate Change 2013: The Physical Science Basis. Contribution of Working Group I to the Fifth Assessment Report of the IPCC.* Cambridge University Press, Cambridge, UK.

IPCC. 2014. Summary for policymakers. In C. B. Field (ed.), *Climate Change 2014: Impacts, Adaptation and Vulnerability. Contribution of Working Group II to the Fourth Assessment Report of the IPCC.* Cambridge University Press, Cambridge, UK.

Janzen, F. J. 1994. Climate change and temperature-dependent sex determination in reptiles. *Proceedings of the National Academy of Sciences of the United States of America* 91:7487–7490.

Keenan, T. F., D. Y. Hollinger, G. Bohrer, D. Dragoni, J. W. Munger, H. P. Schmid, and A. D. Richardson. 2013. Increase in forest water-use efficiency as atmospheric carbon dioxide concentrations rise. *Nature* 499:324–327.

Keller, F., S. Goyette, and M. Beniston. 2005. Sensitivity analysis of snow cover to climate change scenarios and their impact on plant habitats in alpine terrain. *Climatic Change* 72:299–319.

Kelly, A. E. and M. L. Goulden. 2008. Rapid shifts in plant distribution with recent climate change. *Proceedings of the National Academy of Sciences of the United States of America* 105:11823–11826.

Lavoie, L. and L. Sirois, 1998. Vegetation changes caused by recent fires in the northern boreal forest of eastern Canada. *Journal of Vegetation Science* 9:483–492.

Lenoir, J., J. C. Gegout, P. A. Marquet, P. De Ruffray, and H. Brisse. 2008. A significant upward shift in plant species optimum elevation during the 20th century. *Science* 320:1768–1771.

Loarie, S. R., B. E. Carter, K. Hayhoe, S. McMahon, R. Moe, C. A. Knight, and D. D. Ackerly. 2008. Climate change and the future of California's endemic flora. *PLOS ONE* 3(6):e2502. doi:10.1371/journal.pone.0002502.

Loarie, S. R., P. B. Duffy, H. Hamilton, G. P. Asner, C. B. Field, and D. D. Ackerly. 2010. The velocity of climate change. *Nature* 462:1052–1055.

Lund, J., E. Hanak, W. Fleenor, R. Howitt, J. Mount, and P. Moyle. 2007. *Envisioning Futures for the Sacramento-San Joaquin Delta.* Public Policy Institute of California, San Francisco, CA.

Macmynowski, D. P., T. L. Root, G. Ballard, and G. R. Geupel. 2007. Changes in spring arrival of Nearctic-Neotropical migrants attributed to multiscalar climate. *Global Change Biology* 13:2239–2251.

McLachlan, J. S., J. J. Hellmann, and M. W. Schwartz. 2007. A framework for debate of assisted migration in an era of climate change. *Conservation Biology* 21:297–302.

McLaughlin, J. F., J. J. Hellmann, C. L. Boggs, and P. R. Ehrlich. 2002. Climate change hastens population extinctions. *Proceedings of the National Academy of Sciences of the United States of America* 99:6070–6074.

McMenamin, S. K., E. A. Hadly, and C. K. Wright. 2008. Climate change and wetland desiccation cause amphibian decline in Yellowstone National Park. *Proceedings of the National Academy of Sciences the United States of America* 105:16988–16993.

Meehl, G. A., T. F. Stocker, W. D. Collins, P. Friedlingstein, A. T. Gaye, J. M. Gregory, A. Kitoh, R. Knutti, J. M. Murphy, A. Noda et al. 2007. Global climate projections. In S. Solomon, D. Qin, M. Manning, Z. Chen, M. Marquis, K. B. Averyt, M. Tignor, H. L. Miller (eds), *Climate Change 2007: The Scientific Basis. Contribution of Working Group I to the Fourth Assessment Report of the IPCC.* Cambridge University Press, Cambridge, UK. 747–845.

Morgan, J. A., D. E. Pataki, C. Korner, H. Clark, S. J. Del Grosso, J. M. Grünzweig, A. K. Knapp, A. R. Mosier, P. C. D. Newton, P. A. Niklaus et al. 2004. Water relations in grassland and desert ecosystems exposed to elevated atmospheric CO_2. *Oecologia* 140:11–25.

Moritz, C., J. L. Patton, C. J. Conroy, J. L. Patton, C. J. Conroy, J. L. Parra, G. C. White, and S. R. Beissinger. 2008. Impact of a century of climate change on small-mammal communities in Yosemite National Park, USA. *Science* 322:261–264.

Mote, P. W., A. F. Hamlet, M. P. Clark, and D. P. Lettenmaier. 2005. Declining mountain snowpack in western North America. *Bulletin of the American Meteorological Society* 86:39–49.

Myers, N., R. A. Mittermeier, C. G. Mittermeier, G. A. B. de Fonseca, and J. Kent. 2000. Biodiversity hotspots for conservation priorities. *Nature* 403:853–858.

OEHHA (Office of Environmental Health Hazard Assessment). 2004. *Environmental Protection Indicators for California.* California Environmental Protection Agency, Department of Health Services, California Resources Agency, Sacramento, CA.

Ogden, J., L. Basher, and M. McGlone. 1998. Fire, forest regeneration and links with early human habitation: Evidence from New Zealand. *Annals of Botany* 81:687–696.

Oren, R., D. S. Ellsworth, K. H. Johnsen, N. Phillips, B. E. Ewers, C. Maier, K. V. R. Schafer, H. McCarthy, G. Hendrey, S. G. McNulty, and G. G. Katul. 2001. Soil fertility limits carbon sequestration by forest ecosystems in a CO_2-enriched atmosphere. *Nature* 411:469–472.

Orr, J. C., V. J. Fabry, O. Aumont, L. Bopp, S. C. Doney, R. A. Feely, A., Gnanadesikan, N. Gruber, A. Ishida, F. Joos, R. M. Key et al. 2005. Anthropogenic ocean acidification over the twenty-first century and its impact on calcifying organisms. *Nature* 437:681–686.

Parmesan, C. 2006. Ecological and evolutionary responses to recent climate change. *Annual Review in Ecology and Evolution* 37:637–669.

Parmesan, C., T. L. Root, and M. R. Willig. 2000. Impacts of extreme weather and climate on terrestrial biota. *Bulletin of the American Meteorological Society* 81:443–450.

Pauchard, A., C. Kueffer, H. Dietz, C. C. Daehler, J. Alexander, P. J. Edwards, J. R. Arevalo, L. A. Cavieres, A. Guisan, S. Haider et al. 2009. Ain't no mountain high enough: Plant invasions reaching new elevations. *Frontiers in Ecology and the Environment* 7:479–486.

Peñuelas, J., J. G. Canadell, and R. Ogaya. 2011. Increased water-use efficiency during the 20th century did not translate into enhanced tree growth. *Global Ecology and Biogeography* 20:597–608.

Pinsky, M. L., B. Worm, M. J. Fogarty, J. L. Sarmiento, and S. A. Levin. 2013. Marine taxa track local climate velocities. *Science* 341:1239–1242.

Pounds, J. A., M. R. Bustamente, L. A. Coloma, J. A. Consuegra, M. P. Fogden, P. N. Foster, E. La Marca, K. L. Masters, A. Merino-Viteri, R. Puschendorf et al. 2006. Widespread amphibian extinctions from epidemic disease driven by global warming. *Nature* 439:161–167.

Root, T. L. 1988. Energy constraints on avian distributions and abundances. *Ecology* 69:330–339.

Root, T. L., J. T. Price, K. R. Hall, S. H. Schneider, C. Rosenzweig, and J. A. Pounds. 2003. Fingerprints of global warming on wild animals and plants. *Nature* 421:57–60.

Root, T.L., D.P. MacMynowski, M.D. Mastrandrea, and S.H. Schneider. 2005. Human-modified temperatures induce species changes: Joint attribution. *Proceedings of the National Academy of Science of the United States of America* 102:7465–7469.

Rosenzweig, C., G. Casassa, D. J. Karoly, A. Imeson, C. Liu, A. Menzel, S. Rawlins, T. L. Root, B. Seguin, and P. Tryjanowski. 2007. Assessment of observed changes and responses in natural and managed systems. In M. L. Parry, O. F. Canziani, J. P. Palutikof, P. J. van der Linden, and C. E. Hanson (eds), *Climate Change 2007: Impacts, Adaptation and Vulnerability. Contribution of Working Group II to the Fourth Assessment Report of the Intergovernmental Panel on Climate Change.* Cambridge University Press, Cambridge, UK. 79–131.

Sagarin, R. D., J. P. Barry, S. E. Gilman, and C. H. Baxter. 1999. Climate-related change in an intertidal community over short and long time scales. *Ecological Monographs* 69:465–490.

Sala, O. E., I. F. S. Chapin, J. J. Armesto, E. Berlow, J. Bloomfield, R. Dirzo, E. Huber Sanwald, L. F. Huenneke, R. B. Jackson, A. Kinzig et al. 2000. Global biodiversity scenarios for the year 2100. *Science* 287:1770–1774.

Schneider, S. H. and R. Londer. 1984. *The Coevolution of Climate and Life.* Sierra Club Books, San Francisco, CA.

Scholes, R., J. Settele, R. Betts, S. Bunn, P., Leadley, D. Nepstad, J. Overpeck, and M. A. Taboada. 2014. Terrestrial and inland water systems. In C. B. Field and V. Barros (eds), *Climate Change 2014: Impacts, Adaptation and Vulnerability. Contribution of Working Group II to the Fifth Assessment Report of the Intergovernmental Panel on Climate Change.* Cambridge University Press, Cambridge, UK. 79–131.

Shaw, M. R., E. S. Zavaleta, N. R. Chiariello, E. E. Cleland, H. A. Mooney, and C. B. Field. 2002.

Grassland responses to global environmental changes suppressed by elevated CO_2. *Science* 298:1987–1990.

Sinervo, B., F. Mendez-de-la-Cruz, D. B. Miles, B. Heulin, E. Bastiaans, Maricela V.-S. Cruz, R. Lara-Resendiz, N. Martínez-Méndez, M. L. Calderón-Espinosa, R. N. Meza-Lázaro et al. 2010. Erosion of lizard diversity by climate change and altered thermal niches. *Science* 328:894–899.

Still, C. J., P. N. Foster, and S. H. Schneider. 1999. Simulating the effects of climate change on tropical montane cloud forests. *Nature* 398:608–610.

Stein, B. A., L. S. Kutner, G. A. Hammerson, L. L. Master, and L. E. Morse. 2000. State of the states: Geographic patterns of diversity, rarity, and endemism. In B. A. Stein and L. S. Kutner (eds), *Precious Heritage: The Status of Biodiversity in the United States*. Oxford University Press, Cambridge, UK.

Stralberg, D., D. Jongsomjit, C. A. Howell, M. A. Snyder, J. D. Alexander, J. A. Wiens, and T. L. Root. 2009. Re-shuffling of species with climate disruption: A no-analog future for California birds. *PLOS ONE* 4:e6825.

Stralberg, D., B. Matthew, J. C. Callaway, J. K. Wood, L. M. Schile, D. Jongsomjit, M. Kelly, V. T. Parker, and S. Crooks. 2011. Evaluating tidal marsh sustainability in the face of sea-level rise: A hybrid modeling approach applied to San Francisco Bay. *PLOS ONE* 6(11):e27388.

Thomas, C. D., A. Cameron, R. E. Green, M. Bakkenes, L. J. Beaumont, Y. C. Collingham, B. F. N. Erasmus, M. F. De Siqueira, A. Grainger, and L. Hannah. 2004. Extinction risk from climate change. *Nature* 427:145–148.

Thomas, C. D., M. C. Singer, and D. Boughton. 1996. Catastrophic extinction of population sources in a butterfly metapopulation. *American Naturalist* 148:957–975.

Tingley, M. W., M. S. Koo, C. Moritz, A. C. Rush, and S. R. Beissinger. 2012. The push and pull of climate change causes heterogeneous shifts in avian elevational ranges. *Global Change Biology* 18:3279–3290.

Tingley, M. W., W. B. Monahan, S. R. Beissinger, and C. Moritz. 2009. Birds track their Grinnellian niche through a century of climate change.

Proceedings of the National Academy of Science of the United States of America 106:19637–19643.

Trenberth, K. E., P. D. Jones, P. Ambenje, R. Bojariu, D. Easterling, A. Klein Tank, D. Parker, F. Rahimzadeh, J. A. Renwick, M. Rusticucci et al. 2007. Observations: surface and atmospheric climate change. In S. Solomon, D. Qin, M. Manning, Z. Chen, M. Marquis, K. B. Averyt, M. Tignor, and H. L. Miller (eds), *Climate Change 2007: The Scientific Basis. Contribution of Working Group I to the Fourth Assessment Report of the IPCC*. Cambridge University Press, Cambridge, UK. 747–845.

Visser, M. E. and C. Both. 2005. Shifts in phenology due to global climate change: The need for a yardstick. *Proceedings of the Royal Society Series B* 272:2561–2569.

Westerling, A. L., H. G. Hidalgo, D. R. Cayan, and T. W. Swetnam. 2006. Warming and earlier spring increase western US forest wildfire activity. *Science* 313:940–943.

White, M. A., S. W. Running, and P. E. Thornton. 1999. The impact of growing-season length variability on carbon assimilation and evapotranspiration over 88 years in the eastern US deciduous forest. *International Journal of Biometeorology* 42:139–145.

Whittaker, R. H. 1975. *Communities and Ecosystems*. Collier MacMillan, London, UK.

Williams, A. P., C. D. Allen, A. K. Macalady, D. Griffin, C. A. Woodhouse, D. M. Meko, and N. G. McDowell. 2013. Temperature as a potent driver of regional forest drought stress and tree mortality. *Nature Climate Change* 3:292–297.

Williams, J. W. and S. T. Jackson. 2007. Novel climates, no-analog communities, and ecological surprises. *Frontiers in Ecology and the Environment* 5:475–482.

Wolkovich, E. M., B. I. Cook, J. M. Allen, T. M. Crimmins, J. L. Betancourt, S. E. Travers, and E. E. Cleland. 2012. Warming experiments underpredict plant phenological responses to climate change. *Nature* 485:494–497.

Zhao, M. and S. W. Running. 2010. Drought-induced reduction in global terrestrial net primary production from 2000 through 2009. *Science* 329:940–943.

PART II

Learning from Case Studies and Dialogues between Scientists and Resource Managers

CHAPTER 4

Modeling Krill in the California Current

A 2005 CASE STUDY

Jeffrey G. Dorman

Abstract. Examining ecosystem response in coastal upwelling regions to variable atmospheric conditions can help us understand how sensitive these ecosystems may be to climate-driven atmospheric perturbations. A coupled ocean circulation model (ROMS) and a model representing the biology of a prey species (krill) was used to compare predator species during a typical year off northern California (2001) and a year in which the onset of upwelling occurred later and weaker (2005). Decreases in predator survival (salmon) and reproductive success (seabirds) in 2005 were related to poor ocean-feeding (krill) conditions. Hence, these commercially important regions need to be managed from an ecosystem-based approach that depends on a better understanding of the impacts of climate change on atmosphere and biology of the region.

INTRODUCTION

The California Current System (CCS) is one of the four major coastal upwelling regions in the world's oceans. These regions are some of the

---

Key Points

- Anomalous physical conditions (i.e., climate change effects on water temperature, and extent of upwelling) can influence the highest trophic levels within upwelling ecosystems.
- Shifts in the timing and intensity of upwelling can disrupt predator–prey interaction in the California Current.
- Single species fishery management is insufficient in light of the ecosystem variability expected with climate change. Continued implementation of Ecosystem-Based Fishery Management (EBFM) techniques to account for climate-driven ecosystem variability will be important for coastal resource management.
- Marine Protected Areas (MPAs) may not serve their designated purposes in light of changing oceanographic conditions. Maintaining flexibility in the design of MPAs over their lifetime will ensure they are able to serve their intended purpose in light of environmental variability due to climate change.

most productive ecosystems in the ocean, and despite occupying less than 1% of the ocean's surface area, they contribute over 20% of the global commercial fish catch. Within California, there are over 100 commercial fisheries that account for annual revenues of over 100 million dollars (Pacific Fisheries Information Network 1981–2011) and large numbers of coastal jobs. As such, it is of great interest to understand how climate change could impact these important coastal resources.

In most of the world's oceans, primary productivity in surface waters is considered "nutrient-limited," as any nutrients required for photosynthesis and plant growth that enter surface waters are quickly utilized. Waters that are below the depth where light is sufficient for photosynthesis, which can range from 10 to 100 m depending on the water clarity, are typically high in nutrients, and photosynthesis in this zone is said to be "light-limited." The exchange of water (and the nutrients in the water) between deep and surface regions is limited in most regions due to the density difference between warm surface water (less dense) and cold deep water (more dense). The surface waters within upwelling regions are so productive because the alongshore winds provide a means to draw up cold, nutrient-rich water from the deep ocean and essentially fertilize the well-lit surface waters. The upwelled nutrients are quickly utilized by small plant species (phytoplankton), which create "blooms" of phyto plankton in surface waters. The phytoplankton are fed on by small zooplankton (copepods, krill) and small fish (anchovy and herring), which are in turn fed upon by many commercially important species (hake, salmon, rockfish, sablefish, squid). Without the wind-driven upwelling of nutrients to surface waters, the base of the food chain (phytoplankton) would have low abundance and productivity, and the resulting higher trophic level productivity would be greatly reduced. The nutrient-driven phytoplankton blooms and the small number of trophic steps from phytoplankton to commercially important predator species are

the primary reasons that the CCS is one of the most productive regions for commercial fisheries in the world.

Winds are the driver of the productivity of upwelling systems, and any change in atmospheric conditions due to climate change will influence the biological productivity of coastal regions through changes in wind strength, direction, and / or timing. Regional modeling studies of the CCS indicate that changes in atmospheric conditions are likely (Snyder et al. 2003), and there is evidence of increase in wind strength in upwelling regions (Bakun 1990). Changes in wind patterns can affect physical attributes of the water in many different ways (density structure, temperature, currents), which in turn affect individual organism physiology (metabolic rates and duration of the larval life stage), species populations (larval dispersal, size structure, range shifts), and entire communities (changes in predator–prey dynamics). Increasing our understanding of how organisms, populations, and communities will most likely respond to these changes will greatly aid the management of California's coastal resources in a changing climate.

To help assemble information needed to manage ocean resources as climate changes, we are examining the potential climate change impacts on the biological productivity of coastal upwelling regions, with a focus on the krill species *Euphausia pacifica*. *E. pacifica* is a common species of zooplankton throughout the California Current (Brinton 1962) and plays an important role in the regional food web (Field et al. 2006). *E. pacifica* feeds primarily on the upwelling region's diatom-rich phytoplankton blooms, and is preyed upon by a myriad of higher trophic level predators including many commercially important fishes (hake, salmon, rockfish) (Genin et al. 1988, Yamamura et al. 1998, Tanasichuk 1999) and seabirds (Ainley et al. 1996). As the primary productivity in the region is tied to climatic events (e.g., upwelling-favorable winds), and the krill provide a direct path to higher trophic levels, changes in their population biology in response to changing climate events could reverberate up to the higher tropic levels. Decreases in *E. pacifica* abundance due to atmospheric forcing (unfavorable upwelling winds) are hypothesized to be the likely cause of two recent years (2005 and 2006) of reproductive failure in a seabird population (Cassin's auklet; see Box 4.1) on the Farallon Islands (Sydeman et al. 2006), and notable declines in salmon returns in 2008 and 2009 that resulted in the closure of the commercial salmon fishery (see Box 4.1; Lindley et al. 2009). These are pointed examples of how understanding the response of important prey species to climate-driven ocean conditions will enable better management of the organisms that are dependent upon them as a food source.

METHODS

An oceanographic model, a Nutrient-Phytoplankton-Zooplankton-Detritus (NPZD) model, and an individual-based model parameterized for the krill species *E. pacifica* were utilized and linked to model the response of prey species to changing ocean conditions. We ran the models for the years 2001 and 2005 to compare a "normal year" (2001) with the conditions of 2005 that led to anomalously low higher trophic level success.

The coastal ocean was simulated using the Regional Ocean Modeling System (ROMS) (Shchepetkin and McWilliams 2005, Haidvogel et al. 2008) over a region of the eastern Pacific Ocean from Newport, Oregon (~44° 30′ N latitude), to Point Conception, California (~35° N latitude), and up to 450 km offshore. To run ROMS, inputs of atmospheric conditions are required that drive currents (via winds) and heat and cool the ocean (with radiation fluxes, air temperature, humidity, and precipitation). Atmospheric conditions were provided at three-hour intervals from the North American Regional Reanalysis model dataset provided by the National Centers for Environmental Prediction (Mesinger et al. 2006). Ocean conditions at the edges of the modeled region were provided from the global ocean model Estimating the

Coastal Circulation of the Ocean (ECCO2) (Menemenlis et al. 2008). A simple NPZD model (Powell et al. 2006) was incorporated into the ROMS model to provide a food source (phytoplankton) for the simulated *E. pacifica*.

The population biology of *E. pacifica* was simulated using the Individual-Based Model (IBM) POPCYCLE (Batchelder and Miller 1989, Batchelder et al. 2002). The use of an IBM to simulate *E. pacifica's* population biology allows parameters for bioenergetics (ingestion, respiration, assimilation) to vary with attributes of the individual (e.g., size, age, sex), and allows a more accurate representation of discrete events such as reproduction or mortality. As we have incorporated the IBM into the three-dimensional ROMS domain, organism movements and behaviors can be included, along with known variability in those behaviors. For each time-step of the IBM, data on temperature, salinity, current velocity, and phytoplankton abundance are retrieved from the ROMS model output. Each krill individual in the IBM then undergoes growth, stage development, evaluation for reproduction (female only), evaluation for mortality, and spatial position update based on the corresponding ROMS and NPZD data (Figure 4.1).

The bioenergetics of the model were parameterized for the species *E. pacifica* based on extensive laboratory studies on feeding (Ohman 1984), growth, and development (Ross 1982a, Ross 1982b, Feinberg et al. 2006). For every time-step of the IBM, each krill individual is evaluated for growth based on food resources (phytoplankton) and on the physical environment (temperature) from its corresponding location within the ROMS model. Krill were seeded at a weight (40 µg C) representing an early larval stage based on field data from northern California (Dorman et al. 2005). Krill then advance through life stages (based on weight gain), are evaluated for reproduction (based on life stage, sex, and reproductive parameters) and starvation (based on potential weight loss under food limited conditions), and krill location is updated within the model domain based on currents from the ROMS model. Individual krill are removed from the

BOX 4.1 · Two recent examples of dramatic changes to upper trophic level marine populations

Cassin's auklet (*Ptychoramphus aleuticus*) and Chinook salmon (*Oncorhynchus tshawytscha*) have captured public attention and exposed how climate change may be interrupting important tropic interactions. In 2005 and 2006, the Cassin's auklet population at the Farallon Islands experienced near total reproductive failure due to inadequate food resources. In 2007, record low numbers of juvenile salmon returned to the Sacramento River, ultimately resulting in the closure of the 2008 commercial and recreational salmon fishery in California (Box Figure 4.1.1).

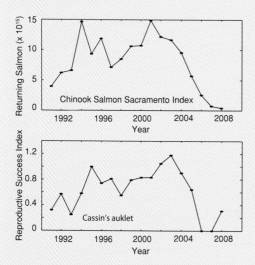

BOX FIGURE 4.1.1: The Chinook Salmon Sacramento Index (sum of escapement, ocean harvest and river harvest) and the reproductive success of Cassin's auklets on the Farallon Islands from 1990 to 2008. Sacramento Index data from O'Farrell et al. (2009) and on Cassin's auklets reproductive success from monitoring conducted by US Fish and Wildlife Service and Point Blue.

Cassin's Auklets

Cassin's auklets are diving marine birds that feed on zooplankton and small fishes over the continental shelf off California. During the breeding season on the Farallon Islands, the adults lay eggs in early April, the eggs hatch in May, and adults feed chicks from May through early July. The breeding and chick rearing period coincides with the onset of upwelling-favorable conditions and

increases in abundances of krill species *Euphausia pacifica* and *Thysanoessa spinifera*. Both of these species are an important part of their diet. In spring of 2005, egg laying by Cassin's auklets was delayed by almost a month and all eggs were abandoned within weeks of laying (Sydeman et al. 2006).

Abandonment of eggs is a survival strategy in long-lived birds to essentially cut losses when reproductive conditions are poor and ensure adult survival so that reproduction will occur in subsequent years. The reproductive failure in 2005 coincided with delayed upwelling and anomalously low primary and secondary productivity early in the year. Events of this type highlight the sensitivity of higher trophic levels to the potential impacts of climate change on the coastal ocean. In 2006, another year of reduced upwelling-favorable winds and warmer surface temperatures (especially during the spring months), a high percentage of Cassin's auklet eggs did not hatch due to abandonment, and only a few chicks survived to fledgling. Reproductive failure of this nature had never been observed in the 35 years of monitoring the Cassin's auklet breeding colony at the Farallon Islands (Sydeman et al. 2006).

Chinook Salmon

The Sacramento River salmon run is one of the largest on the West Coast of the United States and provides a significant portion of the salmon caught commercially and recreationally off Oregon and California. Salmon begin their life in freshwater, spend the majority of their lives feeding in the coastal ocean, and return to freshwater as adults to spawn after 3–4 years (see Quinones and Moyle in this volume). Some juvenile salmon, called "jacks," return to the freshwater environment a year early and their number is considered a good indicator of the number of adults that will return to spawn in the subsequent year. In the fall of 2007, the number of returning jacks (<2000 salmon) was the fewest since records have been kept and well below the average number (~40,000 salmon). This prompted the unprecedented closure of the 2008 commercial and recreational salmon fishery in Oregon and California. As salmon live across such varied habitats, there could be many reasons for the decline in abundance. However, as stock estimates for Oregon and British Columbia rivers were also depressed, it is likely that ocean conditions are at least partially to blame. At least some of the salmon that returned to spawn in the fall of 2008 were juveniles in 2005, and entered the ocean to feed on krill when ocean productivity was very low. It is possible that the reduced productivity that led to reproductive failure in Cassin's auklets also led to increased mortality in juvenile salmon, decreased juvenile returns in 2007, and a closed salmon fishery in 2008.

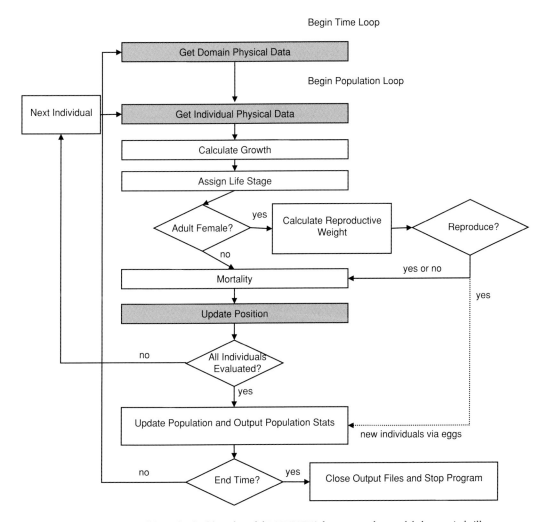

FIGURE 4.1: The structure of the individual-based model POPCYCLE that was used to model changes in krill.

TABLE 4.1

Monthly mean along shore current velocity from locations approximately 30 km offshore.
Poleward flow is represented by positive values and flow toward the equator by
negative values

NOTE THE STRONG POLEWARD FLOW IN JANUARY 2005

	2001					
Latitude	*January*	*February*	*March*	*April*	*May*	*June*
42	0.067	0.069	0.071	0.078	−0.084	−0.052
40	0.004	0.002	0.012	0.073	−0.123	−0.035
38	0.028	0.150	0.085	−0.028	−0.111	0.009
36	0.013	0.087	0.054	0.120	−0.042	−0.024
	2005					
Latitude	*January*	*February*	*March*	*April*	*May*	*June*
42	0.010	0.109	0.062	0.003	0.065	0.051
40	0.210	−0.112	−0.031	−0.016	−0.036	−0.049
38	0.048	0.064	0.014	−0.115	−0.195	−0.045
36	0.169	−0.167	−0.374	−0.272	−0.084	−0.096

population either through starvation or by being transported beyond the domain of the physical model. The IBM was seeded with approximately 5000 krill that were evenly distributed over the continental shelf (within 60 km of the coast) between Point Arena, California (~38 N latitude) to the southern model boundary. Data were collected from the model runs of 200 days, beginning on January 5 of 2001 and 2005.

The use of an IBM that includes stage- or size-specific parameters allows the inclusion of life history traits such as reproduction or growth rates that vary with life stage or size. The modeled krill for this study undergoes variable diel vertical migration based on both size and life stage. This results in differences between the depths occupied by adults and larvae during a 24-hour cycle, thereby exposing various life stages to potentially differing currents, and allowing us to more realistically simulate potential changes in the spatial distribution of krill as ocean conditions change.

RESULTS

We compared modeled alongshore currents between 2001 and 2005 from locations 30 km offshore at 36°, 38°, 40°, and 42° N latitude (Figure 4.2). Model results indicate anomalously strong and northward (poleward, upwelling unfavorable) currents in the coastal region of interest during January 2005 (Table 4.1). Upwelling-favorable (equatorward) currents were fully established during May and June in 2001, and one month earlier in 2005 except in the northernmost regions of the model domain.

Northward krill advection by currents was most evident during January 2005, resulting in 10.6% of simulated krill being advected north of Cape Mendocino (~40° 20′ N latitude), compared with 0.6% in January 2001 (Figure 4.3).

Mean krill weight was significantly greater, for much of the model run time, in 2001 than in 2005 (Figure 4.4). The model results suggest that krill during 2001 and 2005 gained a similar amount of weight until approximately year-day 40, after which krill in 2001 gained 0.35 µg more carbon per day on average than those in 2005. As a result of this difference in weight gain, at the end of the model run, average krill weight was significantly greater in 2001 (mean = 178.7 µg C) than in 2005 (mean = 120.9 µg C) ($t = 18.88$, $df_{2001} = 1088$, $df_{2005} = 545$, $p < 0.001$). An analysis of the 100 modeled krill that gained

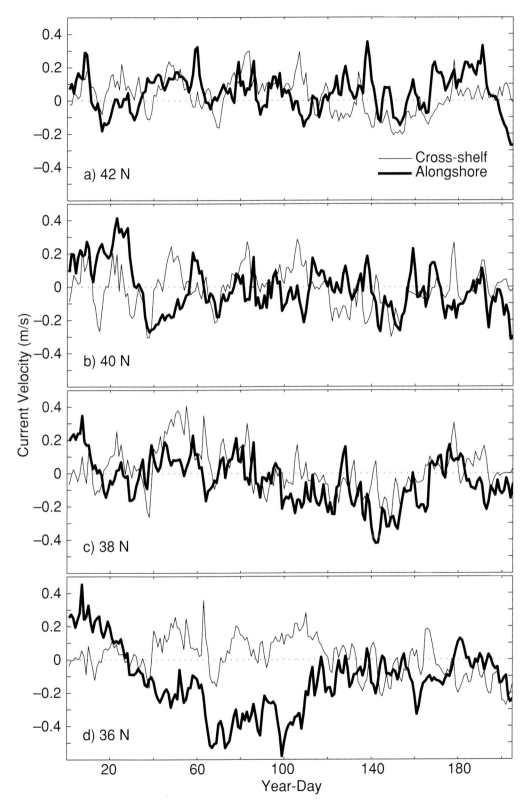

FIGURE 4.2: Alongshore and cross-shelf surface current velocity at 30 km offshore of (a) 42° North, (b) 40° North, (c) 38° North, and (d) 36° North. Poleward and onshore flows are represented by positive values and negative values indicating flow toward the equator and offshore flows.

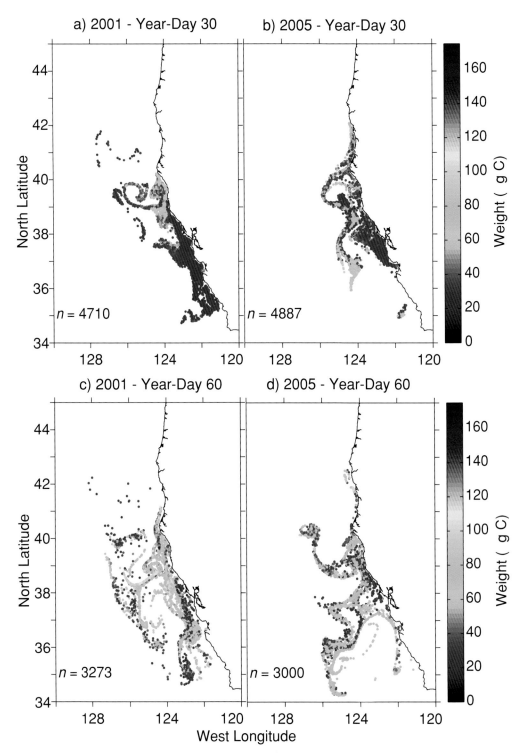

FIGURE 4.3: Krill distribution and weight on year-day 30 of (a) 2001 and (b) 2005, and year-day 60 of (c) 2001 and (d) 2005.

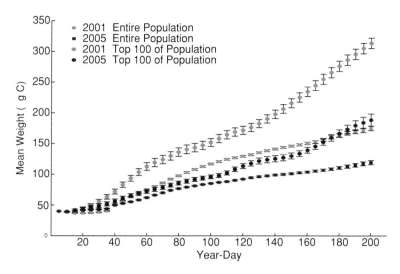

FIGURE 4.4: Mean population weight of all live krill within the model domain during 2001 and 2005, and the mean weight of the 100 individuals that gained the most weight during the model run during 2001 and 2005. Error bars represent 95% confidence intervals of the mean.

the greatest weight over the course of the model run (approximately 2% of the total population) also showed significantly more weight gain during 2001 (mean = 321.7 µg C) than in 2005 (mean = 192.0 µg C) ($t = 19.1$, $df_{2001} = 99$, $df_{2005} = 99$, $p < 0.001$) (Figure 4.5).

The models suggest that food limitation was an important factor limiting population growth. Starvation of the initial seeding of krill was evident during both 2001 and 2005 with 31.9% and 45.2% of individuals starving, respectively, without molting to the next life stage (Figure 4.5). For those individuals that did gain enough weight to molt to the next life stage, furcilia IV, and beyond, starvation was less in 2001 than in 2005, with 3.8% and 21.8% of krill starving, respectively.

DISCUSSION

These simulations indicate that higher starvation mortality of juvenile and adult krill was the primary reason for the lower abundance of *E. pacifica* observed during the spring and summer of 2005 in the central California Current. The greater modeled mortality in 2005 was caused by anomalously low phytoplankton food sources during periods of typical high primary production, and was influenced by northward advection of krill to regions that typically experience lower primary productivity at this time of year.

During both 2001 and 2005, there was high starvation of krill early on in the model run (~year-day 34). High starvation is to be expected during winter months due to the reduced wintertime primary productivity, yet starvation continued on through the spring months in much greater numbers in 2005 than in 2001 due to a delayed onset of upwelling during 2005. Advection also played a role in increasing winter mortality as much of the coastal region experienced strongly poleward currents during January 2005. This resulted in many krill being transported to the north of Cape Mendocino where seasonal upwelling does not typically begin until early summer. Thus, these krill were transported to a region still months away from any significant upwelling of nutrients, and over 90% of the modeled population that was transported to the north died in 2005. Increases in wintertime starvation of *E. pacifica* like the ones suggested by these results would ultimately reduce the abundances of this key prey species during the spring

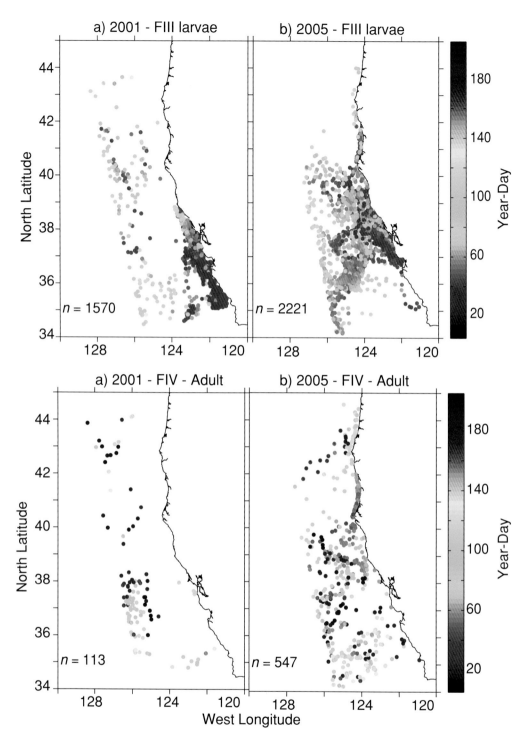

FIGURE 4.5: Starvation location and time of furcilia III larval stage krill during (a) 2001 and (b) 2005, and starvation location and time of later stage krill (furcilia IV through adult) during (c) 2001 and (d) 2005.

months, as there is little observational evidence of any significant reproduction by *E. pacifica* occurring during winter months that would maintain population size. Thus, decreases in the wintertime population would have negative impacts on predators that utilize the krill population in early spring, before reproduction begins again with the onset of upwelling.

In addition to increased starvation, those krill that did survive gained less weight over the course of the model run during 2005. Growth of krill during the early part (January–March) of any year is typically a period of slow growth (or actual weight loss) as this is not a time of strong upwelling in the region. During 2005, the mean weight of the population was significantly less than during 2001, with the greatest difference in weight gain occurring in the spring (April–June). This is due to the unusually late onset of upwelling during 2005. Smaller krill in an ecosystem equates to less krill biomass, and impacts predators through their need to increase the time and energy spent foraging, which also increases their exposure to predators higher up in the food chain.

While the entire population of krill is of interest as a food source for higher trophic levels, only a small percentage of the population will survive to adulthood and reproduce. Those individuals that have gained the most weight during the course of the model run are representative of krill that will reach adulthood the fastest, and provide information on the potential timing of future reproductive events. A comparison of the 100 krill that gained the most weight over the course of the model run revealed a greater disparity in weight gain between 2001 and 2005 than when comparing the entire population, with greater weight gain in 2001. Thus, in addition to suggesting population-level impacts on current krill, these simulations suggest that in 2005 those krill that would have produced the next generation of larvae would have required more time to mature, resulting in greater exposure to potential predators, and would have reproduced later in the year than the 2001 population.

The impacts of the sort of anomalous conditions experienced during 2005 in the California Current have been the source of much interest (see special section of *Geophysical Research Letters* 2006, Volume 33, Issue 2, Jahncke et al. 2008). A season of delayed upwelling to the extent that was observed in 2005 allows a glimpse into how anomalous atmospheric conditions influenced the entire food web of the California Current up to the highest trophic levels (see Box 4.1). Our results mechanistically explore how the spatial and temporal distribution of krill, a key prey species, might have played a role in the low production observed at these higher trophic levels.

SOLUTIONS, ADAPTATIONS, AND LESSONS

The impacts of climate change on zooplankton in the California Current, and specifically on the krill species *E. pacifica*, are far from certain. The anomalous conditions from 2005 that were modeled in this case study are not necessarily indicative of the types of conditions that we should expect under a warmer climate. However, these types of modeling results highlight the possible ramifications of changing ocean conditions on a key prey species of many higher trophic levels in the California Current (see Box 4.1). This interconnection between upwelling-favorable winds and krill, and krill with many top predators, highlights the importance of shifting the management of our coastal resources to a broad ecosystem-based approach.

Traditionally, fisheries management decisions in the United States have considered species-scale variability, with little focus on ecosystem variability and interactions between trophic levels. Single-species techniques are valuable, but are insufficient by themselves in light of the potential ecosystem-wide changes that will likely result from climate change. Ecosystem-scale assessments must also be incorporated into fisheries management to make sense of system variability and the large-scale impacts of climate change. The framework for Ecosystem-Based

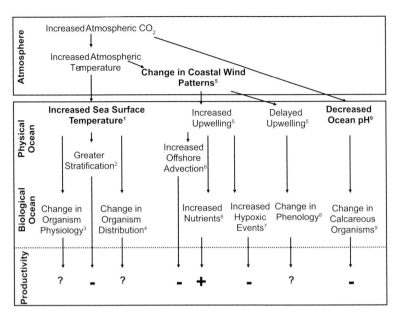

FIGURE 4.6: Three impacts (in bold text) of increasing temperatures and increased atmospheric CO_2. The physical and biological paths through which these changes might impact productivity and the potential impacts on productivity are noted.

1. (Mendelssohn et al. 2003, Palacios et al. 2004, Bograd and Lynn 2003)
2. (Palacios et al. 2004, Capotondi et al. 2012)
3. (O'Connor et al. 2007, Pörtner and Farrell 2008)
4. (Beaugrand et al. 2002, Lindley and Daykin 2005)
5. (Bakun 1990, Schwing and Mendelssohn 1997, Snyder et al. 2003, Bograd et al. 2009, García-Reyes and Largier 2012)
6. (Cury and Roy 1989, Gargett 1997, Botsford et al. 2003, Botsford et al. 2006, García-Reyes and Largier 2012)
7. (Chan et al. 2008, Bograd et al. 2008, Pierce et al. 2012)
8. (Mackas et al. 1998, Edwards and Richardson 2004, Costello et al. 2006, Text Box 4.1)
9. (Fabry et al. 2008, Doney et al. 2009)

Fishery Management (EBFM) was laid out by the Ecosystem Principles Advisory Panel (1999) in response to the Sustainable Fisheries Act (SFA) of 1996, which calls for fisheries to be managed while taking the biological complexities and the overall health of the ecosystem into account. EBFM is a move away from single-species fishery models to a more holistic view of the ecosystem that may incorporate prey-species population dynamics, habitat conditions, and environmental variation. Where data on these parameters are not available, the SFA urges erring on the side of caution. Implementation of EBFM is an ongoing process (Field and Francis 2006) in California and will be essential to managing fisheries in response to climate change.

The primary hindrance to effective implementation of EBFM is the uncertainty regarding the basic ecosystem changes that are expected (Figure 4.6). It is understood how increased atmospheric CO_2 will directly impact the physical ocean characteristics of the California Current (increasing sea surface temperatures, increasing upwelling favorable winds, decreasing oceanic pH). These changes will influence a myriad of other physical characteristics (water column stratification, ocean currents, salinity, dissolved oxygen levels), all of which can impact the biology of the coastal ocean. In many cases, we have a fairly complete understanding of how a change in one single process will impact biological productivity, but our level of understanding is often lacking when considering how changes in multiple physical factors will impact the biology of the region. For example, increased upwelling favorable winds may cause upwelling of more nutrients into surface waters, but increased sea surface temperatures will cause greater thermal stratification (and thus a stronger barrier) between warm nutrient-poor surface water and cold nutrient-rich deep waters. Nutrients will only be upwelling into surface waters if the increase in wind strength is great enough to overcome the increased strength of thermal stratification. Extrapolating the expected changes in physics to the biology of individual organisms and food webs of the California Current introduces even more uncertainty into our understanding. Research to clarify the impacts of climate change on the biological productivity of the coastal ocean is needed in the following areas: Ocean ecosystem response, species-specific response, and food web response.

Ocean Ecosystem Response

Significant work has been accomplished toward understanding the impacts of climate change on the global atmosphere using large-scale climate models. As California's coastal ocean is primarily atmospheric driven, emphasis is also needed on applying these warmer atmospheric conditions to California's upwelling system. Ocean models forced by an atmosphere containing increased CO_2 concentrations will provide a better understanding of the oceanographic conditions necessary (currents, nutrients, temperature, pH, oxygen) to predict lower trophic level biological responses (phytoplankton and zooplankton productivity) using knowledge of ecosystems and individual species. It should be noted that many of these models do not take pH or temperature into consideration, but they can provide a first approximation of the impacts of climate change on primary producers and consumers.

Species-Specific Response

There are many commercially important species in California's coastal ocean and each species will have a unique biological response to climate change. For example, species that have calcareous exoskeletons or shells may be more susceptible to changing ocean pH levels (increased mortality due to a weakening of calcareous structures), while pelagic fish may be affected by increasing temperature (range shifts, changes in physiology). Research on the dominant species responses (physiology, range and phenology shifts, population biology, etc.) to changing ocean conditions needs to be a priority if we are to effectively manage these populations and the pelagic ecosystem with climate change considerations.

Food Web Response

Perhaps the biggest void in our understanding of how climate change may impact coastal ecosystems relates to how food-web dynamics will likely change in response to a warmer climate and more acidic ocean. A greater understanding of changing species interactions (competition and predator–prey dynamics) based on changing abundances and spatial ranges is critical to understanding the flow of trophic energy to the highest levels. Putting together the species-specific research on the major components of the coastal food web will be an important step in understanding the impacts of climate change on California's coastal ocean.

Finally, managers of marine resources will need to maintain flexibility in management plans to incorporate new understandings of the impacts of climate change on the coastal ocean. Marine Protected Areas (MPAs), which are primarily "no take" zones for one or many species, may not serve their designed purpose of promoting population recovery under a warmer ocean (range shifts of organisms, decreased larval dispersal distances due to faster development) or if located in regions of increased hypoxia. Fishery seasons may need to be adjusted later or earlier as shifts in the timing of reproduction or migration occur. Adaptive management of these resources will be crucial to maintaining sustainable fisheries as atmospheric CO_2 increases.

LITERATURE CITED

Ainley, D. G., L. B. Spear, and S. G. Allen. 1996. Variation in the diet of Cassin's auklet reveals spatial, seasonal, and decadal occurrence patterns of euphausiids off California, USA. *Marine Ecology Progress Series* 137:1–10.

Bakun, A. 1990. Global climate change and intensification of coastal ocean upwelling. *Science* 247:198–201.

Batchelder, H. P. and C. B. Miller. 1989. Life history and population dynamics of *Metridia pacifica*: Results from simulation modelling. *Ecological Modelling* 48:113–136.

Batchelder, H. P., C. A. Edwards, and T. M. Powell. 2002. Individual-based models of copepod populations in coastal upwelling regions: Implications of physiologically and environmentally influenced diel vertical migration on demographic success and nearshore retention. *Progress in Oceanography* 53:307–333.

Beaugrand, G., P. C. Reid, F. Ibañez, J. A. Lindley, and M. Edwards. 2002. Reorganization of North Atlantic marine copepod biodiversity and climate. *Science* 296:1692–1694.

Bograd, S. J. and R. J. Lynn. 2003. Long-term variability in the southern California current system. *Deep-Sea Research II* 50:2355–2370.

Bograd, S. J., C. G. Castro, E. DiLorenzo, D. M. Palacios, H. Bailey, W. Gilly, and F. P. Chavez. 2008. Oxygen declines and the shoaling of the hypoxic boundary in the California current. *Geophysical Research Letters* 35, L12607. doi:10.1029/2008GL034185.

Bograd, S. J., I. Schroeder, N. Sarkar, X. Qiu, W. J. Sydeman, and F. B. Schwing. 2009. Phenology of coastal upwelling in the California Current. *Geophysical Research Letters* 36:L01602. doi:10.1029/2008GL035933.

Brinton, E. 1962. The distribution of Pacific euphausiids. *Bulletin of the Scripps Institute of Oceanography* 8:51–269.

Botsford, L. W., C. A. Lawrence, E. P. Dever, A. Hastings, and J. Largier. 2003. Wind strength and biological productivity in upwelling systems: An idealized study. *Fisheries Oceanography* 12(4–5):245–259.

Botsford, L. W., C. A. Lawrence, E. P. Dever, A. Hastings, and J. Largier. 2006. Effects of variable winds on biological productivity on continental shelves in coastal upwelling systems. *Deep-Sea Research Part II* 53:3116–3140.

Capotondi, A., M. A. Alexander, N. A. Bond, E. N. Curchitser, and J. D. Scott. 2012. Enhanced upper ocean stratification with climate change in the CMIP3 models. *Journal of Geophysical Research* 117:C04031. doi:10.1029/2011JC007409.

Chan, F., J. A. Barth, J. Lubchenco, A. Kirincich, H. Weeks, W. T. Peterson, and B. A. Menge. 2008. Emergence of anoxia in the California current large marine ecosystem. *Science* 319:920.

Costello, J. H., B. K. Sullivan, and D. J. Gifford. 2006. A physical-biological interaction underlying variable phonological responses to climate change by coastal zooplankton. *Journal of Plankton Research* 28(11):1099–1105.

Cury, P. and C. Roy. 1989. Optimal environmental window and pelagic fish recruitment success in upwelling areas. *Canadian Journal of Fisheries and Aquatic Sciences* 46:670–680.

Doney, S. C., V. J. Fabry, R. A. Feely, and J. A. Kleypas. 2009. Ocean acidification: The other CO_2 problem. *Annual Review of Marine Science* 1:169–192.

Dorman, J. G., S. M. Bollens, and A. M. Slaughter. 2005. Population biology of euphausiids off northern California and effects of short timescale wind events on *Euphausia pacifica*. *Marine Ecology Progress Series* 288:183–198.

Edwards, M. and A. J. Richardson. 2004. Impact of climate change on marine pelagic phenology and trophic mismatch. *Nature* 430:881–884.

Fabry, V. J., B. A. Seibel, R. A. Feely, and J. C. Orr. 2008. Impacts of ocean acidification on marine fauna and ecosystem processes. *ICES Journal of Marine Science* 65:414–432.

Feinberg, L. R., C. T. Shaw, and W. T. Peterson, 2006. Larval development of *Euphausia pacifica* in the laboratory: Variability in the developmental pathways. *Marine Ecology Progress Series* 316:127–137.

Field, J. C. and R. C. Francis. 2006. Considering ecosystem-based fisheries management in the California Current. *Marine Policy* 30:552–569.

Field, J. C., R. C. Francis, and K. Aydin. 2006. Top-down modeling and bottom-up dynamics: Linking a fisheries-based ecosystem model with climate hypotheses in the Northern California Current. *Progress in Oceanography* 68:238–270.

García-Reyes, M. and J. L. Largier. 2012. Seasonality of coastal upwelling off central and northern California: New insights, including temporal and spatial variability. *Journal of Geophysical Research* 117:C03028. doi:10.1029/2011JC007629.

Gargett, A. E. 1997. The optimal stability "window": a mechanism underlying decadal fluctuations in the North Pacific salmon stocks? *Fisheries Oceanography* 6(2):109–117.

Genin, A., L. Haury, and P. Greenblatt. 1988. Interactions of migrating zooplankton with shallow topography: Predation by rockfishes and intensification of patchiness. *Deep Sea Research* 2:151–175.

Haidvogel, D. B., H. Arango, W. P. Budgell, B. D. Cornuelle, E. Curchitser, E. Di Lorenzo, K. Fennel, W. R. Geyer, A. J. Hermann, L. Lanerolle et al. 2008. Ocean forecasting in terrain-following coordinates: Formulation and skill assessment of the Regional Ocean Modeling System. *Journal of Computational Physics* 227:3595–3624.

Jahncke, J., B. L. Saenz, C. L. Abraham, C. Rintoul, R. W. Bradley, and W. J. Sydeman. 2008. Ecosystem responses to short-term climate variability in the Gulf of the Farallones, California. *Progress in Oceanography* 77:182–193.

Lindley, J. A. and S. Daykin. 2005. Variations in the distributions of *Centropages chierchiae* and *Temora stylifera* (Copepoda: Calanoida) in the north-eastern Atlantic Ocean and western European shelf waters. *ICES Journal of Marine Science* 62:869–877.

Lindley, S. T., C. B. Grimes, M. S. Mohr, W. Peterson, J. Stein, J. T. Anderson, L. W. Botsford, D. L. Bottom, C. A. Busack, T. K. Collier et al. 2009. *What Caused the Sacramento River Fall Chinook Stock Collapse?* Report to the Pacific Fishery Management Council.

Mackas, D. L., R. Goldblatt, and A. G. Lewis. 1998. Interdecadal variation in developmental timing of *Neocalanus plumchrus* populations at Ocean Station P in the subarctic North Pacific. *Canadian Journal of Fisheries and Aquatic Sciences* 55:1878–1893.

Mendelssohn, R., F. B. Schwing, and S. J. Bograd. 2003. Spatial structure of subsurface temperature variability in the California Current, 1950-1993. *Journal of Geophysical Research* 108(C3):3093. doi:10.1029/2002JC001568.

Menemenlis, D., J. Campin, P. Heimbach, C. Hill, T. Lee, A. Nguyen, M. Schodlok, and H. Zhang. 2008. ECCO2: High resolution global ocean and sea ice data synthesis. *Mercator Ocean Quarterly* 31:3–21.

Mesinger, F., G. DiMego, E. Kalnay, K. Mitchell, P. C. Shafran, W. Ebisuzaki, D. Jović, J. Woollen, E. Rogers, E. H. Berbery et al. 2006. North

American regional reanalysis. *Bulletin of the American Meteorological Society* 87:343–360.

O'Connor, M. I., J. F. Bruno, S. D. Gaines, B. S. Halpern, S. E. Lester, B. P. Kinlan, J. M. Weiss. 2007. Temperature control of larval dispersal and the implications for marine ecology, evolution, and conservation. *Proceedings of the National Academy of Sciences* 104(4):1266–1271.

O'Farrell, M. R., M. S. Mohr, M. L. Palmer-Zwahlen, and A. M. Grover. 2009. *The Sacramento Index*. Report to the Pacific Fishery Management Council.

Ohman, M. D. 1984. Omnivory by *Euphausia pacifica*: The role of copepod prey. *Marine Ecology Progress Series* 19:125–131.

Pacific Fisheries Information Network (PacFIN). 1981–2011. *Commercial Landed Catch: Metric Tons, Revenue, Price/Pound*. California All Species Report 308. Pacific States Marine Fisheries Commission. http://pacfin.psmfc .org/pacfin_pub/all_species_pub/woc_r308 .php.

Palacios, D. M., S. J. Bograd, R. Mendelssohn, and F. B. Schwing. 2004. Long-term and seasonal trends in stratification in the California Current, 1950-1993. *Journal of Geophysical Research* 109:C10016. doi:10.1029/2004JC002380.

Pierce, S. D., J. A. Barth, R. K. Shearman, and A. Y. Erofeev. 2012. Declining oxygen in the Northeast Pacific. *Journal of Physical Oceanography* 42:495–501.

Pörtner, H. O. and A. P. Farrell. 2008. Physiology and climate change. *Science* 302:690–692.

Powell, T. M., C. V. W. Lewis, E. N. Curchitser, D. B. Haidvogel, A. J. Hermann, and E. L. Dobbins. 2006. Results from a three-dimensional, nested biological-physical model of the California Current System and comparisons with statistics from satellite imagery. *Journal of Geophysical Research* 111:C07018. doi:10.1029/2004 /JC002506.

Ross, R. M. 1982a. Energetics of *Euphausia pacifica*. I. Effects of body carbon and nitrogen and temperature on measured and predicted production. *Marine Biology* 68:1–13.

Ross, R. M. 1982b. Energetics of *Euphausia pacifica*. II. Complete carbon and nitrogen budgets at 8° and 12°C throughout the life span. *Marine Biology* 68:15–23.

Schwing, F. B. and R. Mendelssohn. 1997. Increased coastal upwelling in the California Current System. *Journal of Geophysical Research* 102(C2):3421–3438.

Shchepetkin, A. F. and J. C. McWilliams. 2005. The regional ocean modeling system (ROMS): a split-explicit, free-surface, topography-following-coordinate oceanic model. *Ocean Modelling* 9:347–404.

Snyder, M. A., L. C. Sloan, N. S. Diffenbaugh, and J. L. Bell. 2003. Future climate change and upwelling in the California Current. *Geophysical Research Letters* 30(15):1823. doi:10.1029/2003GL017647.

Sydeman, W. J., R. W. Bradley, P. Warzybok, C. L. Abraham, J. Jahncke, K. D. Hyrenbach, V. Kousky, J. M. Hipfner, and M. D. Ohman. 2006. Planktivorous auklet *Ptychoramphus aleuticus*

responses to ocean climate, 2005: Unusual atmospheric blocking? *Geophysical Research Letters* 33:L22S09. doi:10.1029/2006GL026736.

Tanasichuk, R. W. 1999. Interannual variation in the availability and utilization of euphausiids as prey for Pacific hake (*Merluccius productus*) along the south-west coast of Vancouver Island. *Fisheries Oceanography* 8:150–156.

Yamamura, O., T. Inada, and K. Shimazaki. 1998. Predation on *Euphausia pacifica* by demersal fishes: Predation impact and influence of physical variability. *Marine Biology* 132:195–208.

Manager Comments

Jeff G. Dorman
in conversation with
Dan Howard

Dorman: Can you provide an introduction to Cordell Bank, and describe how you expect climate change to impact the region?

Howard: Cordell Bank National Marine Sanctuary (CBNMS) was designated in 1989 and is one of the 14 federally protected marine and great lake areas in the National Marine Sanctuary System. Under the authority of the National Marine Sanctuaries Act, CBNMS provides comprehensive and coordinated conservation and management of the marine resources within the sanctuary's 399 square nautical miles. CBNMS is an offshore site located about 43 nautical miles northwest of San Francisco off the coast of Marin and Sonoma County. The sanctuary encompasses an offshore area roughly between Bodega Bay and Point Reyes from six nautical miles offshore to 30 nautical miles from shore. The centerpiece of the sanctuary is a granite bank located 18 nautical miles west of Point Reyes that is 4.5 miles wide by 9.5 miles long. The rocky bank emerges from the soft sediments of the continental shelf, with the upper pinnacles reaching within 115 ft of the ocean's surface, and shelf depths at the base of the Bank in roughly 300–400 ft of water. This is an area of special significance due to its position in the California Current and unique geological and oceanic features that create productive conditions supportive of diverse and abundant marine life.

Dorman: How do you expect climate change to impact CBNMS?

Howard: It is difficult to predict the sequence of change that will occur within the cool temperate California Current Ecosystem including CBNMS. But since these organisms and communities have evolved within an oceanographic system that is driven by an annual productivity cycle that is fueled by upwelling, any long-term change affecting the timing or intensity of upwelling would have significant impacts on early life stages, trophic interactions, and community structure. For example, many fishes and sessile invertebrates inhabit-

ing Cordell Bank release their gametes or larvae into the water column in late winter or early spring. This release is timed to coincide with the onset of the annual productivity cycle (upwelling season) so their young can maximize feeding opportunities and therefore increase their odds of surviving. There would be high mortality in these early life stages if the timing or intensity of the annual productivity cycle changed. There are many similar stories including the Cassin's Auklets dependence on early season productivity that you covered in your chapter. I expect that climate change will restructure the current composition of our regional ocean community. As changes occur in upwelling, sea surface temperature or ocean acidification, those species that cannot adapt will move or die and a new suite of organisms will assume their ecological roles, and what we now consider as our typical ocean community may look very different in the future.

Dorman: Are there any aspects of climate change that stand out to you as a primary threat to conservation and management goals established for the CBNMS?

Howard: One of the reasons Cordell Bank is so productive is the fact that it sits on the very edge of the continental shelf on a little peninsula surrounded by deep water on three sides. This position may also make Cordell Bank susceptible to hypoxic events related to the Oxygen Minimum Zone (OMZ). If increased upwelling related to climate change draws the OMZ up onto the shelf, it could potentially devastate the benthic invertebrates that cover the upper reaches of the bank.

It has also been documented that Humboldt or Jumbo squid (*Dosidicus gigas*) are associated with the OMZ during the day. The area west of Cordell Bank has been a hotspot for catching these squid since they moved into the area several years ago. If climate change shifted the OMZ shallower and closer to the bank, rockfish populations could be in jeopardy from increased predation by the squid. Regulations implemented by the Pacific Fisheries

(continued)

(continued)

Management Council have prohibited all commercial and recreational fishing for rockfish at Cordell Bank since 2003 to allow overfished rockfish populations to recover. It is unclear what kind of impact jumbo squid would have on these recovering populations. The persistent presence of jumbo squid at Cordell Bank may be an indication that ecosystem-level changes are already occurring.

Dorman: How are the staffs at CBNMS responding to the challenge of understanding climate change?

Howard: We have participated in a two-day workshop hosted by the Gulf of the Farallones National Marine Sanctuary (GFNMS) that included CBNMS staff and the broad cross-section of the local scientific community to identify potential climate change drivers and local impacts.

We have also formed a joint working group composed of local scientists, sanctuary staff, and science advisors from the Cordell Bank and Gulf of the Farallones sanctuary advisory councils. This working group will help develop a site scenario for both sanctuaries that will include research and monitoring strategies, management and policy strategies, education and outreach focusing on changing behavior strategies, and operation strategies.

Finally, CBNMS and GFNMS are outfitting our research vessel with instrumentation and developing an at sea sampling plan to monitor ocean acidification in coordination with NOAA Pacific Marine Environmental Laboratory in Seattle, WA.

Dorman: Based on your experiences so far, what do you see as the major hindrances to implementing management strategies that incorporate potential changes in climate?

Howard: I think one of the biggest challenges is the scale of the problem. We manage discreet areas with a limited geography, but many of the drivers associated with climate change are operating on a regional or global scale. So as the ocean gets more acidic and threatens the stability of local food webs, it is hard to implement management actions in local sanctuaries that can effectively address this global issue. Another stumbling block that was identified by both the U.S. Commission on Ocean Policy and the Pew Ocean Commission is the number of different management authorities that have some type of jurisdiction in the ocean. Different missions and overlapping authority can stymie management actions.

CHAPTER 5

Shifts in Marine Biogeographic Ranges

Christopher J. Osovitz and Gretchen E. Hofmann

Abstract. Global climate change brings new threats to wildlife as the environmental conditions of ecosystems rapidly change. Among the most significant of these challenges for California's marine ecosystems is ocean warming, a result of global warming. Ocean warming could impact nearly every marine species, as temperature affects virtually every biological process from cellular reactions to individual behavior, and is believed to be an important factor in setting species distribution limits. In this study, gene expression profiles of purple sea urchins, *Strongylocentrotus purpuratus*, are used as an indicator of the physiological effects of environmental variation along a latitudinal gradient from their southern range boundary in Baja California northward to Oregon, the United States. The purple sea urchin is an ecologically important California species that helps to regulate kelp forest densities, and its sister species, the red sea urchin, *S. franciscanus*, is the subject of one of California's largest fisheries. Gene expression profiles of *S. purpuratus* in Oregon most resembled those from Baja California, suggesting that local thermal variation may

Key Points

- Assuming only poleward movement of purple sea urchins in response to ocean warming is too simplistic. The upwelling patterns along the Pacific Coast create complex temperature patterns that can dictate locations of refugia.
- As part of a long-term conservation strategy, urchin-specific Marine Protected Areas (MPAs) may be appropriate in areas that are currently cooler, such as Northern California and Oregon, where urchins are currently abundant and living substantially below their upper temperature limits. These MPAs should include specific monitoring of sea urchins, as MPAs can also increase urchin predator density.
- In southern California and Baja California, managing *Strongylocentrotus purpuratus* in the cooler refugia created by coastal upwelling will likely be the most efficacious manner of protecting the purple urchin and the closely related red urchin that is the focus of a large California fishery. However,

(continued)

(continued)

persistence of the geographic patterns of upwelling should be monitored.
- Protecting these two species of urchins will likely have the added effect of maintaining kelp forest communities in general.

have a greater physiological impact on this species than differences in temperature seen across large-scale differences in latitude. These data suggest that the small-scale complexity of the marine environment along the California coastline may result in spatial complexity of optimal sea urchin habitat. If true, ocean warming may be more likely to fragment the range of S. purpuratus, rather than shifting it to the predominantly cooler waters of the north Pacific. Therefore, the best management strategy to protect S. purpuratus during a period of ocean warming might be to identify locations where local variation may be optimal for S. purpuratus, instead of managing based on the expectation that each population will migrate poleward.

INTRODUCTION

California's marine ecosystems are already being effected by current ocean warming, and future warming may challenge the survivorship of many members of the California marine ecological community, including fish, sea urchins, mussels, macroalgae, and sea stars, among others (e.g., Harley et al. 2006, Brierley and Kingsford 2009, Hoegh-Guldberg and Bruno 2010). As a result, scientists face a pressing need to understand and predict the nature of current and future biological impacts of climate change in order to best manage and preserve the state's natural marine ecosystems and fisheries. The biological impact of ocean warming on marine species may mirror that of global warming on terrestrial species, that is many species ranges may simply shift either poleward or deeper into the ocean (e.g., Fields et al. 1993, Parmesan and Yohe 2003, Root et al.

2003) or perhaps marine climate change could have a more complex effect on ocean species (e.g., Schiel et al. 2004, Harley et al. 2006). However, very few investigations have undertaken a physiological approach to elucidate the mechanisms of biological change at the cellular level. The central aim of this study is to assay broad-scale patterns of gene expression, which we use as an indicator of variation across the physiological processes within different urchin populations, to examine how ocean warming may impact economically and ecologically important California sea urchin species.

The current prevailing assumption is that ocean warming will force marine species ranges to shift toward the poles or into deeper waters that are historically colder regions (e.g., Fields et al. 1993, Harley et al. 2006). This prediction relies on an often untested assumption that key environmental factors that are important to the well being of a species, such as temperature, vary smoothly and predictably over a geographic scale. A common approach to evaluating risks related to climate change involves linking patterns of key factors to species ranges, based on the idea that species live in areas where such factors are within species' tolerance limits. Plotting the location of these estimated tolerance limits defines what is called the "climate envelope" of a species (Hijmans and Graham 2006), and we can then examine how projected changes in climate may lead to shifts in the location of tolerance limits. The variability of environmental factors across space, however, may appear smooth at a large scale, but when examined on a smaller scale the smoothness often disappears.

When examined at a continent-wide scale, many marine and terrestrial species have been found to have already shifted their ranges poleward (e.g., Sorte et al. 2010, Burrows et al. 2011). An examination of organisms of coastal California's marine ecosystems on a finer scale reveals complex spatial patterns of environmental variation, rather than smooth gradients (e.g., Broitman and Kinlan 2006). Oceanographic properties, such as regional eddies and coastal upwelling, disrupt not only the latitudinal gradient in temperature, but also other factors, such as temporal variability, productivity, and water turbidity. The complexity of these environmental patterns complicates predictions regarding the biological effects of future climate change.

Also undermining our ability to predict the responses of many marine species to climate change is a fundamental lack of understanding of the direct effects of ocean temperature on the physiology of marine species (Helmuth et al. 2005, Osovitz and Hofmann 2007). Stated more simply, we do not yet understand what natural environmental conditions are optimal for many species, largely because it is difficult to measure "contentment" in species. Without knowledge of the physiological mechanisms of environmental tolerance, our ability to predict the biological outcomes of future climate change will be limited to observed ecological patterns of change in response to past climate events (Helmuth et al. 2005). A significant gap in our knowledge regarding the biological impacts of climate change is the extent to which organisms can physiologically tolerate or adjust to environmental change. This knowledge will provide insight into how much environmental change particular species are likely to be able to tolerate, and how successfully the species may respond to novel climate conditions.

Very few studies have measured large-scale patterns of gene expression to study the effects of environmental variation in natural habitats (e.g., Place et al. 2008). The activity, or "expression," of many genes results in the production of a protein that carries out a cellular function. When cells are confronted by stressful or changing conditions, they often respond by expressing different genes that could alleviate the stress of the new environment (Schulte 2001). For example, the freshwater common carp possesses three different genes that each expresses a different version of a single muscle-contraction protein (Turay et al. 1991). Although these three proteins carry out the same function, muscle contraction, they differ in their ability to function at different tempera-

tures (Turay et al. 1991). The cells of the carp express different ratios of these three proteins depending on the temperature of the water in which the carp is swimming (Turay et al. 1991). As a result, the temperature recently experienced by a carp can be inferred simply from the pattern of expression of its muscle-contraction genes. However, we know little about which specific genes marine organisms might activate in order to live within complex fluctuating environmental settings, like those of coastal ecosystems, so investigating this question would benefit from a broader-scale analysis of gene expression.

A molecular tool, called the cDNA (complementary DNA) microarray, represents a powerful technique to assay the expression of thousands of genes simultaneously (Brown and Botstein 1999, Gracey and Cossins 2003). This allows us to investigate which genes vary by location and assumedly play a role in environmental tolerance. Here, custom cDNA microarrays constructed for *Strongylocentrotus purpuratus* (Stimpson), the purple sea urchin, were employed to investigate the association of gene expression and environmental variation among natural populations of *S. purpuratus* at a biogeographic scale among natural populations along the west coast of North America.

The purple sea urchin is an ideal organism to investigate the potential cellular impacts of environmental variation across a species' biogeographic range. One factor that makes it attractive for this type of study is the shape of its range. Its biogeographic range is extensive and essentially one-dimensional (e.g., Sagarin and Gaines 2002a), extending from southeast Alaska to Punta Eugenia in the middle of the Baja California peninsula (Figure 5.1) without extending too far from the shore, reaching a maximum depth of approximately 100 m. Thus, the purple sea urchin's range encompasses thousands of kilometers from the northern range boundary to the southern range boundary, but the east-west range is very narrow. The thermal variation across the full latitudinal extent of the range is substantial, with

Fogarty Creek, Oregon

Point Conception, California
Refugio, California
Puerto Kennedy, Baja California

FIGURE 5.1: A map of the geographic range (dark line) and collection sites (arrows) of the purple sea urchin, *S. purpuratus*.

sustained temperatures spanning from 20°C (68°F) in the south in the summer to 4°C (40°F) in Alaska in winter. However, the range also includes strong variation at much smaller scales: Two locations used in this study, Point Conception and Refugio, California, though separated by less than 100 km, display a mean annual thermal disparity of 3°C (5.4°F) because these two sites bracket a location of ocean currents convergence.

S. purpuratus is an important contributor to the ecology of coastal ecosystems (Pearse 2006), in that algal grazing by this species plays a role in the regulation of the size and abundance of kelp forests (Watanabe and Harrold 1991, Dayton et al. 1992), which host many California marine species including fish and marine mammals. Its sister species *S. franciscanus*, the red sea urchin, is the focus of a large California fishery (Botsford et al. 2004). This urchin is harvested in mass from California waters for sushi. Therefore, the impact of climate change on *S. purpuratus* and *S. franciscanus* could have considerable effects on kelp forest communities as well as the California sea urchin trade. These two sea urchin species are

very similar genetically, and thus, *S. purpuratus* offers an excellent model through which to investigate the potential impacts of climate change on California's sea urchin and kelp communities.

The central approach of this study was to measure gene expression patterns in natural populations of purple sea urchins along their biogeographic range as a proxy measure for their physiological tolerances in these environments. As typically framed on a large spatial scale, "rules of thumb" for how species respond to warming (i.e., based on climate envelope models built for other species, especially in terrestrial systems) would predict that the urchins located closest to each other geographically would display the most similar patterns of gene expression, because gene expression is influenced by environment. Alternatively, a geographic pattern of gene expression where pairs of more distant urchin populations display gene expression patterns just as similar or even more similar than pairs that are near each other could indicate that the typical framing on how species might shift geographically is a poor model for species in complex coastal habitats.

METHODS

1. Custom Microarray Construction

To construct a custom cDNA microarray, we first isolated RNA from several genes that may be of interest from *S. purpuratus* tubefoot tissue (see Osovitz and Hofmann 2005 for details on RNA isolation methods). For this microarray, RNA was isolated from over 200 *S. purpuratus* individuals that were exposed to a broad range of environmental conditions. The intent of this exposure was to collect isolated fragments of all of the genes that may be involved in environmental tolerance across the natural range of *S. purpuratus*. This collection of RNA fragments was then reverse transcribed into DNA (called cDNA). Each cDNA fragment was then stored separately in bacterial cells on plastic plates at −80°C (this process is called gene

TABLE 5.1
Details of the four field collection sites

Collection site	Latitude/ longitude	Collection date (2007)	Collection depth (m)	Temperature at collection (°C)
Fogarty Creek	44° 50′ N 124° 03′ W	Aug 28	1	10.5
Point Conception	34° 31′ N 120° 31′ W	Oct 15	12	10.5
Refugio	34° 27′ N 120° 04′ W	Aug 10	7	17.7
Puerto Kennedy	31°43′ N 116°42′ W	Aug 24	3	10.1

cloning). This type of collection of cDNA fragments is called a cDNA library, and it allows the isolated gene fragments to be stored indefinitely.

A total of 3072 gene clones from this library were sequenced in order to examine how many of the genes were unique, and how many were redundant. Of this total, 564 were found to be non-redundant and were selected for microarray printing (defined below). A total of 227 of these gene clones remain unidentified, meaning that we do not yet know the function of these genes. In addition to the 564 clones, an additional 232 unique cDNA clones, selected from results of preliminary temperature experiments (unpublished data), were added from a separate *S. purpuratus* cDNA library generated from urchin larvae.

A total of 796 (564 + 232) unique *S. purpuratus* genes were then printed in the following manner onto several cDNA microarrays, which were used for all analyses in this study. A single microarray is a regular glass microscope slide chemically coated to allow attachment of DNA segments. The "printing" is a procedure that transfers cDNA gene clones from our plastic plates onto the glass slides by a microarray printing robot. A very similar construction process is described in more detail by Place and Hofmann (2008).

2. Tissue Collection and Custom Microarray Use

Tissue collection and field temperature data

For this experiment, nine *S. purpuratus* individuals were collected by SCUBA from each of four subtidal sites (approximately 10–20 m depths) along the organisms' biogeographic range: (1) Fogarty Creek, Oregon, (2) Point Conception, California, (3) Refugio, California, and (4) Puerto Kennedy, Baja California (Table 5.1, Figure 5.1). Tissue samples (tubefoot) were dissected immediately after collection; tubefoot tissue was flash frozen in liquid nitrogen to preserve natural gene expression levels. During the tissue dissection period, adult sea urchins were suspended in seawater until 100–250 mg portions of tubefoot tissue could be excised from each individual. Field temperatures were acquired by a dive computer at the time of collection, and in addition for Refugio, by tidbit loggers attached to nearby moorings at 9 m depths.

cDNA sample preparation and microarray hybridization

RNA was isolated from the tubefoot tissue using the protocol similar to that used by Place et al. (2008), as described above. Microarray processing and analysis were carried out according to the methods described by Place et al. (2008). Briefly, three pooled samples were generated for each site, by combining equal RNA quantities of three individual *S. purpuratus* RNA preparations. The pooled RNA samples were then transcribed to cDNA and applied to the custom cDNA microarrays to competitively hybridize to the cDNA printed on the slides. This hybridization (which is when DNA strands naturally attach to each other) was done in the presence of fluorescent tags, which enabled later quantification of the RNA for each gene that was present in each sample at the time of collection. This quantification was carried out by a fluorescent microarray scanner and analyzed using scanner-specific software (see Place et al. 2008). After quality screening, data were analyzed using one-way ANOVA by site and Principal Component Analysis (PCA) analysis where groups of genes whose expression were correlated were used as principal components (see Raychaudhuri et al. 2000). The False Discovery Rate (FDR) method (Benjamini and Hochberg 1995) was used to control for error associated with multiple comparisons in the results of the one-way ANOVAs.

RESULTS

Geographic variation in gene expression

Only 195 of the 796 genes were found in sufficient abundance to be analyzed. Of these 195 gene comparisons, 36 were found to vary significantly using one-way ANOVA. Of these genes, 12 were previously identified and cataloged, eight were identified as hypothetical genes, and 16 remain as unidentified DNA segments. A summary of the results of PCA performed on the geographic expression are presented in Table 5.2. In the PCA, 60% of the total variation can be represented among the first three principal components (Table 5.2). However, the second and third components are the focus of this chapter, as each of these two components represented a different suite of genes that covaried to a relatively large degree among the samples. The reason is that there exist some genes that are expressed in all individuals at high levels, among them so-called

TABLE 5.2

Percentage of variance explained by each of the first three principal components, and the cumulative variance of all of the components

Component	Variance (%)	Total (%)
1	40.57	40.57
2	14.86	55.43
3	11.58	60.01

"house-keeping" genes required for basic cell function, and others that are rarely expressed. Since this study was concerned with variation among urchins exposed to different environments, not the gene-to-gene variation within individuals that comprised the first component of the statistical analysis, this component was ignored. Focusing on these components allows the analysis of geographic and environmental patterns of gene expression. Unfortunately, many of the genes represented in these principal components remain unidentified, so we are unable to analyze which biochemical pathways may be most affected by environmental variation. However, we are able to analyze the geographic patterns of expression of these gene suites to investigate which geographic sites have the most similar impact on the function of the cells of these purple sea urchins.

To visualize the relationships among the microarrays, each array was plotted onto components 2 and 3 (Figure 5.2). The individual arrays were generally assorted in groups within their sites (Figure 5.2). For comparisons among sites, Fogarty Creek, Oregon, and Puerto Kennedy, Baja California, overlapped very closely on component 2 and partially overlapped on component 3. The gene expression patterns observed in the urchins collected at Refugio, California, and Point Conception, California, both plotted onto their own coordinates of the graph (Figure 5.2).

Habitat temperatures

The habitat temperatures at the time of collection were similar for all sites [near 10°C (50°F)]

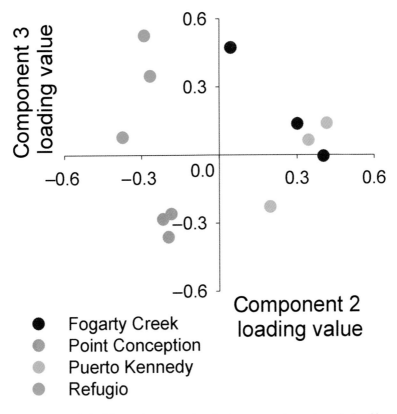

FIGURE 5.2: Each of the 12 microarrays plotted onto components 2 and 3 and colored by their geographic site.

except Refugio, which was approximately 17.7°C at the time of collection (Table 5.1). Over a longer timescale, however, the temperatures of the collection sites decreased from south to north by approximately 2–5°C between each site from Refugio, California, to Fogarty Creek, Oregon, for much of the month prior to tissue collection (Figure 5.3). Gene expression can change in response to short-term as well as long-term environmental conditions, so temperature over both of these timescales could have influenced the expression of genes in these sea urchin populations. Long-term temperature data for the Baja site were not available.

DISCUSSION

This study is one of the few to use functional measures to directly test the cellular state of natural populations across their range (e.g., Sagarin and Somero 2006, Lester et al. 2007, Place et al. 2008), by comparing natural populations of urchins near the middle of their range in Oregon to those closer to their range edge in Baja California. A central aim of this study was to utilize a multi-gene effort in an attempt to assemble environmental "fingerprints" of cellular and organismal state at varying biogeographic positions, and to use this information to infer the influences of environmental variation on natural marine populations. The results of this approach contrasted with the prediction that geographic distance would be negatively correlated with gene expression similarity. Instead, variation in gene expression was at least as pronounced over short distances (100 km) as larger ones (>1500 km). This finding supports the notion that the California coast is a variable complex region, driven by regional

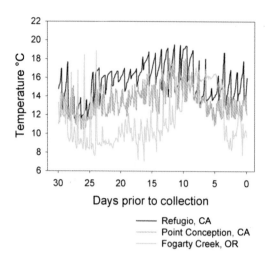

FIGURE 5.3: Thermal profiles near the collection sites of Fogarty Creek, OR; Point Conception, CA; and Refugio, CA, for 30 days prior to collection. Note: Data for the Baja California site were not available.

processes, such as upwelling and variation in coastal topography and ocean currents. These results also suggest that it is unlikely that a simple "climate envelope" model, which can be thought of as a temporal snapshot of a highly variable system, will be able to predict the biological impacts of climate change on California's marine coastal ecosystems. Change in the climate of this habitat could lead to such changes as range contractions and / or fragmentation based on small-scale geographic changes rather than generally predicted range shifts.

The central aim of the research described in this study was to assess the cellular expression states of *S. purpuratus* along the biogeographic range in order to better understand how it might respond to future environmental change. This was addressed by measuring gene expression levels with a custom cDNA microarray in natural populations of *S. purpuratus* at four sites along its range: (1) Fogarty Creek, Oregon, (2) Point Conception, California, (3) Refugio, California, and (4) Puerto Kennedy, Baja California. The central finding of the study is that the gene expression profiles of *S. purpuratus* located at the two most distant sites, Fogarty Creek, Oregon and Puerto Kennedy, Baja California, were the two most similar among all four sites.

These observed geographic patterns of gene expression were inconsistent with the predictions of a generalized climate envelope as typically envisioned on a large (e.g., continent-wide scale) scale, which would predict that urchins located at sites nearest to each other would experience the most similar physiological challenges, and thereby display the more similar gene expression patterns. The reasons for this unexpected geographic pattern are not readily apparent. One possible explanation is that these patterns are associated with the level of temporal environmental variation experienced at each site caused by upwelling, as was found by Stillman and Tagmount (2009) in porcelain crabs on the California coast. Indeed, the local temporal variation in temperature due to the timing of upwelling events was a better predictor of gene expression than mean habitat temperature in natural populations of these crabs (Stillman and Tagmount 2009). The frequency of coastal upwelling is possibly a factor in producing the gene expression profiles we found in the purple urchin, but the patterns of regional coastal upwelling are both spatially and temporally complex (Pickett and Schwing 2006), making this hypothesis difficult to test. Certainly, in theory, gene expression would more likely be affected by the high seawater-nutrient content and / or cool waters brought to the surface by coastal upwelling than by similarities in mean sea-surface water temperature. Complexity in upwelling patterns thus suggests that predicting the future spatial patterns of species associated with upwelling conditions will be particularly challenging.

A similar association between the gene expression profiles of spatially distant populations was found in intertidal *Mytilus californianus* mussels in central Oregon and Northern Baja, California (Place et al. 2008). These similarities between shallow water *S. purpuratus* sea urchins and intertidal *M. californianus* mussels suggest that intertidal communities may respond to changes due to climate change in a manner similar to how shallow subtidal species react, despite the fact that intertidal mussels are expected to respond predominantly to the timing of tides (Helmuth et al. 2002). Instead, the temporal variability in the environment, derived from night / day cycles, upwelling, etc., may be a large driver of these gene expression patterns (e.g., Stillman and Tagmount 2009).

Another example of the physiological effects of the complex California coastal ecosystem is the difference in *S. purpuratus* expression profiles across the area around Point Conception. The Point Conception and Refugio, California sites, though separated by less than 100 km, are also separated by Point Conception itself, which experiences colliding ocean currents. As a result, there is a mean annual thermal disparity of 3°C (5.4°F) across this point. The thermal disparity is similar to that observed between Point Conception and central Oregon. At Point Conception, colliding ocean currents carry waters southward from Alaska and northward from Southern California, producing a broad range of environmental conditions over a small geographic distance. This sharp thermal cline is believed to be a reason why Point Conception acts as a major marine zoogeographic boundary in California (Briggs 1974), again highlighting the spatial complexity of the California near-shore coastal ecosystem. These results suggest that the spatial complexity of environmental factors along the marine coast of California may invalidate the assumptions and thereby the predictions generated by a standard "climate envelope" model for these species.

Another aspect of the study that could have influenced our results is the differences in collection depth among the different sites (Table 5.1). While the overall pattern of gene expression that we observed did not correlate very well with the depth of collection, it cannot be ruled out that the similarity of the gene expression profiles of the Oregon and Baja California sea urchins could be a product of those urchins' relatively shallow collection depth. However, no study of such an effect has ever been conducted, so the potential for this effect is unknown.

One aspect that could confound the results of this study is that some *S. purpuratus* individuals may be more closely related to those living near-est to them, and we would expect more closely related individuals to have more similar gene expression patterns. However, the only examination of the genetic relatedness of *S. purpuratus* looked at the sequence of genes along its range, and found that the gene sequences of the Oregon sea urchins probably differ from those occurring in Baja California (S. Palumbi, pers. comm.), but our study found similarity in gene expression patterns between these two populations. Therefore, it seems that the pattern of gene expression similarity between Oregon sea urchins and those from Baja California found in this study is unlikely a result of genetic relatedness.

SOLUTIONS, ADAPTATION, AND LESSONS LEARNED

In the end, we are still left with the question "Where will *S. purpuratus* and *S. franciscanus* sea urchins go in response to climate change?" The answer to this question lies in the answers to two more basic questions: (1) What are the physiological or ecological limits of environmental change for California's marine species? And (2) how will patterns of environmental change manifest themselves along the California coast? The study described in this chapter was primarily aimed at answering the physiological aspect of the first question. Species' ecologies, for example, range shifts, will certainly play a role in determining the biological fate of California's marine ecosystems. However, while the potential biological impacts of climate change on the marine environment have been considered extensively (e.g., Parmesan and Yohe 2003, Helmuth et al. 2005, Harley et al. 2006), predictions of species responses have not.

On average, the surface seawater of California has warmed 1.3°C (2.3°F) over the last 50 years (Di Lorenzo et al. 2005), and while some California marine species' ranges have recently shifted poleward into historically cooler habitats (e.g., Zacherl et al. 2003, Sorte et al. 2010), the environmental complexity of the California coastal marine ecosystem could instead force species ranges to fragment, species to go extinct, or cause complex ecological shifts (e.g., Schiel et al. 2004, Helmuth et al. 2006). In other words, species may be isolated from historically cooler habitats by virtual blockades of warmer waters to the north of their range. While many marine invertebrates, such as sea urchins, mussels, and snails, do not travel far as adults, their larvae can float in the water column for up to a few weeks often traveling great distances before they morph into their adult form. However, since marine larvae are too small to fight against water currents, their movements are at the mercy of these currents, and current dynamics can also act as a dispersal barrier. Owing to the high levels of environmental variation in the intertidal zone, intertidal species may be more likely to go extinct than subtidal species (Stillman 2003). Shallow subtidal species, on the other hand, may be able to tolerate greater levels of environmental change (Stillman 2003) or migrate into deeper waters in response to climate change. Alternatively, the distributions of some shallow-water species may be regulated by factors other than temperature and pH, such as coastal topography (e.g., Broitman and Kinlan 2006), loss of "ecosystem engineers" (Moore et al. 2007), or a combination of several environmental factors (Scavia et al. 2002, Thompson et al. 2002). Likely, the biological impacts of climate change will be mediated synergistically through both physiological and ecological mechanisms (Clarke 1993, Sanford 1999).

California sea urchin species face not only ocean warming, but also acidification driven by rising CO_2 emissions and commercial harvesting. One of the main management tools at our disposal is the implementation of harvesting regulations, such as those in Marine Protected Areas (MPAs) to limit the fishing stress. However, no-take zones, such as MPAs, also have complex ecological effects and have been shown to limit urchin populations through increases in fish- and lobster-predator abundance (Behrens and Lafferty 2004). Climate change, particularly ocean acidification, which may negatively affect sea urchin growth rates (O'Donnell et al. 2010), in conjunction with harvesting and MPAs, may increase the rate of sea urchin decline in Califor-

nia. In this case, urchin-specific MPAs may be recommended in areas that are currently cooler, such as Northern California and Oregon, where urchins are currently abundant and living substantially below their upper temperature limits.

LITERATURE CITED

Behrens, M. D. and K. D. Lafferty. 2004. Effects of marine reserves and urchin disease on southern Californian rocky reef communities. *Marine Ecology-Progress Series* 279:129–139.

Benjamini, Y. and Y. Hochberg. 1995. Controlling the false discovery rate: A practical and powerful approach to multiple testing. *Journal of the Royal Statistical Society: Series B* 57:289–300.

Botsford, L. W., A. Campbell, and R. Miller. 2004. Biological reference points in the management of North American sea urchin fisheries. *Canadian Journal of Fisheries and Aquatic Sciences* 61:1325–1337.

Brierley, A. S. and M. J. Kingsford. 2009. Impacts of climate change on marine organisms and ecosystems. *Current Biology* 19:R602–R614.

Briggs, J. C. 1974. *Marine Zoogeography*. McGraw Hill, New York, NY.

Broitman, B. R. and B. P. Kinlan. 2006. Spatial scales of benthic and pelagic producer biomass in a coastal upwelling ecosystem. *Marine Ecology-Progress Series* 327:15–25.

Brown, P. O. and D. Botstein. 1999. Exploring the new world of the genome with DNA microarrays. *Nature Genetics* 21:33–37.

Burrows, M. T., D. S. Schoeman, L. B. Buckley, P. Moore, E. S. Poloczanska, K. M. Brander, C. Brown, J. F. Bruno, C. M. Duarte, B. S. Halpern et al. 2011. The pace of shifting climate in marine and terrestrial ecosystems. *Science* 334:652–655.

Clarke, A. 1993. Temperature and extinction in the sea – A physiologist's view. *Paleobiology* 19:499–518.

Dayton, P. K., M. J. Tegner, P. E. Parnell, and P. B. Edwards. 1992. Temporal and spatial patterns of disturbance and recovery in a kelp forest community. *Ecological Monographs* 62:421–445.

Di Lorenzo, E., A. J. Miller, N. Schneider, and J. C. McWilliams. 2005. The warming of the California current system: Dynamics and ecosystem implications. *Journal of Physical Oceanography* 35:336–362.

Fields, P. A., J. B. Graham, R. H. Rosenblatt, and G. N. Somero. 1993. Effects of expected global climate-change on marine faunas. *Trends in Ecology & Evolution* 8:361–367.

Gracey, A. Y. and A. R. Cossins. 2003. Application of microarray technology in environmental and comparative physiology. *Annual Review of Physiology* 65:231–259.

Harley, C. D. G., A. R. Hughes, K. M. Hultgren, B. G. Miner, C. J. B. Sorte, C. S. Thornber, L. F. Rodriguez, L. Tomanek, and S. L. Williams. 2006. The impacts of climate change in coastal marine systems. *Ecology Letters* 9:228–241.

Helmuth, B. E., B. R. Broitman, C. A. Blanchette, S. Gilman, P. Halpin, C. D. G. Harley, M. J. O'Donnell, G. E. Hofmann, B. Menge, and D. Strickland. 2006. Mosaic patterns of thermal stress in the rocky intertidal zone: Implications for climate change. *Ecological Monographs* 76:461–479.

Helmuth, B., C. D. G. Harley, P. M. Halpin, M. O'Donnell, G. E. Hofmann, and C. A. Blanchette. 2002. Climate change and latitudinal patterns of intertidal thermal stress. *Science* 298:1015–1017.

Helmuth, B., J. G. Kingsolver, and E. Carrington. 2005. Biophysics, physiologicalecology, and climate change: Does mechanism matter? *Annual Review of Physiology* 67:177–201.

Hijmans, R. J. and C. H. Graham. 2006. The ability of climate envelope models to predict the effect of climate change on species distributions. *Global Change Biology* 12:2272–2281.

Hoegh-Guldberg, O. and J. F. Bruno. 2010. The impact of climate change on the world's marine ecosystems. *Science* 328:1523–1528.

Lester, S. E., B. P. Kinlan, and S. D. Gaines. 2007. Reproduction on the edge: Large-scale patterns of individual performance in a marine invertebrate. *Ecology* 88:2229–2239.

Moore, P., S. J. Hawkins, and R. C. Thompson. 2007. Role of biological habitat amelioration in altering the relative responses of congeneric species to climate change. *Marine Ecology-Progress Series* 334:11–19.

O'Donnell, M. J., A. E. Todgham, M. A. Sewell, L. M. Hammond, K. Ruggiero, N. A. Fangue, M. L. Zippay, and G. E. Hofmann. 2010. Ocean acidification alters skeletogenesis and gene expression in larval sea urchins. *Marine Ecology Progress Series* 398:157–171.

Osovitz, C. J. and G. E. Hofmann. 2005. Thermal history-dependent expression of the hsp70 gene in purple sea urchins: Biogeographic patterns and the effect of temperature acclimation. *Journal of Experimental Marine Biology and Ecology* 327:134–143.

Osovitz, C. J. and G. E. Hofmann. 2007. Marine macrophysiology: Studying physiological variation across large spatial scales in marine

systems. *Comparative Biochemistry and Physiology A-Molecular & Integrative Physiology* 147:821–827.

Parmesan, C. and G. Yohe. 2003. A globally coherent fingerprint of climate change impacts across natural systems. *Nature* 421:37–42.

Pearse, J. S. 2006. Perspective – Ecological role of purple sea urchins. *Science* 314:940–941.

Pickett, M. H. and F. B. Schwing. 2006. Evaluating upwelling estimates off the west coasts of North and South America. *Fisheries Oceanography* 15:256–269.

Place, S. P., M. J. O'Donnell, and G. E. Hofmann. 2008. Patterns of gene expression in the intertidal mussel, *Mytilus californianus*: Physiological response to environmental factors on a biogeographic scale. *Marine Ecology-Progress Series* 356:1–14.

Raychaudhuri, S., J. Stuart, and R. Altman. 2000. Principal components analysis to summarize microarray experiments: Application to sporulation time series. *Pacific Symposium on Biocomputing* 5:452–463.

Root, T. L., J. T. Price, K. R. Hall, S. H. Schneider, C. Rosenzweig, and J. A. Pounds. 2003. Fingerprints of global warming on wild animals and plants. *Nature* 421:57–60.

Sagarin, R. D. and S. D. Gaines. 2002a. The 'abundant centre' distribution: To what extent is it a biogeographical rule? *Ecology Letters* 5:137–147.

Sagarin, R. D. and G. N. Somero. 2006. Complex patterns of expression of heat-shock protein 70 across the southern biogeographical ranges of the intertidal mussel *Mytilus californianus* and snail *Nucella ostrina*. *Journal of Biogeography* 33:622–630.

Sanford, E. 1999. Regulation of keystone predation by small changes in ocean temperature. *Science* 283:2095–2097.

Scavia, D., J. C. Field, D. F. Boesch, R. W. Buddemeier, V. Burkett, D. R. Cayan, M. Fogarty, M. A. Harwell, R. W. Howarth, C. Mason et al. 2002. Climate change impacts on US coastal and marine ecosystems. *Estuaries* 25:149–164.

Schiel, D. R., J. R. Steinbeck, and M. S. Foster. 2004. Ten years of induced ocean warming causes comprehensive changes in marine benthic communities. *Ecology* 85:1833–1839.

Schulte, P. M. 2001. Environmental adaptations as windows on molecular evolution. *Comparative Biochemistry and Physiology Part B: Biochemistry & Molecular Biology* 128:597–611.

Sorte, C. J. B. and G. E. Hofmann. 2004. Changes in latitudes, changes in aptitudes: *Nucella canaliculata* (Mollusca: Gastropoda) is more stressed at its range edge. *Marine Ecology-Progress Series* 274:263–268.

Sorte, C.J.B., S.L. Williams, and J.T. Carlton. 2010. Marine range shifts and species introductions: Comparative spread rates and community impacts. *Global Ecology and Biogeography* 19:303–316.

Stillman, J.H. 2003. Acclimation capacity underlies susceptibility to climate change. *Science* 301:65.

Stillman, J.H. and A. Tagmount. 2009. Seasonal and latitudinal acclimatization of cardiac transcriptome responses to thermal stress in porcelain crabs, *Petrolisthes cinctipes*. *Molecular Ecology* 18:4206–4226.

Thompson, R.C., T.P. Crowe, and S.J. Hawkins. 2002. Rocky intertidal communities: Past environmental changes, present status and predictions for the next 25 years. *Environmental Conservation* 29:168–191.

Turay, L., G.F. Gerlach, and G. Goldspink. 1991. Changes in myosin HC gene-expression in the common carp during acclimation to warm environmental temperatures. *Journal of Physiology* 435:102.

Watanabe, J.M. and C. Harrold. 1991. Destructive grazing by sea-urchins *Strongylocentrotus* spp in a central California kelp forest – Potential roles of recruitment, depth, and predation. *Marine Ecology-Progress Series* 71:125–141.

Zacherl, D., S.D. Gaines, and S.I. Lonhart. 2003. The limits to biogeographical distributions: Insights from the northward range extension of the marine snail, *Kelletia kelletii* (Forbes, 1852). *Journal of Biogeography* 30:913–924.

Manager Comments

Christopher J. Osovitz
in conversation with
Deborah Aseltine-Neilson

Osovitz: To your knowledge, are temperature gradients considered in locations for marine protected areas, or in other designations that limit fishing in California?

Aseltine-Neilson: Yes, temperature gradients are considered in locations for management of species. Currently, the California Department of Fish and Wildlife has divided the California coast info four management regions, and some biogeographic regions have been identified within those regions. These four areas are used for wildlife management, but fisheries management as well. The understanding is that these regions will likely change as California's climate changes.

Osovitz: To the extent that climate change is currently considered in marine spatial planning and policy, would you say that in general there is an assumption that marine species will "simply" shift toward the north? Are there good examples of plans or policies that have tried to delve into the complexity of factors that may influence species distributions, and potential responses to changes in ocean conditions?

Aseltine-Neilson: This is a relatively new area of research. As of now, there is no scientific consensus in predictions regarding the biogeographic responses of California's marine species. This is partly because the geography of the physical conditions is complicated by oceanographic factors, such as upwelling, currents, and El Nino. As of now, we are not beginning with the prediction that species will simply shift northward. Since we do not have the specifics of all of these factors, we are not making specific predictions regarding future changes in biogeographic ranges. As we learn more, these factors will be considered in future action plans.

Osovitz: To what extent is ocean acidification, and potentially higher vulnerability of species like urchins that produce calcareous shells, figuring into marine conservation?

Aseltine-Neilson: We certainly recognize that ocean acidification is a potentially very serious issue, but there isn't very much information on specifically how ocean acidification might impact California's ecosystems yet. The research is still be compiled, but the California Department of Fish and Wildlife is very interested in seeing the results of these studies. One way that this type of change may impact management strategies is that these strategies may become more precautionary, as there is some data that suggest that ocean acidification could have the largest impacts of any of the changing climate factors. One question left unanswered is whether the current species have the genetic makeup to tolerate acidification. For sea urchins, for example, their ability to produce roe in an acidified climate is the most important factor for fisheries. Do these species currently possess the genes to produce roes in acidified environments? Another system that we are monitoring is marine pteropods, which are tiny shelled mollusks. These tiny creatures play an important role in foodwebs, which support important commercial fisheries, such as salmon. As a result, it is likely that in the future we will integrate an ecosystem management approach with approaches intended to mitigate the biological impacts of climate change. But again, before any of these approaches are included in management action plans in California, we will need to know much more about the potential effects of OA on California's ecosystems.

CHAPTER 6

Integrating Global Climate Change and Conservation

A KLAMATH RIVER CASE STUDY

Rebecca M. Quiñones

Abstract. Climate change is expected to degrade all habitats used by anadromous fishes—those that migrate between rivers and oceans. For anadromous Pacific salmonids (*Oncorhynchus* spp.), the effect of climate change is likely to be severe in California, which is near the southern end of their range in North America. Most wild salmon and steelhead trout populations in the state are already under stress and in decline. Climate change is altering flow patterns in rivers, including the timing and magnitude of droughts and floods. Estuaries are being altered by more frequent and extreme tides and storms, and are experiencing changes in salinity associated with droughts and sea level rise. Water temperatures are predicted to steadily increase throughout the 21st century, likely beyond salmonid tolerances. In the case of the Klamath River basin, salmonid abundances may decrease by more than 50% of current numbers by 2100 unless climate change effects are actively considered in conservation efforts. Owing to the uncertainty and difficulty in addressing climate change effects in the ocean, our discussion focuses on conservation actions that can

Key Points

- Climate change effects are decreasing the suitability of freshwater and estuarine habitats for anadromous salmonids in the Klamath River basin, but the magnitude of effects will differ by species, life history, and location; this variation suggests that we need to think at both local and regional scales as we devise and prioritize conservation actions.
- Anadromous salmonids will respond to climate change effects in many, and perhaps novel, ways. As a result, we need to be prepared to rapidly change our approaches in light of new information.
- The combined effects of climate change and existing stressors are expected to result in declining salmonid abundances and populations with more restricted distributions.
- Salmonids need access to a variety of suitable rearing and spawning habitats if their resiliency is to be enhanced; the prioritization of restoration efforts to increase habitat should include consideration of key climatic factors.

(continued)

(continued)

- Reform of current hatchery practices is needed in order to protect the genetic and life-history diversity necessary for salmonids to adapt to climate change. Conservation actions that protect and restore cold-water habitats, maintain and enhance stream flow, and mitigate impacts of existing stressors (including those from hatcheries) offer the best opportunity for salmonids to endure changing environmental conditions.

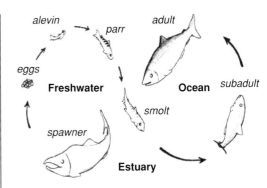

FIGURE 6.1: Salmonid life cycle depicting developmental stages in freshwater (rivers and streams), estuary, and ocean.

improve conditions in streams and estuaries. Salmonids need cold water and access to a variety of habitats, which is why we recommend actions that could be implemented rapidly to increase the likelihood of salmonid persistence in the face of climate change. Recommended conservation actions include the removal of barriers to migration, and reductions in the volume of diverted flows. Conservation of Klamath basin salmonids will also require aggressive strategies that protect genetic and life-history diversity.

INTRODUCTION

The suitability of rivers in the United States for supporting salmon and trout (salmonids) is estimated to decrease by 4–20% by 2030 and by as much as 60% by 2100 (Eaton and Scheller 1996), with the greatest loss projected for California (O'Neal 2002). Because climate change will affect the efficacy of existing and future conservation efforts, resource managers who consider the impacts of climate change on Pacific salmonids (*Oncorhynchus* spp.) are more likely to reach their conservation goals (Battin et al. 2007, Schindler et al. 2008).

In North America, California marks the southern end of the range of six species of anadromous salmonids: Chinook salmon (*Oncorhynchus tshawytscha*), coho salmon (*O. kisutch*), pink salmon (*O. gorbuscha*), chum salmon (*O. keta*), steelhead trout (*O. mykiss*), and coastal cutthroat trout (*O. clarki clarki*)

(Moyle 2002). The general salmonid life cycle is one of adults spawning in streams, eggs incubating in stream gravel, juveniles rearing in rivers and estuaries, and subadults growing in the ocean until sexual maturity (Figure 6.1). However, several species display diverse life histories expressed as differences in the timing and duration of migration and residency between habitats. For example, Chinook salmon can begin upstream spawning migrations in the summer, fall, or winter, at the age of 2–5 years, and rear in freshwater for weeks to months (Groot and Margolis 1991). Life histories specific to each species are largely governed by water temperatures and flow regimes of natal streams (Moyle 2002), but can also be rooted in the genetics of populations (Allendorf and Luikart 2007). Thus, climate change can strongly affect salmonids by changing aquatic habitat conditions (i.e., water temperature, streamflow patterns, stream sediment composition) that influence life-history expression.

The ability of salmonid populations to persist in the face of climate change will largely be determined by their genetic and life-history diversity if suitable habitat is available. A population's adaptive capability (Schindler et al. 2008) has been defined by the magnitude of its genetic variability (Reusch et al. 2005). High genetic and life-history diversity can bolster species viability, even when some populations are lost to adverse environmental conditions (Greene et al. 2010). Exist-

TABLE 6.1
Coho salmon life history, habitat used, and optimal habitat characteristics

Life history stage	Habitat used	Optimal habitat characteristics
Migrating adults (to spawning areas)	Pacific Ocean, estuary, main stem, tributaries	Estuary—open river mouth; increasing flow Tributaries—cold (~8–15°C), clean, flowing water with pools Minimum water depth ~18 cm Maximum water velocity ~2.44 m/s
Spawning adults (spawner)	Tributaries	Clean gravel (~4–14 cm in diameter) Water velocity <1 m/s Minimum water depth of ~10 cm Water temperature <12°C
Incubating embryos (egg)	Tributaries	Intergravel flow Low silt loading Water temperature 8–10°C
Alevin/fry	Tributaries	Low concentration of fine (<3.5 mm) sediment in gravel beds Water temperatures <12°C
Rearing juveniles (parr)	Tributaries	Slow water habitat, especially pools Lots of cover (e.g., fallen trees, overhanging vegetation) Lots of food (e.g., aquatic insects) Water temperature <15°C
Out-migrating juveniles	Tributaries, main stem, estuary	Tributaries—increasing flow and temperature (but <16°C) Main stem—combination of slow and fast moving water
Smoltifying juveniles (smolts)	Main stem, estuary	Main stem—cool water (<16°C), unobstructed access Estuary—cool, productive slough habitats; open river mouth
Rearing subadult	Pacific Ocean	Highly productive coastal habitats (strong upwelling) Sea surface temperature <9°C

SOURCE: Groot and Margolis (1991), Richter and Kolmes (2005), Moyle et al. (2008), NOAA (2011).

ing stressors such as hatchery supplementation and fisheries overharvest can decrease salmonid life-history and genetic diversity (Lackey et al. 2006). Because both climate change effects (Bottom et al. 2011) and loss of life-history diversity (Allendorf and Luikart 2007) increase the likelihood of extinction, we consider their impacts on anadromous salmonids.

Salmonids are adapted to thrive in streams characterized by cold (usually <18–20°C), clear water with high levels of dissolved oxygen (Moyle 2002). Temperature plays an important role during all life stages by affecting oxygen uptake, metabolic rate, growth, and immune function. In general, salmonids can withstand temperatures from 0°C to 25°C, but physiological processes are typically optimal at temperatures between 4°C and 20°C (Moyle 2002, Richter and Kolmes 2005). However, coho and Chinook salmon in some basins (e.g., Klamath River) are able to withstand temperatures 2–3°C higher than those within the conventional tolerance range for each species (Sutton et al. 2007, Strange 2010), perhaps due to their evolution in these warmer environments.

Current and projected changes in climate factors, such as increased temperatures and decreased streamflow, shift the environmental conditions that influence species with complicated life histories (e.g., coho salmon; Table 6.1) and the stressors already acting on them. To conserve salmonids, threats to species and habitats need to be considered concurrently with climate change and specific life-history traits. Here, we briefly discuss threats salmonids face in the Klamath basin, review potential impacts

TABLE 6.2

Taxa (including distinct life histories) of anadromous salmonids in the Klamath River basin

Species / life history	Spawning / rearing habitat	Time in freshwater (F)[1], estuary (E), Pacific Ocean (PO)[2]	Reproductive strategy; peak spawning run
Steelhead trout (*O. mykiss*) Winter run* Summer run**	*Tributary streams **Upper elevation tributaries (migrate farthest upstream)	F: Few months to 2 years E: Days to months PO: 1–2 years	Iteroparous[3]; *November–December **April–June
Chinook salmon (*O. tshawytscha*) Fall run* Spring run**	*Lower elevation tributaries and main stem Klamath River **Tributary streams and upper elevation tributaries	F: Few months to 1 year E: Days to weeks PO: 1–5 years	Semelparous[4]; *October–December **May–July
Coho salmon (*O. kisutch*)	Tributary streams and upper elevation tributaries	F: One year E: Days to months, PO: 1–2 years	Semelparous; October–November
Pink salmon (*O. gorbuscha*)	Main stem Klamath River	F: Days to weeks E: Days to weeks PO: 2–3 years	Semelparous; September–October
Chum salmon (*O. keta*)	Main stem Klamath River	F: Days to weeks E: Weeks to months PO: 2–7 years	Semelparous; October–December
Coastal cutthroat trout (*O. clarki clarki*)	Main stem Klamath River and lower elevation tributaries (migrate shortest distance)	F: 2–3 years E: Weeks to months PO: Months	Iteroparous; August–October

SOURCE: Groot and Margolis (1991), Moyle (2002).

1. Time in freshwater refers to duration of residency by juveniles rearing in freshwater.
2. Time in Pacific Ocean refers to duration of residency by subadults and adults in the ocean.
3. Iteroparous = spawning more than once.
4. Semelparous = spawning only once.

of climate change on salmonid habitats and populations, and propose conservation actions to better manage species as environmental conditions change. We focus our discussion on the Klamath River in northwestern California because it is the second largest river in the state, and currently or previously supported all six anadromous salmonid species, consisting of eight taxonomic units (Table 6.2).

STUDY AREA

The Klamath River flows into the Pacific Ocean about 65 km south of the Oregon / California border, draining approximately 40,400 km² in northern California and southern Oregon (Figure 6.2). The river below Iron Gate Dam was designated in 1981 by the U.S. Congress as a Wild and Scenic River because of the value and diversity of its anadromous fisheries. However, abundances of some species have declined so that they are now a mere fraction of historical levels (Hamilton et al. 2011). Anadromous salmonids are subject to multiple stressors, including water diversions, dam operations, sedimentation, hatchery supplementation, and habitat degradation (NRC 2004, Moyle et al. 2008, Katz et al. 2013). The cumulative impact

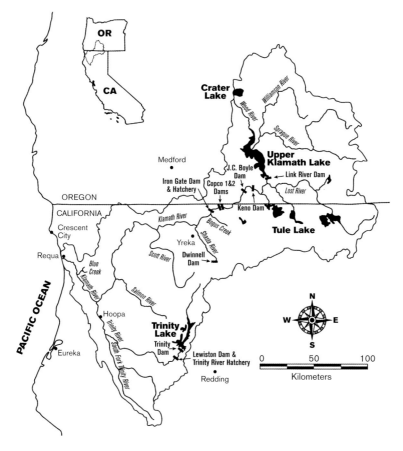

FIGURE 6.2: The Klamath River basin in Oregon and California.

from these stressors has resulted in many once-suitable habitats becoming suboptimal for salmonid survival, and many populations becoming vulnerable to extinction (NRC 2004). Climate change is already intensifying some of these stressors and will likely create new ones (Moyle et al. 2013).

EFFECTS OF CLIMATE CHANGE ON SALMONID HABITATS IN THE KLAMATH RIVER BASIN

Climate change is already stressing, and will further stress salmonids in California and Oregon by additionally degrading aquatic habitats beyond the level of degradation attributable to nonclimate stressors. While indirect effects (e.g., wildfire frequency) and oceanic conditions are also important factors, here, we focus

our discussion on the direct impacts that climate change will likely have on river and estuarine habitats.

Effects on River Habitats

The quality of salmonid habitat in California rivers is already changing as a result of increases in air and water temperatures, decreases in snowpack, and changes in precipitation patterns (Field et al. 1999, Spence and Hall 2010). Based on an estimate of a 2°C increase in air temperatures in the western United States, average annual water temperatures are estimated to have increased by 0.7°C in the last 100 years (based on a conversion factor of 0.35; Eaton and Scheller 1996). Although the magnitude of temperature change may be seen as small, it can result in suboptimal and even lethal conditions

TABLE 6.3

Maximum temperatures (°C) critical to Chinook salmon, coho salmon and steelhead trout survival at different life stages, with a brief description of their effects

Temperatures in bold are reported to result in mortality. Letters (s) and (f) denote Chinook spring and fall run life histories

	Species			Effect of exposure to maximum (near-lethal) temperatures (°C)
Life stage	Chinook	Coho	Steelhead	
Incubation	9–12 **14**	11–12 **13–14**	15–16	Reduction in alevin survival
Juvenile rearing / growth	14–20	15–16.5 **22–26**	20.5–22.5 **24**	Reduction in growth, induction of heat shock proteins. Varies with food availability
Smoltification	12 (s) 17 (f) 18	12–15.5	11–14	Inhibits activity of gill ATPase, reduces ability to osmoregulate, loss of migratory behavior
Adult migration	19–24	15.6 **21–22**	16–22 **21–22**	Stops upstream migration
Spawning	14.5	12–20	13	Reduction in spawning activity and quality of eggs

SOURCE: Adapted from Richter and Kolmes (2005).

to some salmonids already residing in the Klamath River (Table 6.3), where summer water temperatures often exceed 22°C (CDEC 1986, McCullough 1999). Changes in water temperatures, however, vary among watersheds due to location and primary source of water (i.e., snowmelt and rainfall, or groundwater).

Stream and river sections along the coast are usually cooler in summer than inland streams due to the presence of fog. Areas in the fog belt are associated with cooler air temperatures because fog intercepts solar radiation that would otherwise warm stream temperatures (Lewis et al. 2000). In the Klamath River basin, maximum air temperatures in August average 18.9°C for coastal areas within the fog belt, 32.8°C for inland areas at low elevation, and 27.8°C for inland areas at high elevation (Lewis et al. 2000). Streams within the fog belt therefore can offer suitable habitat to cold-water fishes like salmonids during the warmest part of the year. Increased temperatures associated with climate change, however, are changing

wind patterns that influence fog formation (Johnstone and Dawson 2010). Fog frequency since the beginning of the 20th century has decreased by 33%, and this change is already contributing to warming daytime temperatures along the northern California coast (Johnstone and Dawson 2010).

The primary source of water to streams can also exacerbate (i.e., snowmelt and rainfall) or moderate (i.e., groundwater) the impacts of climate change. Snowmelt-fed streams (e.g., Salmon and Scott rivers) are projected to be warmer and drier during the summer and fall months due to reduction in total snowpack and seasonal retention of snow (Hamlet et al. 2005, Stewart et al. 2005). Several monitoring stations in the Klamath basin show that snowpack water content in the last 50 years has significantly declined (Van Kirk and Naman 2008). Lower flows further exacerbate increasing water temperatures, as shallower waters show more of a temperature increase with increasing air temperature (Allan and Castillo 2007). Cli-

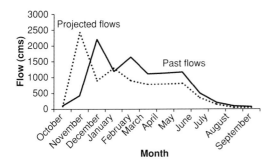

FIGURE 6.3: Illustration of potential changes in flows potential [Cubic Meter Per Second (cms)] of an undammed river (Salmon River) in the Klamath River basin, based on the projection of a 10% increase in winter flow, 30% reduction in spring and summer flows, and a 30-day shift in peak flows (as in the work of Leung et al. 2004, Kim 2005, Stewart et al. 2005).

mate change is also expected to alter streamflow patterns by increasing winter runoff as precipitation increasingly falls as rain rather than snow, likely decreasing spring and summer streamflow, and increasing the occurrence of winter floods and summer droughts (Knox and Scheuring 1991, Field et al. 1999). With increased temperatures causing earlier snowmelt, the timing of peak flows has already shifted earlier by 10–30 days (Stewart et al. 2005), with the largest shifts in more recent decades (Cayan et al. 2001). Flows in undammed snowmelt-fed rivers in the Klamath basin usually peak in winter with a second, smaller peak in spring, and then gradually decrease to lowest levels in summer (Figure 6.3). If changes in flow regimes continue at the current rate, then streamflow in the Klamath River basin is expected to decrease by 10–50% in the spring and summer, while the frequency of extreme high and low flows increases by 15–20% (Leung et al. 2004, Kim 2005).

Streams fed by a steady supply of cool groundwater are generally cooler in the summer and warmer in the winter than those fed by surface runoff (e.g., snowmelt). In the Klamath basin, a significant number of groundwater springs flow into the Shasta River (Nichols 2008), Wood River, lower Williamson River, and along the margin of the Cascade Range

(Gannett et al. 2010). Large springs may be expected to dampen water-temperature changes on the order of decades (Manga 1999) but to a lesser degree as air temperatures continue to increase. Spring-fed rivers are also expected to flow year-round due to large groundwater sources (Thompson 2007, Tague et al. 2008), with less interannual variation in flow (Jefferson et al. 2007). However, groundwater may become depleted as water demands increase with a growing human population and streamflows diminish (Hanak et al. 2011). Water temperatures and flows in streams above Iron Gate Dam are more likely to be more buffered from climate change than those downstream of the dam.

Increases in water temperatures will strongly affect the physiology and behavior of salmonids throughout their life histories. The resulting changes in survival and reproductive success will likely be complex (Table 6.3). Changes in movement patterns are likely to be the most obvious response of salmonid individuals to climate change, particularly as fish are exposed to increases in water temperature and changes in streamflow patterns. Most behavioral responses in salmonids are triggered by temperature thresholds and changes in flow (Groot and Margolis 1991). Because temperature increases will hasten developmental rates, and streamflows will peak earlier in the year, the migration patterns of Klamath salmonids will likely shift to earlier in the year. However, photoperiod (day length) at a given site can also influence the initiation of salmonid migrations, so migration may become unsynchronized with temperature (Feder et al. 2010). Another behavioral response to increased temperatures is the movement of salmonids into colder waters as a method of thermoregulation. Salmonids use cold-water pockets (thermal refugia) in rivers during juvenile rearing and adult migration when water temperatures exceed 22°C (Nielsen et al. 1994, Ebersole et al. 2003, Strange 2010). In summer, use of thermal refugia may make juveniles less susceptible to disease (Foott et al. 1999). Climate change could diminish or eliminate cold-water pockets as temperatures increase. The reduction of suitable freshwater habitat is expected to result in a northward and or higher elevation shift in the range of cold-water fishes where cold-water habitat is accessible (Mohseni et al. 2003, Battin et al. 2007). As a result, salmonids in the Klamath River basin may experience local extinctions, range contractions, and abundance losses of up to 50%, following the predictions of Chatters et al. (1995) and Crozier et al. (2008b) for rivers further north.

Changes in flow patterns due to changes in precipitation may decrease salmonid embryo and juvenile survival by changing suitable habitat conditions. Extreme high flows that move stream gravel can expose eggs by scouring redds, flush juveniles out of the river system before they reach critical size, and desynchronize juvenile arrival into the ocean with the spring phytoplankton bloom (Mote et al. 2003). Fine (<4 mm) sediment washed into streams by intense storm events can smother redds, preventing oxygen from reaching developing embryos or acting as a barrier to fry emergence (Furniss et al. 1991). Decreases in summer and fall flows may increase juvenile mortality through stranding, if flows go subsurface or change faster than juvenile movements. Changes in the timing of spring and fall flows may reduce survival of juveniles migrating from rivers into the ocean (Lawson et al. 2004).

Salmonid survival is simultaneously affected by multiple, often conflicting, factors. For example, survival of juvenile Chinook salmon in the Klamath River may be positively correlated with flow because of increased habitat availability but negatively correlated with high juvenile abundance due to competition between conspecifics for limited resources (Crozier and Zabel 2006). In another case, an increase in high-flow events may reconnect the Klamath River to its floodplain when overflowing river channels spill water to adjacent land, benefitting salmonid populations because highly productive floodplains can provide juvenile salmon with optimal rearing habitat (Jef-

fres et al. 2008). But as discussed earlier, high-flow events that occur too early may scour redds or carry small juveniles into the estuary before they are ready to smolt.

Similar variation in responses to changes in river flow and temperature will occur for adult salmon. Decreases in summer flows may increase adult survival because the metabolic cost of swimming against the current will be reduced (Crozier and Zabel 2006). Yet, high mortality of migrating adults may occur if low, warm flows halt upstream movement. In addition, spawners migrating in warm water use more energy for upstream migration than spawners in cold water, with this increase in stress resulting in higher prespawning mortality from disease, greater vulnerability to predators, and lowered egg quality. Consequently, it is difficult to predict specifically how salmonids will respond to climate change effects due to the many mechanisms for positive and negative impacts, but management practices can aim to provide flexibility in responses that enhance salmonid resilience (Bottom et al. 2011).

Effects on Estuarine Habitats

Productivity in the Klamath River estuary will likely change with altered freshwater, sediment, and nutrient delivery. Decreases in freshwater delivery during the summer may push the estuary toward eutrophication because accumulating organic matter may not be flushed out to sea, increasing the risk of localized anoxic conditions (Moore et al. 1997). Projections are that sea levels may increase by as much as 2 m by 2100 (Allison et al. 2009), worsening the impacts from extreme tidal, storm, and flooding events (Cayan et al. 2008). The rates and magnitude of salt water and freshwater mixing within estuaries will change as temperatures increase, sea level rises, and streamflows become more variable. As a result, salinities are likely to vary among seasons, with concentrations highest in the summer and lowest in the winter (Dettinger and Cayan 1995,

Knowles and Cayan 2004). In addition, sea-level rise and increased rates of erosion caused by more frequent and stronger tides and storms may become other agents to estuarine habitat change (Scavia et al. 2002).

Juvenile salmonids use estuaries to varying degrees (Table 6.1). Chinook salmon and steelhead trout can rear in estuaries for weeks to several months (Healey 1991, Bottom et al. 2005), while coho salmon usually reside in estuaries for shorter periods (Miller and Sadro 2003). Length of estuarine residency can vary according to temperature, streamflow, ocean productivity, and condition of individual fishes (Healey 1982, Kjelson et al. 1982), all factors that can be altered by climate change. Studies of juvenile Chinook in the Klamath River estuary concluded that they did not rear extensively in the estuary (Sullivan 1989, Wallace and Collins 1997). Rearing habitat dimensions, emigration timing, and salmonid condition upon entering the estuary were dependent on streamflow. Warmer temperatures in the river likely forced juvenile Chinook salmon to enter the estuary earlier, resulting in smaller sizes at ocean entry and the potential for lower survival in the ocean (Zabel and Williams 2002).

Saltwater intrusion from the ocean, and therefore salinity, is controlled by river flow into the estuary and sand bar formation at the river mouth. When river flows are high or a sand bar is formed, salinities in the estuary remain low. Increases in storm surges expected with climate change may constrain sand bar formation and increase saltwater intrusion from the ocean, potentially changing estuarine salinity and invertebrate communities (Winder et al. 2011). Migrating juvenile salmonids can benefit from cooler water temperatures and larger foraging base in the estuary (Levings et al. 1991). Salmonids use estuaries more extensively during periods of poor ocean conditions, in order to take advantage of abundant prey. However, salmonids may be at greater risk of predation by marine fish, mammals, and birds if the duration of estuarine residency increases (Chittenden et al. 2010b). In particular, habitats made accessible by an open river mouth may attract seals and sea lions, particularly during high tides (Williamson and Hillemeier 2001).

CONSERVATION STRATEGIES FOR MAINTAINING SALMONIDS

Conservation of Klamath basin salmonids will require aggressive strategies that protect cold-water habitats, and genetic and life-history diversity. Strategies must start with well-defined principles for actions (e.g., improving habitat condition) that maximize use of limited resources in a rapidly changing environment. It may be desirable to conduct a triage exercise (as in Lackey et al. 2006) to determine which populations will likely be lost no matter what, which populations will benefit most from concerted action, and which populations will likely remain stable without intervention. Here, we discuss general strategies that promote salmonid conservation primarily by improving habitat condition, and enhancing genetic and life-history diversity.

Improving Habitat Condition

Salmonids in the Klamath River will face increasingly hostile habitat conditions unless conservation actions are implemented. Owing to the difficulty of addressing climate change effects on oceans, restoration efforts should focus on improving freshwater and estuarine habitats to reestablish resilient salmonid populations. Over the next 50 years, habitat restoration may help moderate the effects of climate change and enhance population growth (Battin et al. 2007). Actions that can rapidly improve salmonid habitat include the removal of Dwinnell Dam (Shasta River), enforcement of water diversions and groundwater pumping allocations (especially Klamath Project, Scott River, Shasta River), and removal of the lower four dams (Iron Gate, Copco 1, Copco 2, and J. C. Boyle dams; Figure 6.2) on the Klamath River (Table 6.4). The removal of all dams should be

TABLE 6.4

Examples of management actions and suggested locations that can improve cold-water conditions for salmonids in the Klamath River basin, California

Conservation action	Suggested locations
Riparian vegetation planting	Valley portion of Scott River main stem Moffett Creek (Scott River)
Reduction of water extraction (surface water diversion and groundwater pumping)	Shasta River Scott River Indian Creek (Klamath River)
Protection of springs	Shasta River Scott River
Closure to mining	Mouth of Scott River tributaries Mouth of Salmon River tributaries Mouth of Trinity River tributaries Mouth of Klamath River tributaries
Closure to grazing (in riparian areas)	Shasta River Scott River headwaters Salmon River headwaters Trinity River headwaters Blue Creek valley
Dam removal	Dwinnell Dam (Shasta River) Diversion dams (Little Shasta River) Main stem Klamath River
Establishment of salmonid parks (management focused on preserving salmon and steelhead)	Blue Creek Salmon River Shasta River Elk Creek (Klamath River)

accomplished jointly with measures for alleviating potential reduction in flows (e.g., from increasing groundwater pumping) so that benefits from dam removal are not diminished.

Once conservation strategies are identified, specific conservation actions can be implemented to address these. Conservation actions should facilitate access to cold-water habitats by increasing flows or removing physical (e.g., dams, shallow water) and physiological (e.g., warm water temperatures) migration barriers. In the Klamath River basin, cold-water habitats in the fog belt, mouth of tributaries (including Blue Creek, Trinity River, Salmon River, Scott River, and Shasta River), and associated with groundwater inputs (e.g., Scott River valley, Shasta River) are particularly important to salmonids in the summer months. If the lower

four dams (Iron Gate, Copco 1, Copco 2, and J.C. Boyle dams; Figure 6.2) on the Klamath River are removed, cold-water habitat will also become available beneath the current location of Klamath Project reservoirs (Hamilton et al. 2011). Nevertheless, as water temperatures continue to increase, smaller streams at higher elevations may become favored by salmonids seeking colder water (Crozier et al. 2008a). By allowing shifts in distribution and thermoregulation, habitat connectivity may be more important than habitat quantity or quality when populations are suppressed and habitat is fragmented (Isaak et al. 2007).

Restoration of riparian habitats in particular can increase shade and help enable ecological adaptation to climate change (Seavy et al. 2009). Riparian vegetation along smaller streams can cool surface waters within cold-water patches by 2–4°C (Ebersole et al. 2003), so to benefit salmonids, riparian vegetation should be reestablished where it has been removed. A comparison between locations of land-use practices and salmonid habitat can help prioritize areas for restoration (Table 6.4). The size and frequency of thermal refugia can be enhanced by reducing water diversions and protecting springs (Ebersole et al. 2003). The effects of groundwater pumping on base flows are often ignored and yet will become increasingly important as the amount of available surface water decreases (Mohseni et al. 2003). Managers should also consider seasonally closing cold-water habitats to recreation, mining, and other uses that can displace salmonids. Furthermore, activities that degrade local riparian condition and water quality, such as grazing, should be limited or eliminated in riparian areas. Also, fine sediment delivery to streams can be curtailed through outsloping (downslope grading of road surface) and decommissioning of roads, and enforcement of water quality protection measures (e.g., Total Maximum Daily Load Limits).

Enhancing Genetic and Life-History Diversity

Conservation actions that improve habitat conditions enhance genetic and life-history diversity by improving salmonid health and survival. For instance, restoration of a natural-flow regime may accelerate juvenile salmonid out-migration, resulting in decreased susceptibility (Foott et al. 2007) and mortality (Bartholomew et al. 2006) to disease, and predation (Melnychuk et al. 2010) in the Klamath River. Improved connectivity to cold-water habitats decreases infection rates because juvenile Chinook salmon are more resistant to disease when water temperature is below 16–17°C (Foott et al. 1999).

Conservation strategies that enhance genetic and life-history diversity offer the best likelihood for salmonid persistence in the face of climate change. Genetic and life-history

diversity closely reflect adaptations of different populations to local conditions. Therefore, habitat diversity must also be maintained in order to buffer against the effects of climate change (Crozier et al. 2008a), with the emphasis of preserving a variety of habitats to foster diversity among populations (Hilborn et al. 2003, Schindler et al. 2008). However, perhaps the most significant conservation strategy that can enhance salmonid genetic and life-history diversity is the reduction of adverse hatchery effects.

Hatcheries can augment commercial fisheries and mitigate declines in salmonid numbers when freshwater habitat quantity and quality decrease (Lichatowich 1999, Brannon et al. 2004). When native fish are used, hatcheries can increase declining population sizes so genetic diversity is maintained (Eldridge and Killebrew 2008, McClure et al. 2008). Large hatcheries, however, can detrimentally affect naturally spawning populations by reducing phenotypic, genetic, and life-history diversity, and reproductive success (Buhle et al. 2009, Kostow 2009). In the Klamath River basin, current hatchery practices may be contributing to the decline of wild salmonid populations (Quiñones et al. 2014). Consequently, careful consideration must be given to hatchery methods that minimize adverse effects on imperiled species (Box 6.1).

CONCLUSIONS

Anadromous salmonids need a variety of cold-water habitats to thrive. Owing to increases in water temperatures and changes to hydrology, all anadromous Pacific salmonids in California already are, and will likely be more, affected by climate change (Katz et al. 2013, Moyle et al. 2013). Climate change is likely to increase water temperatures beyond salmonid thermal tolerances in many areas, resulting in reduced species distribution and abundance. Changes to streamflow will likely change egg and juvenile survival as well as migration patterns of juveniles and adults. Many local populations

Large-scale hatcheries can have adverse impacts on wild salmon and trout populations. One way around this dilemma is to use smaller conservation hatcheries, with the goal of increasing the number of wild adults (Brannon et al. 2004). However, even small supplementation programs can have significant adverse impacts on local populations if not closely monitored. Here we summarize some recommendations to minimize the adverse effects of hatcheries, both large and small.

Conservation actions can minimize adverse hatchery effects at different stages of the salmonid life cycle. Relocation of hatcheries closer to river mouths and the release of smolts on-site can reduce potential breeding between hatchery and wild adults (Quinn 1993, Buhle et al. 2009). Because most salmonids will return to their natal sites to spawn, on-site releases should result in segregating most adults that reared in hatcheries from those in streams. Imprinting hatchery-reared fish to locations near river mouths also reduces the probability of interactions with wild conspecifics spawning further upstream. During spawning, hatcheries should use locally derived brood stock taken from a large proportion of redds (Van Doornik et al. 2010) to enhance genetic and phenotypic diversity (Mobrand et al. 2005). Because salmonids can become domesticated after several generations of artificial propagation, rearing areas for juveniles in hatcheries should closely mimic natural rearing environments to prevent changes to phenotype and behavior (Chittenden et al. 2010a). Upon release to streams, the number of juveniles should not exceed the carrying capacity of the streams (Riley et al. 2009) or marine habitats, particularly in times of low ocean productivity (Beamish et al. 1997, Levin et al. 2001), to limit competition for limited resources. Additionally, hatchery-reared juveniles should be released at low densities and at similar size to wild conspecifics to further minimize competition between the two groups (Riley et al. 2009).

Although conservation actions can alleviate adverse effects from hatchery operations, they may not sufficiently encourage the recovery of wild populations. Where adverse impacts threaten benefits, hatchery practices should be modified to protect wild stock genetic, phenotypic, and life-history diversity to the extent that hatchery closure should be considered in worst-case scenarios (as in Lackey et al. 2006).

may experience range contractions or extirpation as a result of these changes.

Resource managers should focus conservation efforts to restoring and protecting cold-water habitats as well as genetic and life-history diversity of wild populations. Salmonids will need sufficient connectivity between potential spawning and rearing habitats in order to ecologically adapt to climate change. Preservation of genetic and life-history diversity will also improve the likelihood of local adaptation to changing environmental conditions. Although substantial uncertainties exist in predicting the effects of climate change at the watershed level, numerous conservation actions can still be implemented to rapidly alleviate worsening conditions. Such actions include the removal of migration barriers, maintenance of instream flows, reestablishment of riparian vegetation where necessary, and reforming detrimental hatchery practices.

Anadromous salmonids exhibit an amazing ability to adapt, as evident by their continued existence in highly altered systems with multiple stressors. However, the effects of climate change are additional threats that will also heighten stressors already acting on dwindling populations. The health of anadromous salmonid populations in California will depend on aggressive actions that prioritize their recovery above other competing resource demands.

LITERATURE CITED

Allan, J. D. and M. M. Castillo. 2007. *Stream Ecology: Structure and Function of Running Waters.* Springer, Dordrecht.

Allendorf, F. W. and G. Luikart. 2007. *Conservation and the Genetics of Populations.* Blackwell Publishing, Malden, MA.

Allison, I., N. Bindoff, R. Bindschadler, P. Cox, N. de Noblet-Ducoudre, M. England, J. Francis, N. Gruber, A. Haywood, D. Karoly et al. 2009. *The Copenhagen Diagnosis: Updating the World on the Latest Climate Science.* The University of New South Wales Climate Change Research Center, Sydney.

Bartholomew, J. L., S. D. Atkinson, and S. L. Hallett. 2006. Involvement of *Manayunkia speciosa* (Annelida: Polychaeta: Sabellidae) in the life cycle of *Parvicapsula minibicornis*, a myxozoan parasite of Pacific salmon. *Journal of Parasitology* 92:742–748.

Battin, J., M. W. Wiley, M. H. Ruckelshaus, R. N. Palmer, E. Korb, K. K. Bartz, and H. Imaki. 2007. Projected impacts of climate change on salmon habitat restoration. *Proceedings of the National Academy of Sciences of the United States of America* 104:6720–6725.

Beamish, R. J., C. Mahnken, and C. M. Neville. 1997. Hatchery and wild production of Pacific salmon in relation to large-scale, natural shifts in the productivity of the marine environment. *ICES Journal of Marine Science: Journal du Conseil* 54:1200–1215.

Bottom, D. L., K. K. Jones, T. J. Cornwell, A. Gray, and C. A. Simenstad. 2005. Patterns of Chinook salmon migration and residency in the Salmon River estuary (Oregon). *Estuarine, Coastal and Shelf Science* 64:79–93.

Bottom, D. L., K. K. Jones, C. A. Simenstad, C. L. Smith, and R. Cooper. 2011. *Pathways to Resilience: Sustaining Salmon Ecosystems in a Changing World.* Oregon State University, Corvallis, OR.

Brannon, E. L., D. F. Amend, M. A. Cronin, J. E. Lannan, S. LaPatra, W. J. McNeil, R. E. Noble, C. E. Smith, A. Talbot, G. A. Wedemeyer et al. 2004. The controversy about salmon hatcheries. *Fisheries* 29:12–31.

Buhle, E. R., K. K. Holsman, M. D. Scheuerell, and A. Albaugh. 2009. Using an unplanned experiment to evaluate the effects of hatcheries and environmental variation on threatened populations of wild salmon. *Biological Conservation* 142:2449–2455.

Cayan, D., P. Bromirski, K. Hayhoe, M. Tyree, M. Dettinger, and R. Flick. 2008. Climate change projections of sea level extremes along the California coast. *Climatic Change* 87:57–73.

Cayan, D. R., M. D. Dettinger, S. A. Kammerdiener, J. M. Caprio, and D. H. Peterson. 2001. Changes in the onset of spring in the western United States. *Bulletin of the American Meteorological Society* 82:399–415.

CDEC (California Data Exchange Center). 1986. *Shasta / Klamath Rivers Water Quality Study.* California Department of Water Resources, Sacramento, CA.

Chatters, J. C., B. L. Butler, M. J. Scott, D. M. Anderson, and D. A. Neitzel. 1995. A paleoscience approach to estimating the effects of climatic warming on salmonid fisheries of the Columbia River basin. In R. J. Beamish (ed.), *Climate Change and Northern Fish Populations.* National Research Council of Canada, Ottowa. 489–496.

Chittenden, C. M., C. A. Biagi, J. G. Davidsen, A. G. Davidsen, A. Kondo, A. McKnight, O.-P. Pedersen, P. A. Raven, A. H. Rikardsen, J. M. Shrimpton et al. 2010a. Genetic versus rearing-environment effects on phenotype: Hatchery and natural rearing effects on hatchery- and wild-born coho salmon. *PLOS ONE* 5:e12261.

Chittenden, C. M., J. L. A. Jensen, D. Ewart, S. Anderson, S. Balfry, E. Downey, A. Eaves, S. Saksida, B. Smith, S. Vincent et al. 2010b. Recent salmon declines: A result of lost feeding opportunities due to bad timing? *PLOS ONE* 5:e12423.

Crozier, L. G. and R. W. Zabel. 2006. Climate impacts at multiple scales: Evidence for differential population responses in juvenile Chinook salmon. *Journal of Animal Ecology* 75:1100–1109.

Crozier, L. G., A. P. Hendry, P. W. Lawson, T. P. Quinn, N. J. Mantua, J. Battin, R. G. Shaw, and R. B. Huey. 2008a. Potential responses to climate change in organisms with complex life histories: Evolution and plasticity in Pacific salmon. *Evolutionary Applications* 1:252–270.

Crozier, L. G., R. W. Zabel, and A. F. Hamlet. 2008b. Predicting differential effects of climate change at the population level with life-cycle models of spring Chinook salmon. *Global Change Biology* 14:236–249.

Dettinger, M. D. and D. R. Cayan. 1995. Large-scale atmospheric forcing of recent trends toward early snowmelt runoff in California. *Journal of Climate* 8:606–623.

Eaton, J. G. and R. M. Scheller. 1996. Effects of climate warming on fish thermal habitat in streams of the United States. *Limnology and Oceanography* 41:1109–1115.

Ebersole, J. L., W. J. Liss, and C. A. Frissell. 2003. Cold water patches in warm streams: Physico-chemical characteristics and the influence of shading. *Journal of the American Water Resources Association* 39:355–368.

Eldridge, W. and K. Killebrew. 2008. Genetic diversity over multiple generations of supplementation: An example from Chinook salmon using microsatellite and demographic data. *Conservation Genetics* 9:13–28.

Feder, M. E., T. Garland, J. H. Marden, and A. J. Zera. 2010. Locomotion in response to shifting climate zones: Not so fast. *Annual Review of Physiology* 72:167–190.

Field, C. B., G. C. Daily, F. W. Davis, S. Gaines, P. A. Matson, J. Melack, and N. L. Miller. 1999. *Confronting Climate Change in California: Ecological Impacts on the Golden State.* Cambridge University Press, Cambridge, UK.

Foott, J. S., R. Stone, E. Wiseman, K. True, and K. Nichols. 2007. Longevity of *Ceratomyxa shasta* and *Parvicapsula minibicornis* actinospore infectivity in the Klamath River. *Journal of Aquatic Animal Health* 19:77–83.

Foott, J. S., J. D. Williamson, and K. C. True. 1999. *Health, Physiology, and Migration Characteristics of Iron Gate Hatchery Chinook, 1995 Releases.* California-Nevada Fish Health Center, Anderson, CA.

Furniss, M. J., T. D. Roelofs, and C. S. Yee. 1991. Road construction and maintenance. In W. R. Meehan (ed.), *Influences of Forest and Rangeland Management on Salmonid Fishes and Their Habitats.* American Fisheries Society, Bethesda, MD. 297–323.

Gannett, M. W., K. E. Lite Jr., J. L. L. Marche, B. J. Fisher, and D. J. Polette. 2010. *Ground-Water Hydrology of the Upper Klamath Basin, Oregon and California.* U.S. Geological Survey, Reston, VA.

Greene, C. M., J. E. Hall, K. R. Guilbault, and T. P. Quinn. 2010. Improved viability of populations with diverse life-history portfolios. *Biology Letters* 6:382–386.

Groot, C. and L. Margolis (eds). 1991. *Pacific Salmon: Life Histories.* UBC Press, Vancouver.

Hamilton, J., M. Hampton, R. M. Quiñones, D. Rondorf, J. Simondet, and T. Smith. 2011. *Synthesis of the Effects to Fish Species of Two Management Scenarios for the Secretarial Determination on Removal of the Lower Four Dams on the Klamath River.* U.S. Fish and Wildlife Service, Yreka, CA.

Hamlet, A. F., P. W. Mote, M. P. Clark, and D. P. Lettenmaier. 2005. Effects of temperature and precipitation variability on snowpack trends in the western United States. *Journal of Climate* 18:4545–4561.

Hanak, E., J. Lund, A. Dinar, B. Gray, R. Howitt, J. Mount, P. B. Moyle, and B. Thompson. 2011. *Managing California's Water: From Conflict to Reconciliation.* Public Policy Institute of California, San Francisco, CA.

Healey, M. C. 1982. Timing and relative intensity of size-selective mortality of juvenile chum salmon (*Oncorhynchus keta*) during early sea life. *Canadian Journal of Fisheries and Aquatic Sciences* 39:952–957.

Healey, M. C. 1991. Life history of Chinook salmon (*Oncorhynchus tshawytscha*). In C. Groot and L. Margolis (eds), *Pacific Salmon Life Histories.* UBC Press, Vancouver. 313–393.

Hilborn, R., T. P. Quinn, D. E. Schindler, and D. E. Rogers. 2003. Biocomplexity and fisheries sustainability. *Proceedings of the National Academy of Sciences of the United States of America* 100:6564–6568.

Isaak, D. J., R. F. Thurow, B. E. Rieman, and J. B. Dunham. 2007. Chinook salmon use of spawning patches: Relative roles of habitat quality, size, and connectivity. *Ecological Applications* 17:352–364.

Jefferson, A., G. E. Gordon, and S. L. Lewis. 2007. A river runs underneath it: Geological control of spring and channel systems and management implications, Cascade Range, Oregon. In M. Furniss, C. Clifton, and K. Ronnenberg (eds), *Advancing the Fundamental Sciences: Proceedings of the Forest Service National Earth Sciences Conference, 18–22 October 2004, San Diego, CA.* U.S. Department of Agriculture, Forest Service, Pacific Northwest Research Station, Portland, OR.

Jeffres, C., J. Opperman, and P. Moyle. 2008. Ephemeral floodplain habitats provide best growth conditions for juvenile Chinook salmon in a California river. *Environmental Biology of Fishes* 83:449–458.

Johnstone, J. A. and T. E. Dawson. 2010. Climatic content and ecological implications of summer fog decline in the coast redwood region. *Proceedings of the National Academy of Sciences* 107:4533–4538.

Katz, J., P. B. Moyle, R. M. Quiñones, J. Israel, and S. Purdy. 2013. Impending extinction of salmon, steelhead and trout (Salmonidae) in California. *Environmental Biology of Fishes* 96:1169–1186.

Kim, J. 2005. A projection of the effects of the climate change induced by increased CO_2 on extreme hydrologic events in the western U.S. *Climatic Change* 68:153–168.

Kjelson, M. A., P. F. Raquel, and F. W. Fisher. 1982. Life history of fall-run juvenile Chinook salmon, *Oncorhynchus tshawytscha*, in the Sacramento – San Joaquin Estuary, California. In V. S. Kennedy (ed.), *Estuarine Comparisons.* Academic Press, New York. 393–411.

Knowles, N. and D. R. Cayan. 2004. Elevational dependence of projected hydrologic changes in the San Francisco Estuary and watershed. *Climatic Change* 62:319–336.

Knox, J. B. and F. Scheuring (eds). 1991. *Global Climate Change and California: Potential Impacts and Responses*. UC Press, Berkeley, CA.

Kostow, K. 2009. Factors that contribute to the ecological risks of salmon and steelhead hatchery programs and some mitigating strategies. *Reviews in Fish Biology and Fisheries* 19:9–31.

Lackey, R. T., D. H. Lach, and S. L. Duncan (eds). 2006. *Salmon 2100: The Future of Wild Pacific Salmon*. American Fisheries Society, Bethesda, MD.

Lawson, P. W., E. A. Logerwell, N. J. Mantua, R. C. Francis, and V. N. Agostini. 2004. Environmental factors influencing freshwater survival and smolt production in Pacific Northwest coho salmon (*Oncorhynchus kisutch*). *Canadian Journal of Fisheries and Aquatic Sciences* 61:360–373.

Leung, L. R., Y. Qian, X. Bian, W. M. Washington, J. Han, and J. O. Roads. 2004. Mid-century ensemble regional climate change scenarios for the western United States. *Climatic Change* 62:75–113.

Levin, P. S., R. W. Zabel, and J. G. Williams. 2001. The road to extinction is paved with good intentions: Negative association of fish hatcheries with threatened salmon. *Proceedings of the Royal Society of London. Series B: Biological Sciences* 268:1153–1158.

Levings, C. D., K. Conlin, and B. Raymond. 1991. Intertidal habitats used by juvenile Chinook salmon (*Oncorhynchus tshawytscha*) rearing in the North Arm of the Fraser River estuary. *Marine Pollution Bulletin* 22:20–26.

Lewis, T. E., D. W. Lamphear, D. R. McCanne, A. S. Webb, J. P. Krieter, and W. D. Conroy. 2000. *Regional Assessment of Stream Temperatures Across Northern California and Their Relationship to Various Landscape-Level and Site-Specific Attributes*. Forest Service Project, Humboldt State University Foundation, Arcata, CA.

Lichatowich, J. A. 1999. *Salmon without Rivers: A History of the Pacific Salmon Crisis*. Island Press, Washington, DC.

Manga, M. 1999. On the timescales characterizing groundwater discharge at springs. *Journal of Hydrology* 219:56–69.

McClure, M. M., F. M. Utter, C. Baldwin, R. W. Carmichael, P. F. Hassemer, P. J. Howell, P. Spruell, T. D. Cooney, H. A. Schaller, and C. E. Petrosky. 2008. Evolutionary effects of alternative artificial propagation programs: Implications for viability of endangered anadromous salmonids. *Evolutionary Applications* 1:356–375.

McCullough, D. A. 1999. *A Review and Synthesis of Effects of Alterations to the Water Temperature Regime on Freshwater Life Stages of Salmonids, with Special Reference to Chinook Salmon*. United States Environmental Protection Agency, Seattle, WA.

Melnychuk, M. C., D. W. Welch, and C. J. Walters. 2010. Spatio-temporal migration patterns of Pacific salmon smolts in rivers and coastal marine waters. *PLOS ONE* 5:e12916.

Miller, B. A. and S. Sadro. 2003. Residence time and seasonal movements of juvenile coho salmon in the ecotone and lower estuary of Winchester Creek, South Slough, Oregon. *Transactions of the American Fisheries Society* 132:546–559.

Mobrand, L. E., J. Barr, L. Blankenship, D. E. Campton, T. T. P. Evelyn, T. A. Flagg, C. V. W. Mahnken, L. W. Seeb, P. R. Seidel, and W. W. Smoker. 2005. Hatchery reform in Washington State: Principles and emerging issues. *Fisheries* 30:11–23.

Mohseni, O., H. G. Stefan, and J. G. Eaton. 2003. Global warming and potential changes in fish habitat in U.S. streams. *Climatic Change* 59:389–409.

Moore, K. A., R. L. Wetzel, and R. J. Orth. 1997. Seasonal pulses of turbidity and their relations to eelgrass (*Zostera marina* L.) survival in an estuary. *Journal of Experimental Marine Biology and Ecology* 215:115–134.

Mote, P. W., E. A. Parson, A. F. Hamlet, W. S. Keeton, D. Lettenmaier, N. Mantua, E. L. Miles, D. W. Peterson, D. L. Peterson, R. Slaughter et al. 2003. Preparing for climatic change: The water, salmon, and forests of the Pacific Northwest. *Climatic Change* 61:45–88.

Moyle, P. B. 2002. *Inland Fishes of California*, 2nd ed. UC Press, Berkeley, CA.

Moyle, P. B., J. A. Israel, and S. E. Purdy. 2008. *Salmon, Steelhead, and Trout in California: Status of an Emblematic Fauna*. Center for Watershed Sciences, U.C. Davis, San Francisco, CA.

Moyle, P. B., J. D. Kiernan, P. K. Crain, and R. M. Quiñones. 2013. Climate change vulnerability of native and alien freshwater fishes of California: A systematic assessment approach. *PLOS ONE* 8:e63883.

Nichols, A. L. 2008. *Geological Mediation of Hydrologic Process, Channel Morphology and Resultant Planform Response to Closure of Dwinnell Dam, Shasta River, California*. University of California, Davis, CA.

Nielsen, J. L., T. E. Lisle, and V. Ozaki. 1994. Thermally stratified pools and their use by steelhead in northern California streams. *Transactions of the American Fisheries Society* 123:613–626.

NOAA (National Oceanic and Atmospheric Administration). 2011. *Forecast of Adult Returns for Coho in 2010 and Chinook Salmon in 2011.* Northwest Fisheries Science Center, Seattle, WA.

NRC (National Research Council). 2004. *Endangered and Threatened Fishes in the Klamath River Basin.* The National Academies Press, Washington, DC.

O'Neal, K. 2002. *Effects of Global Warming on Trout and Salmon in U.S. Streams.* Defenders of Wildlife, Washington, DC.

Quinn, T. P. 1993. A review of homing and straying of wild and hatchery-produced salmon. *Fisheries Research* 18:29–44.

Quiñones, R. M., M. L. Johnson, and P. M. Moyle. 2014. Hatchery practices may result in replacement of wild salmonids: Adult trends in the Klamath basin, California. *Environmental Biology of Fishes* 97:233–246.

Reusch, T. B. H., A. Ehlers, A. Hämmerli, and B. Worm. 2005. Ecosystem recovery after climatic extremes enhanced by genotypic diversity. *Proceedings of the National Academy of Sciences of the United States of America* 102:2826–2831.

Richter, A. and S. Kolmes. 2005. Maximum temperature limits for Chinook, coho, and chum salmon, and steelhead trout in the Pacific Northwest. *Reviews in Fisheries Science* 13:23–49.

Riley, S. C., C. P. Tatara, B. A. Berejikian, and T. A. Flagg. 2009. Behavior of steelhead fry in a laboratory stream is affected by fish density but not rearing environment. *North American Journal of Fisheries Management* 29:1806–1818.

Scavia, D., J. Field, D. Boesch, R. Buddemeier, V. Burkett, D. Cayan, M. Fogarty, M. Harwell, R. Howarth, C. Mason et al. 2002. Climate change impacts on U.S. coastal and marine ecosystems. *Estuaries and Coasts* 25:149–164.

Schindler, D. E., X. Augerot, E. Fleishman, N. J. Mantua, B. Riddell, M. Ruckelshaus, J. Seeb, and M. Webster. 2008. Climate change, ecosystem impacts, and management for Pacific salmon. *Fisheries* 33:502–506.

Seavy, N. E., T. Gardali, G. H. Golet, F. T. Griggs, C. A. Howell, R. Kelsey, S. L. Small, J. H. Viers, and J. F. Weigand. 2009. Why climate change makes riparian restoration more important than ever: Recommendations for practice and research. *Ecological Restoration* 27:330–338.

Spence, B. C. and J. D. Hall. 2010. Spatiotemporal patterns in migration timing of coho salmon (*Oncorhynchus kisutch*) smolts in North America.

Canadian Journal of Fisheries and Aquatic Sciences 67:1316–1334.

Stewart, I. T., D. R. Cayan, and M. D. Dettinger. 2005. Changes toward earlier streamflow timing across western North America. *Journal of Climate* 18:1136–1155.

Strange, J. S. 2010. Upper thermal limits to migration in adult Chinook salmon: Evidence from the Klamath River basin. *Transactions of the American Fisheries Society* 139:1091–1108.

Sullivan, C. M. 1989. *Juvenile Life History and Age Composition of Mature Fall Chinook Salmon Returning to the Klamath River, 1984-1986.* Humboldt State University, Arcata, CA.

Sutton, R. J., M. L. Deas, S. K. Tanaka, T. Soto, and R. A. Corum. 2007. Salmonid observations at a Klamath River thermal refuge under various hydrologic and meteorological conditions. *River Research and Applications* 23:775–785.

Tague, C., G. Grant, M. Farrell, J. Choate, and A. Jefferson. 2008. Deep groundwater mediates streamflow response to climate warming in the Oregon Cascades. *Climatic Change* 86:189–210.

Thompson, J. 2007. Running dry: Where will the west get its water? *Science Findings* 97:1–5. U.S. Department of Agriculture, Pacific Northwest Research Station.

Van Doornik, D. M., B. A. Berejikian, L. A. Campbell, and E. C. Volk. 2010. The effect of a supplementation program on the genetic and life history characteristics of an *Oncorhynchus mykiss* population. *Canadian Journal of Fisheries and Aquatic Sciences* 67:1449–1458.

Van Kirk, R. W. and S. W. Naman. 2008. Relative effects of climate and water use on base-flow trends in the Lower Klamath basin. *Journal of the American Water Resources Association* 44:1035–1052.

Wallace, M. and B. W. Collins. 1997. Variation in use of the Klamath River estuary by juvenile Chinook salmon. *California Fish and Game* 83:132–143.

Williamson, K. and D. Hillemeier. 2001. An assessment of pinniped predation on upon fall-run Chinook salmon in the Klamath River estuary, California, 1999. Yurok Tribe Fisheries Department, Klamath, CA.

Winder, M., A. D. Jassby, and R. Mac Nally. 2011. Synergies between climate anomalies and hydrological modifications facilitate estuarine biotic invasions. *Ecology Letters* 14:749–757.

Zabel, R. W. and J. G. Williams. 2002. Selective mortality in Chinook salmon: What is the role of human disturbance? *Ecological Applications* 12:173–183.

Manager Comments

Rebecca M. Quiñones
in conversation with
Julie Perrochet

Quiñones: Have you observed any changes in fish communities or habitats that seem strongly tied to climatic changes?

Perrochet: No, because we don't have site-specific data that show changed conditions that we can strongly link to climate change. However, the Forest Service is concerned with potential warming throughout the region that could result in loss of salmon and trout through reduced habitat quality, immigration of new species, and changed wildfire behavior. Management actions are planned and reviewed to meet regulations such as ESA and the Clean Water Act, and one of the critical assessments of proposed actions is impacts on water temperature.

Quiñones: When you think about the ability of the management community to change management approaches or shift conservation priorities in this system in responses to changes in climate, what do you see as the main constraints?

Perrochet: Coordination of efforts is a major challenge. The good news is that the Klamath Basin Restoration Act was agreed to in 2010, setting up the Klamath Basin Coordinating Council. Before that, the US Congress passed the "Klamath Act" (Public Law 99–552), which allocated $20,000,000 to provide 20 years (1986–2006) of restoration, education, and creation of local watershed community groups in the Klamath basin (U.S. Fish and Wildlife Service 2006). The other good news is that there are strong restoration interests at the federal and state levels, only constrained by funding. Coordination between the research community and these efforts is more important than ever. In addition to the "resource management" side of the federal and state agencies, there is also a need to better coordinate with research entities, such as the Pacific Southwest Research Station of the Forest Service and the USGS research efforts, and university research groups.

While we have come a long way on collaboration in this region, competition among stakeholders to get as much funding as possible to work in their specific area of interest at times reduces the effectiveness of restoration efforts. There is also a need for funding entities to coordinate—sometimes one group will deny funding to a project because of technical problems, while another group will fund it because they don't recognize the technical problems. The funding entities (state, federal, private foundations, and other non-profit groups such as the National Fish and Wildlife Foundation) need to understand and fund actions based on best science and according to a prioritized system of what type of actions are needed in what locations. This also brings up the need for clearer goals and priorities. A prioritized list of where various types of conservation should take place—to the fifth- and sixth-field watershed scales—is needed for the Klamath River basin.

Managers are also constrained by the availability of climate change projections at appropriate scales. Specifically, data that show where the specific effects of climate change override or exacerbate other limiting factors are needed. Your chapter does a good job of identifying genetics, hatchery, and water use as areas of concern. Fishing practices, the role of changing ocean temperatures, species' immigration (through adaptation), and the risk of large-scale wildfires are other areas of concern. Loss of riparian shading and ground stabilization from high severity fires is a concern in forested areas such as the Klamath River basin.

Quiñones: One focus of this book is to help "bridge" the resource management and research communities. With that in mind, are there particular areas of study (ecological science, climate science, policy, data management, etc.) that could help overcome any of the constraints you mention?

Perrochet: Efforts like the Template for Assessing Climate Change Impacts and Management Options (TACCIMO; www.forestthreats.org/taccimotool), and outreach by individuals like Hugh

(continued)

(continued)

Safford of the Forest Service represent a big step forward for creating a bridge between resource management and research communities. In terms of specific research questions, I'd like to know which species are capable of using habitat above the dams. A past study of the energetics needed by salmon to spawn above the dams suggested fall Chinook (the dominant salmonid run in the basin) could not survive long enough in the system to use habitat above Iron Gate. Has this been studied, validated, and responded to by the research community? As far as policy is concerned, I think we already have the needed policies for federal agencies. Clear results from science should be incorporated into the current policies for successful restoration. In terms of bridging the gap, I do wonder if the research communities fully understand the current policies for restoration. As I suggested above, research communities also compete for funds to conduct studies in the Klamath basin and that can create a disjointed mix of efforts. Their efforts should be coordinated with and communicated to the stakeholder groups.

Quiñones: This chapter focuses on actions and priorities for resource managers, but this group is only one part of the set of stakeholders that make key decisions that influence the health and long-term viability of salmonids in light of climate change. Can you comment on other key players and their roles?

Perrochet: How is "Resource Managers" defined? In your question, and the chapter, it seems that "resource managers" is meant to infer state and federal agencies. I think the term is antiquated when applied to the Klamath basin. The basin stakeholders might all be considered resource managers because tribes, water user groups, and watershed council community members are recognized in the basin as "resource managers". Perhaps the research community is applying an artificial delineation. I like Hugh Safford's term, "restoration practitioners." For example, there are the federal and state and county-level agencies. But the Klamath Water Users group may also be considered as resource managers, as also each of the Watershed Restoration groups and tribes. The Klamath Basin Monitoring Program is another key group. Monitoring is critical to determine the success of habitat restoration and population recovery.

Quiñones: Any observations from your work on what factors really help important projects get implemented?

Perrochet: Good science. The Northwest Forest Plan (FEMAT, chapter 5) provided us with the current scientific understanding of the issues and we have been responding to that data since 1994, switching from in-stream short-term benefits to upslope long-term restoration benefits on federally managed land. The issues of hatchery practices, fishing regulations, and water management are largely out of the control of federal land managers in the mid and lower basin . . . so the next emphasis for restoration may be different from the past three decades. Beyond science, agreement and acceptance of priority places and actions is critical. Availability of funds is important but is always an unknown. A key next step is articulating measureable indicators for habitat restoration and population recovery at the basin scale so that we can know what success looks like.

LITERATURE CITED

U.S. Fish and Wildlife Service. 2006. *Klamath River Basin Conservation Area Restoration Program Activities (1986–2006)*. Yreka Fish and Wildlife Office, U.S. Fish and Wildlife Service, Yreka, CA. http://www. fws. gov / yreka / PDF / KRBCARP_Activities. pdf.

CHAPTER 7

Pollinators and Meadow Restoration

Brendan Colloran, Gretchen LeBuhn, and Mark Reynolds

Abstract. There are clear indications that climate change is altering plant and animal populations in ways that confound conservation efforts. The consequences of climate change may be particularly dire for montane and alpine bumble bee communities, which include species that are at the upper elevational and northern limits of their habitat range. We modeled the community dynamics within meadows (modeled aspatially) and migration among meadows (modeled spatially) for a set of 50 meadows in the Sierra Nevada over a span of 100 years to study the effects of meadow condition and climate change on montane bumble bee population dynamics in the Sierra Nevada. These results suggest that increasing the scale of current meadow restoration efforts may be an effective approach to slowing biodiversity loss in montane pollinator communities—and the plants that depend on their pollinator services—in the face of climate change.

INTRODUCTION

Meadows, which are wetlands or semi-wetlands in the subalpine and alpine zone, are some of

Key Points

- Continued restoration of alpine meadows is key to increasing persistence of bumble bee species.
- Our models suggest that persistence of bumble bee species would be higher in undisturbed / restored meadows than in degraded ones in all climate change scenarios considered.
- For the range of climate change scenarios considered, species that emerge earlier appear to gain a competitive advantage from the earlier warming, and generally persist on the order of two to three decades longer than later-emergent species in undisturbed / restored meadows.
- In degraded meadows, drier conditions lead to a shorter resource-rich period when population can grow, which reduces persistence of species and removes the influence of the emergence timing.
- In the Sierra Nevada mountain range where altitude generally decreases toward the north, restoration will be critical in higher elevation meadows as the opportunity for poleward or upslope migration are limited because such sites are lacking.

the most threatened habitats in the Sierra Nevada, with over 80% of the meadows degraded (McKelvey et al. 1996). Meadow systems comprise less than 10% of the land area, and over the last 150 years, the hydrology of meadow systems has been dramatically changed by stream incision resulting from overgrazing (Odion et al. 1988, Kirchner et al. 1998, Blank et al. 2006), erosion (Micheli and Kirchner 2002a, Micheli and Kirchner 2002b), logging, housing, railroad, or road development (Ffolliott et al. 2004, Loheide and Gorelick 2007). The resulting stream incision leads to a lowering of the water table and a drying out of soil in the meadow, leading to changes in the composition of the vegetation and the timing of availability of floral resources (Loheide and Gorelick 2005).

Recently, there has been a move to restore some of these montane meadows. Restoration of meadow hydrology occurs primarily through the building of dams, called "pond and plug" restoration (Loheide and Gorelick 2005, Loheide and Gorelick 2007). In this type of restoration, dams are built at various points across the stream channel and water pools above the dam, rerouting flow, decreasing erosion, and increasing saturation upstream of the dam. This results in an increase in the water table depth and restoration of wet meadow vegetation (Loheide and Gorelick 2005, Loheide and Gorelick 2007). Meadows with restored hydrology have a larger area influenced by surface water and groundwater, and maintain a more diverse plant community with greater vegetation structure, particularly of riparian, emergent, and wet meadow species (e.g., Dobkin et al. 1998, Lang and Halpern 2007).

Bumble bees (genus *Bombus*, approximately 250 species worldwide) are one of the most important groups of pollinators for both native plants and crop plants, and their populations are declining worldwide (Williams 1986, Thorp and Shepard 2005, Cameron et al. 2011). Because they are typically found at higher latitudes and elevations than other bees, bumble bees are the primary pollinators in alpine and arctic communities, including the Sierra Nevada. Climate-driven changes in bumble bee communities, including phenology, may lead to decreased pollinator service for native plants and possibly shifts in the composition of both the plant and the pollinator communities (Williams 1986). In montane and alpine environments, these changes may have devastating effects on native plants and ecosystem processes, potentially creating an "extinction vortex" (Gilpin and Soule 1986, Hegland et al. 2009) in which declining bee populations could result in a reduction in pollinator services leading to the subsequent decline of plant populations, which in turn results in less foraging habitat for bees, reinforcing bee population decline.

Twenty species of bumble bees are known to inhabit California's Sierra Nevada mountain range. These bees are sorted spatially by elevation, and since each bumble bee species has a unique seasonal emergence pattern, they are also sorted temporally by emergence time (Thorp et al. 1983). In the mid-elevations (1500–2500 m) of the Sierra, bumble bees emerge over an 8- to 12-week period (Nordby 2010). *Bombus bifarius* is the earliest species to appear after annual snowmelt (Nordby 2010). The next group to emerge consists of *B. vosnesenskii, B. flavifrons, B. nevadensis, B. fervidus, B. centralis, B. vandykei,* and *B. psithyrus insularis.* The late group includes *B. appositus, B. californicus, B. griseocollis,* and *B. occidentalis.* Californian bumble bees compete for nest sites in a limited number of abandoned rodent holes, and it has been shown that the availability of nest sites plays an important role in determining the structure of these bumble bee communities (McFrederick and LeBuhn 2006). This limit on the number of nest sites suggests that there is the potential for priority effects, if specific bee species have different physiological tolerances that influence their ability to compete for sites (Morin 1999, Fukami and Morin 2003, Price and Morin 2004).

In years where snowmelt is early and there are no late frosts, the early species may have a distinct advantage in securing nest sites and establishing colonies—and since colony growth is exponential after the nest is established, nests that are established early produce more workers and are able to garner more resources than nests that are established later in the season. On the other hand, species that emerge later face fewer risks, as they are less exposed to the cold weather and poor resource availability that is present early in the season. This sensitivity to climatic conditions, particularly early in the season, suggests that bees may be very vulnerable to changes in the phenology of alpine meadows (e. g., season length, timing of extreme events, and availability of resources), and that changes in climate may have dramatic effects on both community dynamics at any given elevation and the distribution of species across elevations.

Given the strong links between meadow community dynamics and climatic drivers, costs and benefits of conservation efforts are best assessed in the context of both current and future conditions. There are clear indications that climate change is altering plant and animal populations in ways that confound conservation efforts. Recent comprehensive meta-analyses (Parmesan and Yohe 2003, Root et al. 2003, Bartomeus et al. 2011) and a multi-taxa synthesis (Parmesan 2006) of changes in plant and animal populations suggest that recent patterns of global warming are already affecting the range, behavior, phenology, and other attributes of many species. In North America, species ranges are expected to respond to temperature increases by shifting north and moving to higher elevations (Parmesan 2006). However, the consequences may be dire for montane and alpine communities, some of which are already at the tops of mountains or the northernmost parts of North America, and which therefore have little opportunity to migrate. Particularly when combined with competition for resources, changes in climate may have dramatic effects on both community dynamics at any given elevation and the distribution of species across elevations (Bowers 1985). Moreover, the consequences

of climate change may be intensified by habitat degradation—in the Sierra Nevada, for example, changes in meadow hydrology affect the availability of resources, possibly exacerbating climate-driven changes in bumble bee communities.

The combination of meadow degradation and climate change may have serious effects on bumble bee communities by changing the quantity and phenology of resources. Fortunately, recent work has suggested that meadow restoration can improve the hydrology of degraded meadows (Loheide and Gorelick 2005). In order to explore the connections between climate change and the benefits of restoring meadow hydrology, we developed a model to examine how meadow restoration might influence bumble bee population dynamics in the Sierra Nevada under various climate change scenarios.

This research is a first step to understanding how these pollinators may respond to climate change. We know of no other work that has modeled the effects of climate change on bees in these environments.

METHODS

Overview of the Model

Our model simulates community dynamics within meadows (modeled aspatially) and migration among meadows (modeled spatially) for a set of 50 meadows in the Sierra Nevada over a span of 100 years. We compare the persistence of bumble bee species in meadows with degraded hydrology to that of meadows with restored hydrology under four climate scenarios. Persistence is defined as the number of years individuals of that species are found in a meadow. Loheide and Gorelick (2007) found restored meadows and those that have not been degraded in the Northern Sierra have similar hydrology, so we lumped them into a single category ("restored"). Each scenario was simulated 500 times under both hydrology regimes (degraded and restored).

Meadows and Species Pool

To generate the meadows in our model, we started with size, position, and elevation data for a set of 61 meadows in the Northern Sierra that ranged in elevation from 1750 to 2400 m (Hatfield and LeBuhn 2007). To simplify our spatial migration model, we treated these meadows as disks; simplifying the geometry of the meadows in this way caused some to overlap, which we merged into a single larger meadow with a total area equal to the sum of the areas of the two meadows merged. The final elevation of the merged meadow was the average of the elevations of the original meadows weighted by the meadow areas. This procedure left us with a set of 50 meadows. We assigned each meadow a maximum carrying capacity (the number of bees it is able to support) based on its area, and a fixed, nonrandom number of nest sites based on its area (1–50 ha^{-1}). All nest sites are identical in our model, and may be occupied by any species. We do not consider any possible effects of meadow hydrology on availability of nest sites.

The hydrological quality of each meadow affects resource availability within that meadow. We used field capacity, a measure of the amount of moisture that soil can hold, to capture the hydrological quality of the meadows. For restored meadows, the model assumed a starting field capacity of 27 cm; for degraded meadows, this is reduced by 40% to 16.5 cm. In any run of our simulation, all of the meadows are either restored or degraded—we do not consider scenarios in which some meadows are restored and others are degraded.

We consider 10 species, each of which was assigned a distinct "degree days" requirement that triggered its emergence from overwinter hibernation (a degree day is a standardized measure of heating and cooling calculated as the integral of a function of time that varies with temperature). For the purpose of this model, the species are otherwise identical, and are not intended to represent particular species. Since their emergence timing is their only distinguishing feature, we may refer to the species by "species number" or "emergence number" interchangeably. Initially, we seeded the meadows with hibernating queens of all 10 species. The number of queens of each species in each meadow was set by the meadow size, and thus was the same for each simulation. Each time a queen was added, its species type was drawn at random from a pool of all potential species, each of which had different emergence times.

Community and Population Dynamics

Each day of the simulation, all meadows accumulated degree days based on the regional temperature (determined by the climate model described below), which is common to all meadows, and adjusted for each meadow's specific elevation. If a meadow has accumulated enough degree days to trigger the emergence of one of the species, then all the queens of that species hibernating in that meadow emerge and begin their search for a nest. These queens either nest in the meadow in which they hibernated or migrate to another meadow if they cannot locate a nest in their initial meadow.

Our simple spatial model of migration assumes that queens travel in a straight line on a random heading from their meadow of origin. If this flight path comes within 200 m of a meadow, we assume that the queen successfully detects the meadow and searches it for a nest site; if there are no meadows within 200 m of this flight path, the queen dies.

A queen that successfully migrates to a new meadow searches for a nest; her probability of finding a nest in any given meadow is determined by how many unoccupied nest sites are available. If she cannot locate an empty nest in this new meadow, she again attempts to migrate to another meadow, repeating these steps until she has either successfully found a nest or died during migration. Our model assumes that the entire process of emergence, nest location, and migration occurs during one day; once all queens have either found nests or died, the simulation proceeds to the next day, repeating these steps until the hibernating queens of each species have emerged.

If a queen successfully locates a nest in a meadow, she enters a 21-day foraging period, during which she may die based on the temperature in the meadow—as the temperature decreases, the probability of mortality increases. If she survives the foraging period, her first brood of 15 workers become active, and begins gathering resources for the nest, allowing the queen to stop foraging and produce more workers. At this point, the size of the active nest increases according to the simple Lotka–Volterra model

$$N_{t+1} = N_t + (r * N_t (1 - N_t/K_t))$$

where N_t is the number of workers in a nest, r is the growth rate of the number of workers in the nest (which is the same for all species), and K_t is the resources currently available in the meadow.

The above model of nest growth applies to all species and all meadows, but resources available in the meadow (K_t) vary based on meadow size, degree days accumulated, and water availability in the meadow (which is in turn determined by the hydrological quality of the meadow). We modeled water availability by using a modified Hargreaves equation to estimate evapotranspiration (Droogers and Allen 2002) and the Thornthwaite–Mather procedure for calculating soil-moisture balance (Steenhuis and Van der Molen 1986).

For simplicity, we assume that each season of colony life ends on the 280th day of the year (which is in early October). At that point, the queens of all existing nests die, all nest sites are vacated and become available for colonization the following season, and nests produce new queens in direct proportion to the number of workers in the nest (one new queen for every 10 workers). These new queens hibernate in the meadow in which they were born. To model overwinter mortality, the model assumed a standard overwinter mortality rate of 60% for queens of all species.

Climate and Weather Model

To capture large-scale temperature and precipitation patterns in the Northern Sierra under a range of climate scenarios, we used projected monthly precipitation and temperature data for the years 2000–2099 under the A2 ("higher") emissions, A1B ("middle") emissions, and B1 ("lower") emissions scenarios described in the IPCC Special Report on Emissions Scenarios (Nakicenovic and Swart 2000). These data were obtained from the Lawrence Livermore National Laboratory, Bureau of Reclamation, and Santa Clara University downscaled climate projections derived from the World Climate Research Programme's Coupled Model Intercomparison Project phase 3 multi-model dataset. Temperature trends for the Northern Sierra under these emissions scenarios are depicted in Figure 7.1. None of these scenarios display any significant trends in precipitation quantity over time, and the seasonal timing of precipitation is essentially the same under all three scenarios.

These climate projections gave us monthly temperature and precipitation averages for each year of the simulation run, but did not provide the daily weather information required by our model. To generate these data in a way that would realistically capture natural variability and extreme events, we based daily temperature and precipitation amounts for each year of the simulation on a randomly selected year of observed temperature and precipitation data from the Sierraville Ranger Station from 1957 to 2007. In order to make these daily observations conform to the monthly averages given by the climate projections, we performed a histopolation procedure, adding a smooth trend to the observed daily time series data that adjusted the monthly averages up or down as needed to match the average for the climate projection.

As a null climate model, we chose a year of historical weather data at random for each

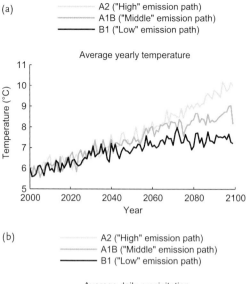

(a)

A2 ("High" emission path)
A1B ("Middle" emission path)
B1 ("Low" emission path)

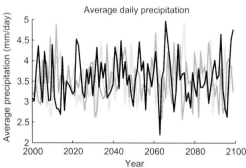

(b)

A2 ("High" emission path)
A1B ("Middle" emission path)
B1 ("Low" emission path)

FIGURE 7.1: (a) Projected annual average temperature for the area surrounding the Sierraville Ranger Station for 2000–2100 under the A2, A1B, and B1 emission scenarios. (b) Projected average daily precipitation for the area surrounding the Sierraville Ranger Station for 2000–2100 under the A2, A1B, and B1 emission scenarios.

year of the 100 years in a model run, simulating a climate scenario with no interannual trend.

RESULTS

Community Structure

While most species decrease in their probability of persistence under climate change, the earlier-emerging species are more likely to persist. Under both the restored and degraded meadow hydrology regimes, all three climate change scenarios favor earlier-emerging species relative to our null climate scenario (Figure 7.2). Under the null scenario, all species fare worse in degraded meadows. This also holds in all three climate change scenarios for all but the last species to emerge. Surprisingly, in degraded meadows the null climate scenario results in the lowest mean persistence times for every species. In restored meadows, we see that some species persist for a much greater duration than others. These differences in persistence times are attenuated in degraded meadows. This flattening of persistence times indicates that those species that are most likely to persist in restored meadows are the most severely affected by meadow degradation.

Community Size

In both the degraded and restored meadows, the A2, A1B, and B1 climate projections result in approximately the same number of species persisting at any given year. The median number of species persisting under the null model follows a trajectory similar to that of the climate change scenarios in restored meadows, but is lower in degraded meadows (Figure 7.3). The rate of species loss is much greater in degraded meadows than restored meadows, and by year 100 the expected number of species persisting in degraded meadows has fallen to around one under all climate scenarios, whereas in restored meadows the expected number of species persisting at year 100 is around two for all climate scenarios. In particular, under all three climate change scenarios, about one additional species persists to year 100 in restored meadows, and under the null scenario, about two more species persist in restored meadows.

DISCUSSION

Meadows with restored hydrology remain wet later into the summer drought, which in our model means that floral resources remain available later into the summer, and hence bumble bee populations are able to continue to

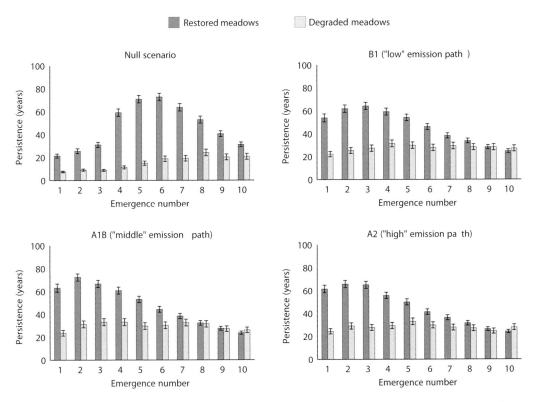

FIGURE 7.2: Mean number of years (starting from 2000) that each species persists in both hydrologic regimes. Each panel depicts one climate scenario. The sample mean is computed across all 500 model runs; error bars are 95% confidence intervals. Persistence times are greater in restored meadows.

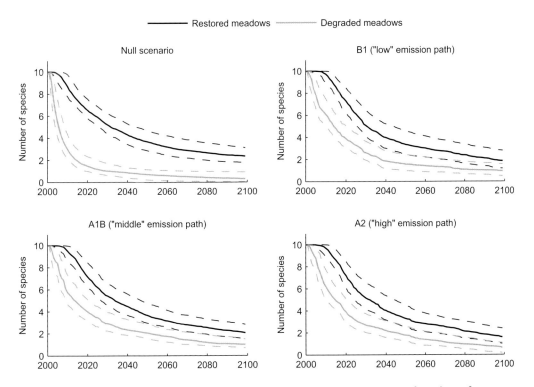

FIGURE 7.3: Resampled estimates of the median number of species persisting in this region for each year from 2000 to 2100 in both hydrologic regimes. Each panel depicts one climate scenario. The dashed lines are the upper and lower limits of the interquartile range. Under all climate scenarios, more species persist in restored meadows at every year of the simulation.

increase in size for a longer period of time. Consequently, restored meadows in our model larger communities of bumble bees for longer periods of time than degraded meadows (Figure 7.3). In all cases, early-emerging species do better under climate change than under current climate conditions because their peak nest sizes are larger (Figure 7.2). In restored meadows, late-emerging species have lower persistence under climate change scenarios than under current climate conditions because their peak nest sizes are smaller, resulting in decreases in new queen production. The same lower persistence time is not observed in degraded meadows. The magnitude of the effect of climate change is less for degraded meadows because these meadows already have diminished ecological function and the time for nest population growth is already reduced because floral resources are reduced. While the magnitude of effect of climate change is less, the loss of bumble bee species in degraded meadows is dramatically higher than that in restored meadows under all scenarios.

Because of their social structure, bumble bees are susceptible to losses in genetic variation when populations are highly fragmented (Chapman and Bourke 2001, Darvill et al. 2006, Ellis et al. 2006). Sociality causes a decreased effective population size because most individuals are sterile workers (Packer and Owen 2001). This reduced genetic diversity may hamper their ability to respond to environmental change. Data on gene flow in bumble bees suggest that for many species long-range dispersal is unlikely (Darvill et al. 2006) and species that have become isolated have rapidly gone extinct even when their habitat persists (Goulson 2003). In the Sierra Nevada, isolation of higher elevation bumble bee species is a potential threat because there are no high-elevation meadows in the Northern Sierra and, therefore, no routes for migration to suitable high-elevation habitats.

A key to understanding the effects of climate change on bumble bee communities will be the changes in the plant community, particularly changes in plant phenology. In this model, the increase in early season temperatures and change in rainfall patterns lead to a shift in both the peak of mean resources and the date at which resources no longer become available in a meadow. The increase in early season temperatures and resources decreases the probability of early season mortality for those species that emerge first. This increase in success of early season species is coupled with a decline in late-emerging species. This decline may simply be a numerical effect where early population growth leads to exploitation of available resources and carrying capacity is reached prior to the emergence of the later species (Louette and De Meester 2007). While the probability of mortality decreases in the early spring under the climate change scenarios, resource availability declines in the early summer, leading to a shorter length of time available for bee worker production. Since the reproductive success of colonies in this model is linearly related to the number of workers produced, the reduced size of late-emerging colonies has a significant effect on their persistence. These results are consistent with a meta-analysis of bumble bees on three continents, which showed that bumble bees with later colony initiation dates were more likely to have declined (Williams et al. 2009)

The model suggests that extending the length of the growing season by improving a meadow's ability to retain moisture may have significant positive effects on the bumble bee community. However, the changes in temperature and season length may cause unexpected problems for some of the high-elevation habitat specialists that are not explored in our model. These problems may be caused directly by a physiological inability of these species to cope with changes to their habitat, or less directly by species adapted to lower elevations migrating upward and increasing competition for resources in high-elevation habitats (Theurillat and Guisan 2001, Halloy and Mark 2003, Pauli et al. 2007).

The benefits to plant species of maintaining larger pollinator communities may be substantial. In the Andes, bees pollinate over 50% of the species in the Andean zone (60% of the plant species between 2200 and 2600 m, over 40% of the species between 2700 and 3100 m, and 13% between 3200 and 3600 m) (Arroyo et al. 1982). For primarily outcrossing plant species, a decline in gene flow may increase the probability of extinction. If migration rates vary across either plant or animal species, there may be significant changes in both plant and animal communities and in ecosystem function (Root et al. 2003). If pollinators and plant phenologies become uncoupled, there may be a negative feedback between the loss of bumble bee species and the maintenance of plant communities (Inouye 2008).

Coadaptation of patterns of emergence, arrival or flowering, or other resources in alpine systems is common. For example, numerous migratory birds' arrival time coincides with peak resources, the emergence time of many insect taxa match the emergence or flowering time of their host plants, and there is a clear sequence of flowering phenologies. While the responses of individual taxa to changes in climate are unpredictable, the model suggests that late emerging or arriving species may be at risk. Consistent with these results, data on pollen records from subalpine meadows suggest that plants in those meadows with shorter growing seasons are more sensitive to climate change (Gavin and Brubaker 1999).

SOLUTIONS, ADAPTION, AND LESSONS

Meadows are some of the most threatened ecosystems in North America (McKelvey et al. 1996). Our results indicate that one of the most effective strategies for managing climate change impacts for montane and alpine meadow pollinator communities may be doing what conservation practitioners and resource managers are already doing—restoring meadows. This model suggests that meadow restoration will be paramount to maintaining diverse bumble bee communities in the face of climate change.

Late-emerging bee species that have lower probabilities of persisting under climate change will respond to climate change in one of three ways: migration, adaptation, or extinction. In the Sierra, the opportunity for northward migration may be limited because the amount of available high-elevation habitat is markedly less toward the northern end of the Sierra Nevada range. This suggests that the potential for finding equivalent habitat by migrating northward may not exist. This is especially true for high-elevation species like B. appositus, B. sylvicola, and B. balteatus. While some species may be able to adapt to changes in their habitats, there currently are no data on what determines the elevational stratification of bumble bee species—it may be that bumble bee distributions are determined by thermal tolerance (Arroyo et al. 1982). Alternatively, distributions may be determined by constraints on flight physiology and the outcome of competitive interactions at different air densities as suggested for hummingbirds (Altshuler and Dudley 2006). Depending on the cause of this elevational stratification, the opportunity for adaptation may be limited.

Conservation practitioners and resource managers are understandably challenged by how to effectively protect, restore, and manage species, natural communities, and ecosystems in the face of climate change. The consequences of climate change for montane and alpine communities are especially vexing, because some of these communities are already at their upper elevational and northern limits. Conservationists will justifiably need to perform habitat "triage" focusing efforts on conservation of habitats most likely to persist under climate change scenarios and leaving species and communities dependent on declining habitats to adapt, go extinct, or be rescued by interventions such as managed relocation.

This model of the effects of meadow condition and climate change on montane bumble bee population dynamics in the Sierra Nevada suggests that meadow restoration is an effective strategy for ameliorating climate change threats to montane bee communities. Restoring montane meadows is an opportunity to increase multiple ecosystem service values. By stabilizing the hydrology in these meadows, there should be less and slower loss of water from the system, decreased flooding, and improved habitat for plants and other wildlife. Rather than "writing these ecosystems off," increasing the pace and scale of current meadow restoration efforts may be an effective approach to conserving montane pollinator communities, and the plants that depend on their pollinator services, in the face of climate change. It has been said that the essential ecological consequence of climate change is that it changes everything. Conserving montane pollinator communities through restoration may be one instance where continuing to do what we are doing already in terms of restoration activities is a wise response to climate change and particularly important in high- and mid-elevation meadows.

LITERATURE CITED

Altshuler, D. and R. Dudley. 2006. The physiology and mechanics of avian flight at high altitude. *Integrative and Comparative Biology* 46:6–71.

Arroyo, M. T. K., R. Primack, and J. Armesto. 1982. Community studies in pollination ecology in the high temperate Andes of central Chile. I. Pollination mechanisms and altitudinal variation. *American Journal of Botany* 69:82–97.

Bartomeus, I., J. Ascher, D. Wagner, B. N. Danforth, S. Colla, S. Kornbluth, and R. Winfree. 2011. Climate-associated phenological advances in bee pollinators and bee-pollinated plants. *Proceedings of the National Academy of Sciences of the United States of America* 108(51):20645–20649.

Blank, R. R., T. Svejcar, and G. Riegel. 2006. Soil attributes in a Sierra Nevada riparian meadow as influenced by grazing. *Rangeland Ecology and Management* 59(3):321–329.

Bowers, M. A. 1985. Experimental analyses of competition between two species of bumble bees (Hymenoptera: Apidae). *Oecologia* 67:224–230.

Cameron, S. A., J. D. Lozier, J. P. Strange, J. B. Koch, N. Cordes, L. F. Solter, and T. L. Griswold. 2011. Patterns of widespread decline in North American bumble bees. *Proceedings of the National Academy of Sciences of the United States of America* 108:662–667.

Chapman, R. E. and A. F. Bourke. 2001. The influence of sociality on the conservation biology of social insects. *Ecology Letters* 4(6):650–662.

Darvill, B., J. Ellis, G. Lye, and D. Goulson. 2006. Population structure and inbreeding in a rare and declining bumble bee, *Bombus muscorum* (Hymenoptera: Apidae). *Molecular Ecology* 15:601–611.

Dobkin, D. S., A. C. Rich, and W. H. Pyle. 1998. Habitat and avifaunal recovery from livestock grazing in a riparian meadow system of the northwestern Great Basin. *Conservation Biology* 12(1):209–221.

Droogers, P. and R. Allen 2002. Estimating reference evapotranspiration under inaccurate data conditions. *Irrigation and Drainage Systems* 16:33–45.

Ellis, J. S., M. E. Knight, B. Darvill, and D. Goulson. 2006. Extremely low effective population sizes, genetic structuring and reduced genetic diversity in a threatened bumble bee species, *Bombus sylvarum* (Hymenoptera: Apidae). *Molecular Ecology* 15:4375–4386.

Ffolliott, P. F., L. F. DeBano, M. B. Baker, D. G. Neary, and K. N. Brooks. 2004. Hydrology and impacts of disturbances on hydrologic function. In D. G. Neary, L. F. DeBano, M. B. Baker, and P. F. Ffolliott (eds), *Riparian Areas of the Southwestern United States: Hydrology, Ecology, and Management.* CRC Press, Boca Raton, FL.

Fukami, T. and P. J. Morin. 2003. Productivity-biodiversity relationships depend on the history of community assembly. *Nature* 424:423–426.

Gavin, D. and L. Brubaker. 1999. A 6000-year soil pollen record of subalpine meadow vegetation in the Olympic Mountains, Washington, USA. *Journal of Ecology* 87:106–122.

Gilpin, M. E. and M. E. Soulé. 1986. Minimum viable populations: processes of species extinction. In M. E. Soule (ed.), *Conservation Biology: The Science of Scarcity and Diversity.* Sinauer Press, Sunderland, MA. 19–34.

Goulson, D. 2003. *Bumblebees: Their Behavior and Ecology.* Oxford University Press, Oxford, UK.

Halloy, S. R. P. and A. F. Mark. 2003. Climate-change effects on alpine plant biodiversity: A New Zealand perspective on quantifying the threat. *Arctic, Antarctic, and Alpine Research* 35:248–254.

Hatfield, R. and G. LeBuhn. 2007. Patch and landscape factors shape community assemblage of bumble bees, *Bombus* spp. (Hymenoptera:

Apidae), in montane meadows. *Biological Conservation* 139:150–158.

Hegland, S. J., A. Nielsen, A. Lázaro, A.-L. Bjerknes, and Ø. Totland. 2009. How does climate warming affect plant-pollinator interactions? *Ecology Letters* 12:184–195.

Inouye, D. 2008. Effects of climate change on phenology, frost damage and floral abundance of montane wildflowers. *Ecology* 89:35–362.

Kirchner, J. W., L. Micheli, and J. D. Farrington. 1998. *Effects of Herbaceous Riparian Vegetation on Streambank Stability*. Technical Completion Report, Project W-872. University of California Water Resources Center, Berkeley, CA, USA.

Knapp, R. A. and K. R. Matthews. 1996. Livestock grazing, golden trout, and streams in the Golden Trout Wilderness, California: Impacts and management implications. *North American Journal of Fisheries Management* 16:805–820.

Lang, N. L. and C. B. Halpern. 2007. The soil seed bank of a montane meadow: Consequences of conifer encroachment and implications for restoration. *Canadian Journal of Botany* 85:557–569.

Loheide, S. and S. Gorelick. 2005. A high resolution evapotranspiration mapping algorithm (ETMA) with hydroecological applications at riparian restoration sites. *Remote Sensing Environment* 98:182–200.

Loheide, S. and S. Gorelick. 2007. Riparian hydroecology: A coupled model of the observed interactions between groundwater flow and meadow vegetation patterning. *Water Resources Research* 43:W07414. doi: 10.1029/2006WR005233.

Louette, G. and L. De Meester. 2007. Predation and priority effects in experimental zooplankton communities. *Oikos* 116:419–426.

McFrederick, Q. and G. LeBuhn. 2006. Are urban parks refuges for bumble bees? *Bombus* spp. (Hymenoptera: Apidae). *Biological Conservation* 129:372–382.

Mckelvey, K. S., C. N. Skinner, C. Chang, D. C. Erman, S. J. Husari, D. J. Parsons, J. W. van Wagendonk, and C. P. Weatherspoon. 1996. *Sierra Nevada Ecosystem Project: Final Report to Congress, Vol. II. Assessments and Scientific Basis for Management Options*. Centers for Water and Wildland Resources, University of California, Davis, CA.

Micheli, E. R. and J. W. Kirchner. 2002a. Effects of wet meadow riparian vegetation on stream-bank erosion. 1. Remote sensing measurements of streambank migration and erodibility. *Earth Surface Processes and Landforms* 27:627–639.

Micheli, E. R. and J. W. Kirchner. 2002b. Effects of wet meadow riparian vegetation on streambank erosion. 2. Measurements of vegetated bank strength and consequences for failure mechanics. *Earth Surface Processes and Landforms* 27:687–697.

Morin, P. J. 1999. *Community Ecology*, 2nd ed. Blackwell Science, Malden, MA.

Nakicenovic, N. and R. Swart (eds). 2000. *IPCC Special Report on Emissions Scenarios*. Cambridge University Press, Cambridge, UK.

Nordby, A. C. 2010. *Influence of Emergence Phenology and Elevation on Community Structure of Bumble Bees*. Master's Thesis, San Francisco State University.

Odion, D. C., T. L. Dudley, and C. M. D'Antonio. 1988. Cattle grazing in southeastern Sierran meadows: Ecosystem change and prospects for recovery. In C. A. Hall and V. Doyle-Jones (eds), *Plant Biology of Eastern California, Mary DeDecker Symposium*. White Mountain Research Station, Los Angeles, CA, USA.

Packer, L. and R. Owen. 2001. Population genetic aspects of pollinator decline. *Conservation Ecology* 5:4.

Parmesan, C. 2006. Ecological and evolutionary responses to recent climate change. *Annual Review of Ecology, Evolution, and Systematics* 37:637–669.

Parmesan, C., N. Ryrholm, C. Stefanescu, J. K. Hill, C. D. Thomas, H. Descimon, B. Huntley, L. Kaila, J. Kullberg, T. Tammaru et al. 1999. Poleward shift of butterfly species' ranges associated with regional warming. *Nature* 399:579–583.

Parmesan, C. and G. Yohe, 2003. A globally coherent fingerprint of climate change impacts across natural systems. *Nature* 421:37–42.

Pauli, H., M. Gottfied, I. K. Reiter, C. Klettner, and G. Grabherr. 2007. Signals of range expansions and contractions of vascular plants in the high Alps: Observations (1994-2004) at the GLORIA master site Schrankogel, Tyrol, Austria. *Global Change Biology* 13:147–156.

Price, J. E. and P. J. Morin. 2004. Colonization history determines alternate community states in a food web of intraguild predators. *Ecology* 85:1017–1028.

Root, T. L., J. T. Price, K. R. Hall, S. H. Schneider, C. Rosenzweig, and J. A. Pounds. 2003. Finger-prints of global warming on wild animals and plants. *Nature* 421:57–60.

Steenhuis, T. and W. Van der Molen. 1986. Thornthwaite-Mather procedure as a simple engineering method to predict recharge. *Journal of Hydrology* 84:221–229.

Theurillat, J. and A. Guisan. 2001. Potential impact of climate change on vegetation in the European Alps: A review. *Climate Change* 50:77–109.

Thorp, R., D. S. Horning, and L. L. Dunning. 1983. *Bumble Bees and Cuckoo Bumble Bees of California.* University of California Press, Berkeley, CA.

Thorp, R. W., and M. D. Shepherd. 2005. Profile: Subgenus Bombus. In *Red List of Pollinator Insects of North America CD-ROM Version 1 (May 2005).* The Xerces Society for Invertebrate Conservation, Portland, OR.

Williams, P. H. 1986. Environmental change and the distributions of British bumble bees (*Bombus* Latr.). *Bee World* 67:50–61.

Williams, P., S. Colla, and Z. Xie. 2009. Bumblebee vulnerability: Common correlates of winners and losers across three continents. *Conservation Biology* 23(4):931–940.

Manager Comments

Brendan Colloran
in conversation with
Tina Mark

Colloran: Have you observed changes in the meadow ecosystems that you manage that seem linked to changes in climate? As climate continues to change, what ecological changes do you expect to see?

Mark: We're seeing changes but at this point climate change does not seem to be the biggest driver of change. The most significant changes we are seeing in montane meadows of the Tahoe area are conifer encroachment likely related to lack of fire and also a drop in water table from historic stream incision. In the long term, these factors may be influenced by climate change. There are also positive changes (rewetting) in meadows related to reduced grazing pressure and meadow engineering (restoration).

Colloran: How do you see this information fitting in with the work that you do to manage / conserve montane meadow systems in California? What kinds of decisions could this type of research be used to inform?

Mark: This information can help the Forest Service in developing effective approaches to managing use of meadows by livestock, which meadows need hydrologic restoration. Also, it may benefit meadow restoration including conifer removal. It can also help understanding the effects of climate change on meadow size and hydrology. There is some work already underway on prescribed fire in meadows of Sierraville Ranger District on the Tahoe National Forest and on the Inyo National Forest.

Colloran: Is this an area where stronger or continued partnership with academic researchers is desired / valued? Can you give an example of an ecological study that would be of particular interest to you, and that would have the potential to influence management actions?

Mark: Yes, partnerships with academic researchers can help identify new issues and approaches for management. Some examples of areas of ecological study that would have the potential to influence

management include design of meadow restoration and conservation of species of management concern (e. g., state and federal listed endangered species such as willow flycatcher); experiments on factors affecting rate of conifer encroachment in meadows; and long-term monitoring of climate change effects on meadows. Additional needs include public education and citizen science on wildflowers and pollinators.

Colloran: When you think about your ability to change management approaches or shift conservation priorities in this system, what do you see as the main constraints? Are there particular areas of study (ecological science, climate science, policy, data management, etc.) that could help overcome this constraint?

Mark: An important issue (or constraint in many cases) for conservation and management priorities is the concept of managing for single species needs versus overall ecological function and resilience. Society and current legal protection are most concerned with rare species (e. g., mountain yellow-legged frog), yet the greatest long-term benefits to species and ecosystems could be achieved in managing for resilient and healthy ecosystems. The other constraints include limited funding, lack of public awareness of the issues, and high turnover (and low replacement) of experienced staff. Areas of study that would help overcome constraints include studies that incorporate species functional groups as indicators of ecological resilience and function rather than studies that incorporate only single species or special status species.

Colloran: What are the key factors that are currently used to guide restoration priorities for alpine meadow communities? Have these changed at all in light of observed and projected changes in climate?

Mark: Stream degradation factors such as stream incision and altered hydrology, along with restoration feasibility, guide priorities. The role of these

(continued)

(continued)

factors in guiding restoration priorities has not changed in light of climate change.

Colloran: Are benefits to pollinators currently being considered, and / or is any monitoring of bees underway or being considered?

Mark: Benefits to pollinators are considered and are a big part of public awareness campaigns. Bees are being considered to being added to sensitive / management species list for Tahoe area National Forests. There is no systematic bee monitoring currently underway. Plant species composition of meadows is being extensively monitored, but no guidelines are being given as to which plant species / pollinator relationships are important to monitor.

Colloran: Are there other key ecological services that are provided by meadows that could be modeled in a similar way to help understand the vulnerabilities and priorities for meadow restoration?

Mark: Modeling of hydrologic and biological responses to restoration could help understanding of vulnerabilities and priorities for meadow restoration. A new meadow guide has been developed which classifies meadow hydrologic types (Weixelman et al. 2011). This new publication has the potential to help guide meadow restoration efforts and to clarify the different types of meadows and restoration possibilities to fit certain meadow types.

Colloran: Understanding hydrologic impacts of climate change, especially in alpine systems where we are concerned with both precipitation in general and the depth and timing of snowpack, is a major challenge. In this system, can you think of restoration approaches or factors to consider in conservation planning that future research could focus on that would help develop adaptation strategies that are robust to a variety of different possible futures?

Mark: Future research could focus on coupling hydrologic and biological restoration goals for resilience, management, and adaptation to various climate change scenarios. A major question is the overall effect on hydrology due to lack of fire, overly dense forested areas, and subsequent effects on water availability as compared to effects on hydrology due to climate change. Teasing out these effects is crucial to management.

LITERATURE CITED

Weixelman, D. A., B. Hill, D. J. Cooper, E. L. Berlow, J. H. Viers, S. E. Purdy, A. G. Merrill, and S. E. Gross. 2011. *A Field Key to Meadow Hydrogeomorphic Types for the Sierra Nevada and Southern Cascade Ranges in California*. General Technical Report R5-TP-034, U.S. Department of Agriculture, Forest Service, Pacific Southwest Region, Vallejo, CA, 34 pp.

CHAPTER 8

Elevational Shifts in Breeding Birds in the Southern California Desert Region

Lori Hargrove and John T. Rotenberry

Abstract. The biogeographical distribution of a species is generally limited by the set of environmental conditions (ecological "niche," including climate and habitat) to which the species is best adapted. Distribution limits often occur along ecological gradients such as temperature isotherms or transition zones between habitat types. If limiting environmental conditions change, then we expect populations at distribution margins to show evidence of expansion or retraction in association with that change. In this study, we test for distributional shifts in breeding birds along an arid elevation gradient in the Santa Rosa Mountains of southern California that is undergoing rapid climate change (locally, mean maximum temperature in spring at the low end of this gradient has increased by 5.0°C since 1961). Increasing temperatures and aridity in this system are expected to cause upward shifts in the elevational distributions of species. Over the past 26 years, five bird species (out of 28 tested) showed statistically significant shifts, all upward in elevation (average increase 496 m). The average shift for all 28 species was upward (average

---

Key Points

- Long-term monitoring is necessary, especially along gradients, and often straddling multiple management boundaries.
- Upper limits of species are expected to expand upward in elevation, while lower limits of species are expected to contract upward in elevation.
- Maintain habitat connectivity along elevation gradients.
- Not all species will shift at the same rate, due to differences in physiology, demography, site tenacity, dispersal, and local differences in climatic change and other environmental factors.
- Monitor fitness measures of marginal populations that are both shifting and not shifting to determine stressors and mechanisms of distribution shifts.

---

increase 116 m), but desert species had stronger upward shifts (average increase 171 m) compared to montane species (average increase 60 m). Our results reveal rapid shifts in avian distributions as predicted by climate warming, which has profound implications for this arid ecosystem, which include potential altered community structure and novel species–habitat associations. Management strategies designed to conserve biodiversity will need to incorporate this rapid dynamism into future conservation efforts.

INTRODUCTION

Biogeography, or the spatial and temporal distribution of species, is a fundamental aspect of ecology and evolution, and understanding distributional change is valuable to conservation and management of species. Across the range of a species, locations with higher abundances tend to be associated with more favorable conditions, while abundances tend to taper off toward range limits where conditions become unfavorable (Brown et al. 1995, Brown et al. 1996). In vagile organisms such as birds, the behaviors of habitat selection and territoriality largely determine spatial distribution, and are expected to link adaptive traits of individuals to suitable habitat. When there are no barriers to dispersal, range limits are often linked to physiological or reproductive constraints directly or indirectly associated with climate. Consequently, species often show strong distributional patterns when surveyed along latitudinal and, especially, elevational gradients because of climatic differences. In birds, climate may be a direct limiting factor on distributions when associated with a species' physiological temperature tolerance limits, and it may also affect species' distributions indirectly through influence on patterns in vegetation and resource availability. Distributional studies of birds often identify climate as an important range boundary predictor (e. g., Root 1988, Bohning-Gaese and Lemoine 2004), while vegetation structure and composition are often identified as important predictors of bird distribution and abundance locally (Cody 1985, Wiens 1989, Block and Brennan 1993).

To the extent that species' distributions reflect climatic limitations, distributions are expected to shift in response to climate change unless these shifts are constrained by some form of barrier or resource limitation. For example, if average temperatures increase, then distributions are expected to expand where a species is cold-limited, and retract where it is heat-limited. Many observed changes in species' distributions are consistent with the hypothesis that they represent a response to climate change. Poleward latitudinal range expansions are well documented and demonstrate a "fingerprint" of global warming (Walther et al. 2002, Parmesan and Yohe 2003, Root et al. 2003, Parmesan 2006, Chen et al. 2011). However, range retractions and elevational shifts are less well known (Shoo et al. 2006, Thomas et al. 2006). Strong climatic gradients can be found with changes in elevation over a relatively short distance, and thus provide a potentially sensitive system for detecting distributional shifts in association with climate change. Examples of recent upward elevational shifts in species assemblages consistent with climate change effects include those for vascular plants (Walther et al. 2005, Kelly and Goulden 2008, le Roux and McGeouch 2008, Lenoir et al. 2008, Brusca et al. 2013, Kopp and Cleland 2014), insects (Konvicka et al. 2003, Franco et al. 2006, Wilson et al. 2007, Baessler et al. 2013), reptiles and amphibians (Raxworthy et al. 2008), small mammals (Moritz et al. 2008), and birds (Reif and Flousek 2012, Freeman and Freeman 2014), but there have been many inconsistencies as well, especially for birds (Archaux 2004, Brommer and Møller 2010, Lenoir et al. 2010, Forero-Medina et al. 2011, Tingley et al. 2012). Although birds are expected to be responsive to climate change and there is evidence of close tracking of ranges with climate (Tingley et al. 2009), fine-scale studies of elevational shifts in birds are greatly needed but are few (Sekercioglu et al. 2008). Ideally, tests of elevational shifts should also be linked to mechanistic studies that determine how climate change affects the demography of local populations, and how that varies among regions and species.

Elevation-based studies usually address the effects of climate change on high-elevation montane species rather than low-elevation desert species, and predictions tend to emphasize extinction risk due to warmer temperatures at high-elevation sites (e. g., Thomas et al. 2006). Although it is often thought that global warming will cause relatively minor impacts on desert species (e. g., Sala et al. 2000, Thomas et al. 2004), little is known about biological responses to climate change in arid ecosystems. Desert species are strongly responsive to variation in precipitation, and may be particularly sensitive to changes in both temperature and precipitation (e. g., McKechnie and Wolf 2010, Sinervo et al. 2010). Yet, how climate parameters change and how ecosystems respond to those changes may vary widely among regions. In the Chihuahuan Desert, the replacement of grassland by desert scrub has been attributed to an increase in winter precipitation (Brown et al. 1997). The desert regions of southern California are generally predicted to become warmer and drier, and variation in extreme events such as floods and droughts is expected to increase (e. g., Hayhoe et al. 2004, Seager et al. 2007). In deserts, although interannual variance in productivity associated with variation in rainfall tends to be high, increasing variance in annual rainfall can reduce population viability (survival and reproduction) by intensifying droughts (Saltz et al. 2006). Changes in drought periodicity and intensity may have even stronger impacts on desert populations than gradually shifting temperature and precipitation means (Albright et al. 2010, Barrows et al. 2010). Although a high rate of turnover is predicted for birds in the southern California deserts due to rapid temperature increases (Stralberg et al. 2009), it is unclear if desert species will be able to simply shift into higher-elevation areas (Hargrove and Rotenberry 2011a).

Along a desert-to-mountain transition, distributional shifts of both desert and montane species can be tested simultaneously. To test for distributional shifts in association with climate change, we quantified current distributions of breeding birds along a desert-to-mountain elevational gradient in southern California that is undergoing rapid climate change. These distributions were then compared to those recorded 26 years earlier using identical methods (Mayhew 1981, Weathers 1983). A warmer, more arid climate is predicted to cause an upward elevational shift in distributions of both desert and montane species. For desert species, the upper elevational limit is expected to advance upward, while for montane species, the lower elevational limit is expected to retract upward. Here, the results of this comparison are followed by a discussion of the possible implications of climate change for desert ecosystems, with suggestions for research needs and updates to management strategies.

METHODS

We conducted our study along the "Deep Canyon Transect" located at the Philip L. Boyd Deep Canyon Desert Research Center, part of the University of California Natural Reserve System, and extending into the San Bernardino National Forest on the north- and east-facing slopes of the Santa Rosa Mountains in central Riverside County, California (Figure 8.1). The Deep Canyon Transect spans an elevational range from near sea level to 2600 m over a distance of 35 km along the transition between the Peninsular Ranges and Colorado Desert. The Peninsular Ranges, which include the Santa Rosa Mountains, run north–south and form a rain shadow for the Colorado Desert to the east. The Colorado Desert is an extension of the Sonoran Desert, and includes areas that rank among the hottest, driest places on earth (Meigs 1953, UNEP 1997). The vegetation varies from Sonoran desert scrub at lower elevations to chaparral and pinyon-juniper woodland at mid-elevations and mixed coniferous woodland at upper elevations (Figure 8.2).

In 1979, a series of plot transects were established along the Deep Canyon Transect, each of varying length and orientation, but typically

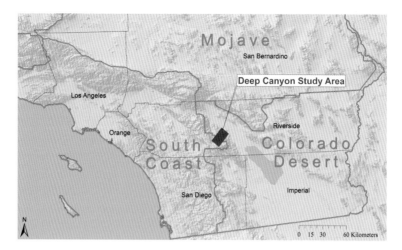

FIGURE 8.1: Location of the Deep Canyon Transect. Located at the transition between the South Coast Peninsular Ranges and Colorado Desert, Southern California. Basemap source: ESRI.

FIGURE 8.2: The range of vegetation types surveyed along the Deep Canyon Transect. A. Chaparral and oak-conifer woodland (elevation 1800 m). B. Chaparral and pinyon-juniper woodland (elevation 1200 m). C. Mid-elevation desert scrub (elevation 900 m). D. Sonoran desert scrub and Palo Verde Wash (elevation 200 m).

1-km long (Mayhew 1981). These transects were systematically surveyed multiple times each year (typically 2–3 times each spring), and all observations of vertebrates were recorded (Mayhew 1981, Weathers 1983). In the spring of 2005–2007, using the same methods, we conducted repeat surveys at 15 of the existing plot transects that were most consistently surveyed and that were stratified along an elevation range of 200–2400 m. We compared our results to data collected at these same locations during the first three years of the original surveys, 1979–1981, using only morning surveys during spring months with good weather conditions. The analyses were restricted to 28 breeding bird species. We excluded species that

BOX 8.1 · Elevational Shifts in the Mountain Quail Distribution

To test for elevational shifts in distributions, we compared observations of 28 bird species along an elevational gradient between two three-year time periods at Deep Canyon, 26 years apart. We quantified patterns of bird abundance along this gradient using the data collected during the breeding season at 15 sites ranging in elevation from 200 to 2400 m, using the same survey methods during both time periods.

In this example for mountain quail (*Oreortyx pictus*), the elevational distributions are shown for the two time periods (1979–1981 in green vs. 2005–2007 in orange; Box Figure 8.1.1). For each time period, the graph shows the cumulative abundance of mountain quail observed at each of the 15 sites as elevation increases from 200 to 2400 m. Thus, the cumulative frequency distribution curves show the sum of the species'

proportional abundance from its lowest to highest altitudinal occurrence, such that the cumulative frequency at the highest elevation site is always 100%.

In 2005–2007, we found fewer mountain quail at lower-elevation sites compared to 1979–1981 (shown by the rightward shift in the distribution), indicating that there was an upward shift in the elevational distribution of abundance for this species. A Kolmogorov–Smirnov test is a nonparametric statistical test used to compare differences between two cumulative distributions. For the mountain quail, this test indicated a difference between the two distributions that was statistically significant ($p < 0.05$). Of the 28 species tested in this way, five showed a statistically significant difference. All of these five species had a strong upward shift in weighted mean elevation.

BOX FIGURE 8.1.1: Upward shift in the elevational breeding distribution of mountain quail (*Oreortyx pictus*) over a 26-year time period, Boyd Deep Canyon Desert Research Center, Riverside County, California. Elevational distributions are shown for the two time periods [1979–1981 (green squares) vs. 2005–2007 (orange triangles); K-S test: $p < 0.05$]. Photo credit: Peter LaTourrette / birdphotography.com.

were rare, nonbreeding migrants, wide-ranging foragers (e. g., swifts overhead), not always identified to species (e. g., hummingbirds), or completely absent during one of the two survey periods. Although the elevational ranges and distributional limits varied widely by species, we categorized species as "desert" if they tended to be more restricted to lower-elevation sites in this study area (with an upper limit distributional margin falling within the study area) or "montane" if they tended to be more restricted to higher-elevation sites (with a lower limit distributional margin falling within the study area).

To provide context for our bird observation data, we analyzed temperature and precipitation data from two local weather stations, one on the desert floor at an elevation of 292 m and one in the mountains at an elevation of 1640 m (WRCC 2007, M. Fisher, pers. comm.). We tested long-term trends for maximum temperature (yearly and spring averages of mean monthly maximum temperatures), minimum temperature (yearly and spring averages of mean monthly minimum temperatures), and precipitation (July–June rain-year). To compare cumulative precipitation between the two survey periods, we summed precipitation over a five-year period (each period containing the three survey years plus two preceding years).

To test for distributional shifts, we used methods that were robust to any differences in absolute abundance between the two time periods that might have arisen, for example, due to observer differences. For each species, we calculated abundance as the average number of birds detected over a 1-km transect for each of the 15 sites, for each three-year period. We calculated a weighted mean elevation for each species for each three-year period as the sum of elevations at which the species was present, each multiplied by the abundance of that species at that site, and divided by the total abundance for that species over all sites. We constructed cumulative frequency distribution curves for each species by summing its proportional abundance from its lowest to highest altitudinal occurrence (Box 8.1). Thus, cumulative frequency at the highest elevation site is always 100%. A Kolmogorov–Smirnov (K-S) test was used to test for any difference in the cumulative elevational distribution of individual species between the two periods (Box 8.1), and a paired-sample t-test was used to test for upward elevational shifts in weighted mean elevations at the community level. Three community groups were considered: All species, desert species alone, and montane species alone.

RESULTS

In the desert (Boyd Deep Canyon Desert Research Center, elevation 292 m), mean maximum temperature increased by 3.8°C from 1962 to 2006, while there was no change in mean minimum temperature. The increase in mean maximum temperature was even greater during the main breeding season when surveys were conducted: 5.0°C since 1961 ($r^2 = 0.48$, March–June, Figure 8.3). In contrast, at a montane site near the upper end of the transect (Idyllwild, elevation 1640 m), there was no trend in the mean maximum temperature, but the mean minimum temperature increased by 1.7°C from 1960 to 2006, with no difference between annual ($r^2 = 0.39$) and spring ($r^2 = 0.16$, Figure 8.3) averages. In the desert, there was no long-term trend in precipitation over the 45 years from 1962 to 2007, but the most recent survey period, 2003–2007, had 44% less cumulative precipitation than the survey period 26 years ago, 1977–1981 (each period containing the three survey years plus two preceding years, using July–June rain-years).

Five individual species (out of 28 tested) showed statistically significant differences in their cumulative elevational distributions (K-S test, $p < 0.05$, $n = 15$; Box 8.1, Table 8.1). All five had an upward shift in their weighted mean elevation, with an average increase of 496 m. Moreover, there was an upward elevational shift in the avian community as a whole (paired-sample t-test, $p < 0.01$, $n = 28$ species). The average weighted mean elevation for all 28 species increased by 116 m. Species categorized as "desert" showed a statistically significant average elevational shift of 171 m upward (paired-sample t-test, $p < 0.05$, $n = 14$), whereas the shift in "montane" species, although upward an average of 60 m, was not significant (paired-sample t-test, $p = 0.12$, $n = 14$).

DISCUSSION

These results suggest that significant elevational shifts in breeding bird distributions are

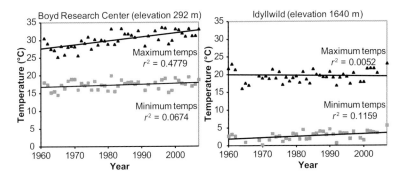

FIGURE 8.3: Spring mean maximum and minimum temperature trends in the desert (left) and mountains (right), 1960–2007. Values are daily maximum and minimum temperatures, averaged for each spring (March–June), and r^2 denotes proportion of variance in temperature statistically explained by linear regression on year. (Boyd Deep Canyon Desert Research Center, elevation 292 m, and Idyllwild, elevation 1640 m.)

possible over a relatively short time. The upward shifts we observed are consistent with expectations given the warmer, drier conditions now versus 26 years ago. Compared to montane species, desert species were more likely to show upward shifts. These shifts in distribution occurred in a diverse group of breeding bird species, each likely to have a different set of direct and indirect ecological links to changes in climate. Response to temperature and precipitation can be direct (e. g., mortality and nest timing), or mediated through biotic factors linked to climate (e. g., habitat and food availability).

In desert birds, physiological adaptations to heat and aridity likely involve decreases in metabolic rates and water loss (Tieleman and Williams 2000). Increasing maximum temperatures at desert sites can exceed tolerance limits, while increasing minimum temperatures at montane sites can improve suitability. Drought is known to cause reproductive failure in many bird species, even in relatively arid environments (e. g., Bolger et al. 2005, Albright et al. 2010). Many bird species are associated strongly with certain habitats, and there is a suggestion that vegetation is similarly shifting upward in this area (Kelly and Goulden 2008). This is likely associated with recent die-off of desert shrubs at low elevations (Miriti et al. 2007). Thus, although changing climate can

have direct effects on species' distributions, biotic interactions (such as species–habitat relationships) can amplify the effects of climate change if one component of the relationship is more sensitive than another. Therefore, biotic interactions should be an important consideration in predictive distribution modeling too (Preston et al. 2008, Angert et al. 2013).

In this study system, desert species were especially likely to show upward elevational shifts. Although it is often thought that desert species will exhibit relatively little negative impact of global warming because desertification will create more desert habitat (e. g., Sala et al. 2000, Thomas et al. 2004), desert species are likely to be closer to their limits of temperature and precipitation tolerance than species of more humid habitats. Even though some desert species may be able to adapt to further increases in temperature and aridity within their current range, it is unlikely that adaptive evolutionary processes can keep pace with the high rate of temperature increase in this system. Results from the North American Breeding Bird Survey indicate decreasing population trends overall for the Sonoran and Mojave deserts from 1966 to 2007 (Sauer et al. 2008), but more study is needed on the causes of these population declines. With increasing desertification throughout the region, relatively mobile species such as birds may be able to shift into new areas

TABLE 8.1

Species in order of weighted mean elevation (m) in 1979–1981 at 15 sites, Boyd Deep Canyon Desert Research Center, Riverside County

Species	Desert /montane	Weighted mean elevation, 1979–1981 (m)	Elevational shift, 2005–2007 (m)	K-S test
Black-tailed gnatcatcher (*Polioptila melanura*)	D	245	+6	NS
Verdin (*Auriparus flaviceps*)	D	248	+81	NS
Phainopepla (*Phainopepla nitens*)	D	269	+562	$p < 0.05$
Northern mockingbird (*Mimus polyglottos*)	D	270	+293	NS
Say's phoebe (*Sayornis saya*)	D	270	+260	$p < 0.05$
American kestrel (*Falco sparverius*)	D	334	+786	$p < 0.05$
California-Gambel's quail (*Callipepla californica-gambelii*)[a]	D	487	+93	NS
Cactus wren (*Campylorhynchus brunneicapillus*)	D	530	+226	NS
Loggerhead shrike (*Lanius ludovicianus*)	D	616	+29	NS
Rock wren (*Salpinctes obsoletus*)	D	638	+65	NS
Ladder-backed woodpecker (*Picoides scalaris*)	D	820	+307	$p < 0.05$
Black-throated sparrowm (*Amphispiza bilineata*)	D	914	−251	NS
Scott's Oriole (*Icterus parisorum*)	D	929	−44	NS
Bushtit (*Psaltriparus minimus*)	M	1165	+309	NS
Greater roadrunner (*Geococcyx californianus*)	D	1216	+9	NS
California rowhee (*Pipilo crissalis*)	M	1262	−10	NS
Western scrub jay (*Aphelocoma californica*)	M	1290	+48	NS
Pinyon jay[b] (*Gymnorhinus cyanocephalus*)	M	1298	−2	NS
Bewick's wren (*Thryomanes bewickii*)	M	1351	+64	NS
Mountain quail (*Oreortyx pictus*)	M	1370	+594	$p < 0.05$
California thrasher (*Toxostoma redivivum*)	M	1424	−88	NS
Oak titmouse (*Baeolophus inornatus*)	M	1509	−38	NS
Spotted Towhee (*Pipilo maculatus*)	M	1597	−53	NS
Wrentit (*Chamaea fasciata*)	M	1598	+47	NS
Black-chinned sparrow (*Spizella atrogularis*)	M	1739	−14	NS
Northern flicker (*Colaptes auratus*)	M	2215	−59	NS
House wren (*Troglodytes aedon*)	M	2314	+66	NS
Mountain chickadee (*Poecile gambeli*)	M	2336	−25	NS

NOTE: "D" (desert) indicates that the species was more common at low-elevation sites in this study system, while "M" (montane) indicates that the species was more common at higher-elevation sites. Elevational shift is the difference in weighted mean elevation between the two time periods (1979–1981 vs. 2005–2007). Positive values indicate upward elevational shifts. The Kolmogorov–Smirnov test was used to compare the cumulative elevational distributions between the two time periods.

a Data combined for two hybridizing species (California and Gambel's quail).
b Pinyon jays were highly localized at intermediate elevation sites.

as they become habitable. However, in southern California, higher elevations are limited in extent, and in the more mesic coastal areas, natural habitats that might otherwise undergo conversion to desert vegetation types are highly fragmented by urbanization. In addition, there is unlikely to be any "backfill" or immigration by new species if lower elevations become increasingly inhospitable to current desert species—the Colorado Desert is already one of the hottest and driest areas on earth. Although this study excluded rare species that were absent during one of the two time periods, no obvious trends were apparent with respect to species additions and extirpations at low versus high elevations.

SOLUTIONS, ADAPTATION, AND LESSONS

Predictions or forecasts from most climate-change models are averaged over large geographical spaces and long periods of time. However, organisms and their habitats are affected by local temperature, precipitation, and short-term extreme events (e.g., McKechnie and Wolf 2010). The distributional shifts observed in this study system over a 26-year period could be partly due to short-term climate fluctuations, but this rapid dynamism suggests that longer-term effects could be equally pronounced. Elevational distributions of birds were found to be relatively static over a three-year period (Hargrove and Rotenberry 2011b), which further suggests that these 26-year shifts are due to longer-term changes. The strong differences in temperature within this study system with opposing temperature trends only a short distance apart are likely to have very different biological effects. Thus, climate-change predictions at a much finer scale are needed to manage biodiversity for these challenges. Furthermore, although trends in temperature and precipitation extremes are less well known, they are even more likely to be biologically relevant than annual averages.

Understanding distributional change is valuable to conservation management. As species may shift across management boundaries, management perspectives must evolve to encompass broader spatial and temporal scales. However, little is known about the rates and mechanisms of range shifts, and there is great uncertainty as to the differences in responsiveness among species. Additional monitoring is needed along elevational gradients and at other transition zones where distribution limits occur. Ideally, monitoring should be linked to mechanistic studies as well. For example, during drought years in this study system, breeding success of desert species tends to be greater at higher elevations, while reproductive failure is more likely on the desert floor, which may be exacerbated by warmer temperatures (Hargrove 2010). However, the link between demographic processes and range shifts remains weak, and both ecological and evolutionary studies of distribution limits are greatly needed (Hoffmann and Blows 1994, Parmesan et al. 2005, Travis et al. 2013). Nevertheless, it is clear that management strategies will need to incorporate dynamic processes into future conservation efforts and encompass broader spatial and temporal scales across management boundaries.

ACKNOWLEDGMENTS

We are grateful to Al Muth and Mark Fisher for assistance with field access and historical data, and to Myung-Bok Lee for data entry. We thank the Biological Impacts of Climate Change in California group and Phil Unitt for comments on the research and manuscript. Field work was supported by a Biological Impacts of Climate Change in California Research grant from the California Energy Commission's Public Interest Energy Research Environmental Area and Point Reyes Bird Observatory Conservation Science, the California Department of Parks and Recreation, a California Desert Research Fund from the Community Foundation Serving Riverside and San Bernardino Counties, a Mildred E. Mathias Graduate Student Research grant from the University of California Natural Reserve System, a

Mewaldt-King Student Research grant from the Cooper Ornithological Society, and a Ralph W. Schreiber Ornithology Research Award from the Los Angeles Audubon Society.

LITERATURE CITED

Albright, T. P., A. M. Pidgeon, C. D. Rittenhouse, M. K. Clayton, C. H. Flather, P. D. Culbert, B. D. Wardlow, and V. C. Radeloff. 2010. Effects of drought on avian community structure. *Global Change Biology* 16:2158–2170.

Angert, A. L., S. L. LaDeau, and R. S. Ostfeld. 2013. Climate change and species interactions: Ways forward. *Annals of the New York Academy of Sciences* 1297:1–7.

Archaux, F. 2004. Breeding upwards when climate is becoming warmer: No bird response in the French Alps. *Ibis* 146:138–144.

Baessler, C., T. Hothorn, R. Brandl, and J. Muller. 2013. Insects overshoot the expected upslope shift caused by climate warming. *PLOS ONE* 8:e65842.

Barrows, C. W., J. T. Rotenberry, and M. F. Allen. 2010. Assessing sensitivity to climate change and drought variability of a sand dune endemic lizard. *Biological Conservation* 143:731–736.

Block, W. M. and L. A. Brennan. 1993. The habitat concept in ornithology: Theory and applications. *Current Ornithology* 11:35–91.

Bohning-Gaese, K. and N. Lemoine. 2004. Importance of climate change for the ranges, communities and conservation of birds. *Advances in Ecological Research* 35:211–236.

Bolger, D. T., M. A. Patten, and D. C. Bostock. 2005. Avian reproductive failure in response to an extreme climatic event. *Oecologia* 142:398–406.

Brommer, J. E. and A. P. Møller. 2010. Range margins, climate change, and ecology. In A. P. Møller, W. Fiedler, and P. Berthold (eds), *Effects of Climate Change on Birds*. Oxford University Press, Oxford. 249–274.

Brown, J. H., D. W. Mehlman, and G. C. Stevens. 1995. Spatial variation in abundance. *Ecology* 76:2028–2043.

Brown, J. H., G. C. Stevens, and D. M. Kaufman. 1996. The geographic range: Size, shape, boundaries, and internal structure. *Annual Reviews of Ecology and Systematics.* 27:597–623.

Brown, J. H., T. J. Valone, and C. G. Curtin. 1997. Reorganization of an arid ecosystem in response to recent climate change. *Proceedings of the National Academies of Science of the United States of America* 94:9729–9733.

Brusca, R. C., J. F. Wiens, W. M. Meyer, J. Eble, K. Franklin, J. T. Overpeck, and W. Moore. 2013. Dramatic response to climate change in the Southwest: Robert Whittaker's 1963 Arizona Mountain plant transect revisited. *Ecology and Evolution* 3:3307–3319.

Chen, I. C., J. K. Hill, R. Ohlemuller, D. B. Roy, and C. D. Thomas. 2011. Rapid range shifts of species associated with high levels of climate warming. *Science* 333:1024–1026.

Cody, M. L. (ed.). 1985. *Habitat Selection in Birds*. Academic Press, Orlando, FL.

Forero-Medina, G., J. Terborgh, S. J. Socolar, and S. L. Pimm. 2011. Elevational ranges of birds on a tropical montane gradient lag behind warming temperatures. *PLOS ONE* 6:e28535.

Franco, A. M. A., J. K. Hill, C. Kitschke, Y. C. Collingham, D. B. Roy, R. Fox, B. Huntley, and C. D. Thomas. 2006. Impacts of climate warming and habitat loss on extinctions at species' low-latitude range boundaries. *Global Change Biology* 12:1545–1553.

Freeman, B. G. and A. M. C. Freeman. 2014. Rapid upslope shifts in New Guinean birds illustrate strong distributional responses of tropical montane species to global warming. *Proceedings of the National Academies of Science of the United States of America* 111:4490–4494.

Hargrove, L. 2010. Limits to species' distributions: Spatial structure and dynamics of breeding bird populations along an ecological gradient. Dissertation. University of California, Riverside, CA. http://proquest.umi.com/pqdweb?did=2019822781&sid=1&Fmt=2&clientId=48051&RQT=309&VName=PQD.

Hargrove, L. and J. T. Rotenberry. 2011a. Breeding success at the range margin of a desert species: Implications for a climate-induced elevational shift. *Oikos* 120:1568–1576.

Hargrove, L. and J. T. Rotenberry. 2011b. Spatial structure and dynamics of breeding bird populations at a distribution margin, southern California. *Journal of Biogeography* 38:1708–1716.

Hayhoe, K., D. Cayan, C. B. Field, P. C. Frumhoff, E. P. Maurer, N. L. Miller, S. C. Moser, S. H. Schneider, K. N. Cahill, E. E. Cleland et al. 2004. Emissions pathways, climate change, and impacts on California. *Proceedings of the National Academies of Science of the United States of America* 101:12422–12427.

Hoffmann, A. A. and M. W. Blows. 1994. Species borders: Ecological and evolutionary perspectives. *Trends in Ecology and Evolution* 9:223–227.

Kelly, A. E. and M. L. Goulden. 2008. Rapid shifts in plant distribution with recent climate change.

Proceedings of the National Academies of Science of the United States of America 105:11823–11826.

Konvicka, M., M. Mardova, J. Benes, Z. Fric, and P. Kepka. 2003. Uphill shifts in distribution of butterflies in the Czech Republic: Effects of changing climate detected on a regional scale. *Global Ecology & Biogeography* 12:403–410.

Kopp, C. W. and E. E. Cleland. 2014. Shifts in plant species elevational range limits and abundances observed over nearly five decades in a western North America mountain range. *Journal of Vegetation Science* 25:135–146.

le Roux, P. C. and M. A. McGeoch. 2008. Rapid range expansion and community reorganization in response to warming. *Global Change Biology* 14:2950–2962.

Lenoir, J., J. C. Gegout, P. A. Marquet, P. de Ruffray, and H. Brisse. 2008. A significant upward shift in plant species optimum elevation during the 20th century. *Science* 320:1768–1771.

Lenoir, J., J. C. Gegout, A. Guisan, P. Vittoz, T. Wohlgemuth, N. E. Zimmermann, S. Dullinger, H. Pauli, W. Willner, and J. C. Svenning. 2010. Going against the flow: Potential mechanisms for unexpected downslope range shifts in a warming climate. *Ecography* 33:295–303.

Mayhew, W. W. 1981. *Line Transects and Photo Sites on the Deep Canyon Transect.* University of California Natural Land and Water Reserves System, Philip L. Boyd Deep Canyon Desert Research Center, Riverside, CA.

McKechnie, A. E. and B. O. Wolf. 2010. Climate change increases the likelihood of catastrophic avian mortality events during extreme heat waves. *Biology Letters* 6:253–256.

Meigs, P. 1953. World distribution of arid and semi-arid homoclimates. *Reviews of Research on Arid Zone Hydrology.* UNESCO, Paris. 203–209.

Miriti, M. N., S. Rodriguez-Buritica, S. J. Wright, and H. F. Howe. 2007. Episodic death across species of desert shrubs. *Ecology* 88:32–36.

Moritz, C., J. L. Patton, C. J. Conroy, J. L. Parra, G. C. White, and S. R. Beissinger. 2008. Impact of a century of climate change on small-mammal communities in Yosemite National Park, USA. *Science* 322:261–264.

Parmesan, C. 2006. Ecological and evolutionary responses to recent climate change. *Annual Review of Ecology Evolution and Systematics* 37:637–669.

Parmesan, C. and G. Yohe. 2003. A globally coherent fingerprint of climate change impacts across natural systems. *Nature* 421:37–42.

Parmesan, C., S. Gaines, L. Gonzalez, D. M. Kaufman, J. Kingsolver, A. T. Peterson, and R.

Sagarin. 2005. Empirical perspectives on species borders: From traditional biogeography to global change. *Oikos* 1:58–75.

Preston, K. L., J. T. Rotenberry, R. A. Redak, and M. F. Allen. 2008. Habitat shifts of endangered species under altered climate conditions: Importance of biotic interactions. *Global Change Biology* 14:1–15.

Raxworthy, C. J., R. G. Pearson, N. Rabibisoa, A. M. Rakotondrazafy, J. Ramanamanjato, A. P. Raselimanana, S. Wu, R. A. Nussbaum, and D. A. Stone. 2008. Extinction vulnerability of tropical montane endemism from warming and upslope displacement: A preliminary appraisal for the highest massif in Madagascar. *Global Change Biology* 14:1703–1720.

Reif, J. and J. Flousek. 2012. The role of species' ecological traits in climatically driven altitudinal range shifts of central European birds. *Oikos* 121:1053–1060.

Root, T. 1988. Environmental factors associated with avian distributional boundaries. *Journal of Biogeography* 15:489–505.

Root, T. L., J. T. Price, K. R. Hall, S. H. Schneider, C. Rosenzweig, and J. A. Pounds. 2003. Fingerprints of global warming on wild animals and plants. *Nature* 421:57–60.

Sala, O. E., F. S. Chapin III, J. J. Armesto, E. Barlow, J. Bloomfield, R. Dirzo, E. Huber-Sanwald, L. Huenneke, R. B. Jackson, A. P. Kinzig et al. 2000. Global biodiversity scenarios for the year 2100. *Science* 287:1770–1774.

Saltz, D., D. I. Rubenstein, and G. C. White. 2006. The impact of increased environmental stochasticity due to climate change on the dynamics of Asiatic wild ass. *Conservation Biology* 20:1402–1409.

Sauer, J. R., J. E. Hines, and J. Fallon. 2008. *The North American Breeding Bird Survey, Results and Analysis 1966 - 2007. Version 5.15.2008.* USGS Patuxent Wildlife Research Center, Laurel, MD.

Seager, R., M. Ting, I. Held, Y. Kushnir, J. Lu, G. Vecchi, H. P. Huang, N. Harnik, A. Leetmaa, N. C. Lau et al. 2007. Model projections of an imminent transition to a more arid climate in southwestern North America. *Science* 316:1181–1184.

Sekercioglu, C. H., S. H. Schneider, J. P. Fay, and S. R. Loarie. 2008. Climate change, elevational range shifts, and bird extinctions. *Conservation Biology* 22:140–150.

Shoo, L. P., S. E. Williams, and J. M. Hero. 2006. Detecting climate change induced range shifts: Where and how should we be looking? *Austral Ecology* 31:22–29.

Sinervo, B., F. Méndez-de-la-Cruz, D. B. Miles, B. Heulin, E. Bastiaans, M. Villagrán-Santa Cruz, R. Lara-Resendiz, N. Martínez-Méndez, M. L. Calderón-Espinosa, R. N. Meza-Lázaro et al. 2010. Erosion of lizard diversity by climate change and altered thermal niches. *Science* 328:894–899.

Stralberg, D., D. Jongsomjit, C. A. Howell, M. A. Snyder, J. D. Alexander, J. A. Wiens, and T. L. Root. 2009. Re-shuffling of species with climate disruption: A no-analog future for California birds? *PLOS ONE* 4:1–8.

Thomas, C. D., A. Cameron, R. E. Green, M. Bakkenes, L. J. Beaumont, Y. C. Collingham, B. F. N. Erasmus, M. Ferreira de Siqueira, A. Grainger, L. Hannah et al. 2004. Extinction risk from climate change. *Nature* 427:145–148.

Thomas, C. D., A. M. A. Franco, and J. K. Hill. 2006. Range retractions and extinction in the face of climate warming. *Trends in Ecology and Evolution* 21:415–416.

Tieleman, B. I. and J. B. Williams. 2000. The adjustment of avian metabolic rates and water fluxes to desert environments. *Physiological and Biochemical Zoology* 73:461–479.

Tingley, M. W., W. B. Monahan, S. R. Beissinger, and C. Moritz. 2009. Birds track their Grinnellian niche through a century of climate change. *Proceedings of the National Academies of Science of the United States of America* 106:19637–19643.

Tingley, M. W., M. S. Koo, C. Moritz, A. C. Rush, and S. R. Beissinger. 2012. The push and pull of climate change causes heterogeneous shifts in avian elevational ranges. *Global Change Biology* 18:3279–3290.

Travis, J. M. J., M. Delgado, G. Bocedi, M. Baguette, K. Bartoń, D. Bonte, I. Boulangeat, J. A. Hodgson, A. Kubisch, V. Penteriani et al. 2013. Dispersal and species' responses to climate change. *Oikos* 122:1532–1540.

UNEP. 1997. *World Atlas of Desertification*. United Nations Environment Programme, Edward Arnold Publishers, London.

Walther, G., E. Post, P. Convey, A. Menzel, C. Parmesan, T. J. C. Beebee, J. Fromentin, O. Hoegh-Guldberg, and F. Bairlein. 2002. Ecological responses to recent climate change. *Nature* 416:389–395.

Walther, G., S. Beibner, and C. A. Burga. 2005. Trends in the upward shift of alpine plants. *Journal of Vegetation Science* 16:541–548.

Weathers, W. W. 1983. *Birds of Southern California's Deep Canyon*. University of California Press, Berkeley, CA.

Wiens, J. A. 1989. *The Ecology of Bird Communities*. Cambridge University Press, Cambridge, UK.

Wilson, R. J., D. Gutierrez, J. Gutierrez, and V. J. Monserrat. 2007. An elevational shift in butterfly species richness and composition accompanying recent climate change. *Global Change Biology* 13:1837–1884.

WRCC (Western Regional Climate Center). 2007. Idyllwild Fire Department, California. http://www.wrcc.dri.edu (accessed on October 1).

Manager Comments

Lori Hargrove
in conversation with
Mark Fisher, Allan Muth,
Jenny Rechel, and Anne Poopatanapong

Hargrove: As climate continues to change, what ecological changes do you expect to see in southern California's ecosystems?

Fisher and Muth: At Boyd Deep Canyon, we have observed changes in the distribution of plants, the timing of flowering, and mortality of plant species that seem linked to a warming and drying climate. For example, the perennial plants *Encelia farinosa* and *Ambrosia dumosa* are missing from large parts of the creosote bush scrub where they were once among the top 10 dominant species. The cactus *Opuntia basilaris* begins flowering about a month earlier than when we both came here in the early 1980s. A prolonged drought resulted in over 75% mortality of the cactus *Cylindropuntia bigelovii* at one site. We expect increased mortality of species at the hotter and drier margins of their distributions in conjunction with an overall decrease in precipitation and increase in temperature. Since the hotter, drier margins are typically at lower elevation, there should be a net uphill movement in species distributions as demonstrated with birds in this case study.

Rechel: I work in the San Jacinto Mountains on the San Jacinto Ranger District. In our area, I expect to see the density of shrubs increasing in the mid-elevations and in the area dominated currently by oak woodlands, and expect to record increased density of *Quercus* species in the upper elevations with conifer species. More generally, I expect to observe an overall drier climate, with more common spikes in storm events, thereby resulting in increased erosion in stream channels (I have already observed this in several areas). I also expect to observe angiosperms flowering earlier in the spring and expect to see different angiosperm species and grasses flowering based on soil moisture resulting from too dry or too wet winter precipitation conditions.

Poopatanapong: In the San Bernardino National Forest (SBNF), we are already seeing changes, such as shorter spring flowering periods. We are

also seeing Quino checkerspot butterflies in places where they have not been documented scientifically before—we are seeing them at higher elevations and farther to the east. In 2010, we saw significant increase in the individuals sighted, which may suggest that the species is actually breeding on the forest that we have not seen before. Also, on the topic of possible range shifts, we haven't seen species that we expect to see on the district, such as the northern flying squirrel. The SBNF is not actively searching for them, but no one seems to see them anymore on the mountain range. We have also seen desert species such as great-tailed grackle in areas that one would not typically see them such as Lake Hemet. It appears that they are expanding their distribution westward for some reason.

Hargrove: In your experience, what do you see as key vulnerabilities of the ecosystems you manage? Are they viewed as highly vulnerable to climate change?

Fisher and Muth: The increased variability in weather that is associated with climate change can make the marginal ecosystems most susceptible to extirpations of endemic species, especially those species that are unable to disperse upward because of specific habitat preferences. The Colorado Desert division of the Sonoran Desert is the driest, hottest desert in the United States, and is thus extremely vulnerable to the effects of climate change in comparison with more insulated ecosystems.

We have recently documented extreme variation in reproductive output in the endangered lizard, *Uma inornata*, that might lead to localized extirpation during prolonged droughts. This should be replicated by many other species that are likewise restricted to a specific substrate type or other limited habitat feature.

Rechel: In terms of how systems in this region are viewed, it depends on the definition of "highly

(continued)

(continued)

vulnerable" and the answer varies by species groups. But I would say that the upper elevations of conifer forests are the most vulnerable. Additionally, avian species that occur in the mid-elevation oak woodlands are expected to move up in elevation and potentially compete with existing conifer forest species.

For the most part, I do believe the forest personnel view the San Bernardino as vulnerable to climate change. The biologists do, and local researchers do, but resource managers likely view vulnerable ecosystems as part of the overall resource management complex of issues. Fire personnel may or may not; if they do, I suspect this vulnerability relates to increases in temperatures and longer droughts and drier winters, thus increasing fire risk in the chaparral; especially at lower elevations.

Hargrove: How do you see the information from this case study and other similar research fitting in with the work that you do in management?

Fisher and Muth: The Deep Canyon Transect, including this UC Reserve and its surrounding matrix comprised primarily of National Forest and BLM lands, contains an elevational gradient that includes plant communities ranging from low-elevation creosote bush scrub to high-elevation yellow pine forest, with chaparral and pinyon-juniper woodland plant communities at mid-elevations.

Many plant and animal taxa are restricted to each of these communities and come into contact or near contact at the transition zones. An example of a management decision made from what was learned from this study would be to restrict disturbance and perturbation (e.g., no exclusionary fencing or removal experiments) so as to avoid interference with the future uphill movement of species in response to climate change.

Poopatanapong: For the San Bernardino, information from researchers is used to help complete the biological assessments and evaluations needed for the NEPA process. Since the Forest Service does not actively survey for many wildlife species, any information gathered by partners is extremely helpful. More generally, having research on the national forest is always desired. At each national forest, the Forest Service does not have funding, staff, and ability to actively conduct research—nor is it the mission of the agency. Having the opportunity to use the data gathered by students and other researchers really helps all groups work together to better manage the land.

Rechel: For the San Bernardino and Angeles National Forests, climate change information is important to use in conjunction with existing long-term (20+ years) datasets of avian population and habitat use questions, especially related to changes in fire regimes in Mediterranean ecosystems.

CHAPTER 9

Conserving California Grasslands into an Uncertain Future

K. Blake Suttle, Erika S. Zavaleta, and Sasha Gennet

Abstract. This chapter focuses on a grassland experiment in Mendocino County, California, to examine the predictability of climate change impacts, and the lessons our work suggests for how we prepare for climate change. Grasslands provide a number of important ecosystem services to the public and are strongly shaped by climate in both composition and function. From experimental data describing multiple forms of ecological response to changing precipitation regimes through time, we show that the complexity of responses increased as the focus of attention narrowed from ecosystem- and community-level metrics down to individual species. Complexity emerged largely from changing interactions among species, which makes aggregate measures like ecosystem function and community diversity more reliable than predictions of responses by individual species. Effects of changing interactions were often lagged relative to the direct effects of physical conditions on species; this means that responses to annual variation in weather may poorly predict responses to longer-term changes in climate. The complexity apparent in species-level

Key Points

- Length of the rainy season is a major factor in the production and diversity of California grasslands, much more so than the total amount of precipitation that falls. A key aspect of understanding how systems might respond is identifying what forms of climate information are most relevant, and understanding both the trends and uncertainty in those aspects of change.
- In highly dynamic systems such as arid grasslands, on-the-ground variation in species composition and diversity in response to annual variation in climate may be a poor predictor of longer-term trajectories under decadal and longer climate change. When short-term research and past observations are used to inform management, particularly at the level of individual species, this should be done with caution and an eye toward potential complexities that could derail expectations.
- Uncertainties surrounding how climate will change and how ecosystems will respond complicate efforts to adapt conservation

(continued)

(continued)

and management practices, but do not preclude effective measures being taken. Approaches like prioritizing conservation and restoration based on abiotic factors and geographic gradients allow actions to take place without reliance on understanding species-specific response.

- Given the complexities that emerge as our geographic focus narrows to the scale of individual sites and our ecological focus narrows to the scale of individual species, the goal of maintaining a given mix of species at a given site may be less practical and less sustainable over the long term than that of maintaining native plant cover and functional and phenological group diversity
- Restoration of specific habitats should be coordinated with and complement landscape-scale conservation actions such as establishment of new protected areas, particularly in light of ongoing grassland habitat loss and conversion in California.

responses indicates that predictions for local plant and animal populations under different climate change scenarios could be more difficult than the sophistication of our modeling capabilities would otherwise suggest. In developing on-the-ground adaptation plans, it will be prudent for scientists and managers to allow for uncertainty and expect surprises, particularly where the focus is on individual species rather than broader patterns of diversity or function. Precautionary and flexible management actions should form the starting point in climate change adaptation, with more interventionist measures accompanied by careful monitoring to limit the consequences of inevitable surprises.

INTRODUCTION

To update management and conservation approaches in an era of climate change, we need to frame our expectations of what the future might look like, and acknowledge which aspects of this future are highly likely and which are most uncertain. Ecological responses to climate change are a challenge to predict, because they may combine many different kinds of direct and indirect effects (Walther 2010, Post 2013). Direct effects include changes in growth, reproduction, and survival rates based on physiological responses to climate itself (i.e., whether a change in physical conditions favors or disfavors an organism, such as by allowing its establishment or preventing its persistence). Indirect effects are those mediated by changes in the biological community around that organism (i.e., how responses to climate change by consumers, competitors, resources, and pathogens further affect its growth, reproduction, or survival). Species do not exist independently of each other, but are enmeshed in intricate webs of interrelations. As a result of the network structure of nature, any one species' experience of climate change can influence and be influenced by the experience of many other species. We consider here how ecological interactions complicate our understanding of future climate change impacts in grasslands, and how scientists and managers can account for such complexities in developing adaptation plans.

Grasslands are widely distributed across California and host a sizeable fraction of its native biodiversity, including numerous threatened and endangered plants and animals (Stromberg et al. 2007). They are of high economic and cultural importance in the state, particularly with respect to ranching, providing critical habitat and food for wildlife and livestock, and places of recreation for the public. Grasslands vary widely in their specific composition across the state, based on geographic, climatic, and edaphic characteristics, but almost all share two important characteristics. First, grasslands are among the most heavily invaded ecosystems in California, hosting a diverse and abundant flora of European species (Mack 1989, D'Antonio et al. 2007). Annual grasses from the Mediterranean have been particularly successful, dominating millions of hectares throughout the state (Huenneke 1989). Most grasslands in California retain native plants, though these are typically scattered amidst abundant exotic plant cover (Bartolome et al. 1986, Huenneke 1989, Hamilton 1997). Second, grasslands in California are distinct from those in many other regions in that they do not follow clear patterns of succession, with composition and production fluctuating from year to year instead of shifting through a typical set of stages over time. Specifically, standing crop and species composition vary with annual variation in weather, and particularly in the timing and amount of precipitation that falls each year (Murphy 1970, Pitt and Heady 1978, Stromberg and Griffin 1996, Hobbs et al. 2007).

Grasslands, therefore, may be particularly susceptible to changes in rainfall levels that accompany long-term climate change in California. Extended drought in the late 1800s is thought to have contributed to the speed and extent of conversion of California grasslands into exotic-dominated systems (Burcham 1957, Burcham 1961, Major 1988, Corbin and D'Antonio 2003), exacerbating concomitant effects of land-use change, altered fire regimes, and repeated introductions of exotic species. Where abundant native grasses persist in northern coastal regions, their success has been attributed in part to the longer winter rainy seasons and less severe summer droughts that characterize the region (Hektner and Foin 1977, Hayes and Holl 2003). Directional changes in annual precipitation regimes could have important bearing not just for the structure and diversity of these systems but also for how we use them as a result. Conservation of threatened or valuable species, restoration of degraded habitats, and maintenance of the many services provided by California grasslands will all benefit from advance knowledge of what the future holds for these ecosystems.

In this chapter, we consider lessons for grassland management under climate change from a large-scale experiment directly manipulating climate variables over natural grassland plots in northern California. In 2001, Blake Suttle and colleagues at the University of California in Berkeley set up an experiment modi-fying the intensity and duration of the annual rainy season over replicate plots of grassland in a Mendocino County nature reserve. Leading climate projections at the time (National Assessment Synthesis Team 2000) called for considerable increases in annual precipitation across much of the state, with disagreement as to the specific seasonal timing of those increases. The experiment was originally set up to test how alternative scenarios of rainfall change could affect native–exotic balances among grassland plants in these already heavily invaded systems (Thomsen et al. 2006, Suttle and Thomsen 2008). Knowledge of basic life history differences among the main constituents of California grasslands suggested contrasting expectations for how major groups of native and exotic species should be affected. The most prominent group of invaders, annual species that evolved in the Mediterranean, would be more likely to benefit from additional rainfall during the winter rainy season than from increases that came during the summer drought, as these plants typically complete their life cycle in spring and persist through the summer as seed (Pitt and Heady 1978, Jackson and Roy 1986). Native perennial plants, on the other hand, could benefit from increases in rainfall during the summer drought, as this is a period of low competitive pressure from most exotic species, allowing natives to take advantage of additional precipitation to increase vegetative growth and produce more and larger seeds.

Initial research in the system showed that when plant groups were considered in isolation, these physiology-based mechanisms of response were largely borne out, but that interactions between different plant groups could moderate these responses in important ways (Thomsen et al. 2006, Suttle and Thomsen 2008). The contrasting phenologies of different plant groups in the system, and the differing physiological responses to seasonal rainfall they entail, therefore provide a basis for comparisons of direct and indirect effects in long-term responses; this became the focus of later work in the system (Suttle et al. 2007, Suttle et al., in review), and this work is part of our focus here in considering opportunities for prediction and preemptive management in grasslands under climate change. We compare initial and decadal trajectories in net primary production, biological diversity, and individual species abundances to highlight potential differences in predictability among different levels of ecological response, and to explore the implications of these for grassland management under climate change.

Several features of this experiment make it well suited for addressing questions of predictability. The design emphasized "naturalness"—minimizing infrastructure and observer interference so that grassland species experienced imposed changes in rainfall amid the full complexity of their biological and physical surroundings. Plots consisted of open communities in a protected grassland with no recent history of livestock grazing or anthropogenic disturbance (>75 years). Water was manipulated in a manner that did not change other environmental conditions, such as wind or shade, that would introduce other changes that could potentially confound our ability to understand responses to the treatment itself. Treatments themselves did not represent realistic approximations of climate change, but highly simplified changes in a single climatic variable, setting up straightforward expectations and hypothesis tests. By minimizing complexities introduced from the physical side, the experiment focused on biological complexities such as species interactions, response thresholds, and other nonlinearities. The work also encompassed large enough spatial scales (plots >70 m²) and long enough temporal scales (>10 years) for a rich suite of life history processes to play out in organism responses, but all within a fully controlled, randomized, and replicated framework that allowed formal hypothesis testing.

METHODS

Beginning in January 2001, 36 plots of grassland (~70 m² each) at the Angelo Coast Range Reserve in Mendocino County, California (39° 44′ 17.7″ N, 123° 37′ 48.4″ W) have been subjected to one of three annual precipitation treatments: A winter addition of water simulating an intensification of the rainy season (January through March), a spring addition of water simulating an extension of the rainy season (April through June), and an unmanipulated ambient control (Figure 9.1). Treatments simulate predictions for the region from climate models developed at the *Hadley Centre for Climate Prediction and Research* (HadCM2) and the *Canadian Centre for Climate Modeling and Analysis* (CCM1) (National Assessment Synthesis Team 2000). Both models projected substantial increases in precipitation throughout northern California over the next century, but the seasonality of these increases differed. The *Hadley* model called for increases during the current winter rainy season, while the *Canadian* model predicted increased rainfall extending into the summer, when rain is typically scarce and acts as a key limiting resource.

Each watered plot receives approximately 44 cm of supplementary water *over ambient* rainfall per year, roughly a 20% increase over mean annual precipitation (216 cm) but within the range of natural variability in both amount and timing at the study site (details in Suttle et al. 2007). Water is collected from a natural spring on a forested mountain southeast of the grassland. The spring drains to the south of the grassland, and a portion of its flow is diverted via pipe to an 1100-L irrigation tank placed on the mountainside approximately 40 vertical meters above the grassland immediately to its east. The tank is continually replenished via gravity feed from the spring, and water has been tested and found to contain nitrogen concentrations within the range present naturally in rainwater at the study site (Suttle et al. 2007).

Water is delivered evenly over the surface of each plot from a RainBird® RainCurtain™ sprinkler specially designed to simulate natural rainfall. Experimental rain additions begin approximately two hours after dawn every third day during the three-month watering period of

FIGURE 9.1: The experiment as seen from the air in summer 2001. The spring-addition plots, in which the rainy season is extended into summer, can be recognized from the green coloration imparted by the late flush of nitrogen-fixing forbs in these systems.

each addition treatment, and last for one hour. The watering radius is 5 m, and all samples are collected at least 0.5 m inside of the outer edge of the watered area (Figure 9.2).

The grassland contained a well-mixed assemblage of grasses and forbs of both native and exotic origin on an abandoned terrace of the South Fork Eel River. Approximately 50 plant species are present in the system in a given year, dominated by the exotic annual grasses *Bromus hordeaceus, Vulpia myuros, Aira caryophyllea,* and *Bromus diandrus*; the exotic annual forbs *Erodium cicutarium, Gallium parisiense,* and *Hypochaeris glabra*; the native annual forbs *Madia gracilis, Trichostemma lanceolatum,* and *Eremocarpus setigerus*; and the native perennial forbs *Eschscholzia californica* and *Sanicula bipinnatifida*. Relative abundances among these species varied year to year. The three native perennial bunchgrass species *Danthonia californica, Elymus glaucus,* and *Elymus multisetus* were also present scattered throughout the system at generally low densities.

In order to understand how adding rainfall to the system would affect three different forms of biological response—productivity, biological diversity, and population abundances—the experiment was designed to meet the complementary goals of obtaining long-term observations of minimally disturbed plots and allowing follow-on experimentation in other, more heavy-use plots. Prior to any work or manipulation in the system, the 36 plots were divided randomly into "Pattern" plots and "Process" plots (18 Pattern plots and 18 Process plots, comprising six replicates each of three rainfall treatments). The Pattern plots served as the template on which the precipitation manipulations could play out over the long term and are the data presented and discussed in this chapter. These plots provide a window into the emergent result of numerous processes playing out in experimental communities in response to the altered precipitation regimes. The Process plots have been used for more intensive and potentially destructive research aimed at understanding causal mechanisms underlying the dynamics observed in the Pattern plots, providing more specific insights into the implications of altered precipitation regimes for restoration of native bunchgrasses (Suttle and Thomsen 2008), biological invasion (Thomsen

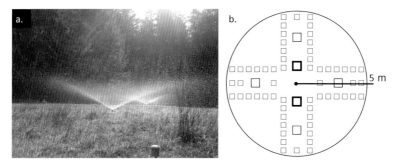

FIGURE 9.2: Manipulation and sampling of experimental plots. (a) Water is delivered evenly over the surface of each 5 m radius plot by a sprinkler. (b) Plant production was measured in two predesignated 900 cm² subplots (small squares) on May 25, July 1, and August 25, targeting peak biomass of different phenological groups. Richness was measured across the growing season by visual inspection in two central 2500 cm² subplots (bolded large squares). Abundances were measured by point-frame in six permanent 2500 cm² subplots.

et al. 2006), carbon and nutrient cycles (Hawkes et al. 2010, Berhe et al. 2012), and dynamics within the soil microbial community (Cruz-Martinez et al. 2009, Hawkes et al. 2010).

In each replicate Pattern plot (3 treatments × 6 replicates), plant production was measured every year in two predesignated 900 cm² subplots on May 25, July 1, and August 25, in order to target the peak biomass of different phenological groups of plants present in the study system (Figure 9.2). Each subplot was harvested once and then eliminated from the sampling regime. Subplots were positioned at regular intervals along a single transect laid out in a randomly selected cardinal direction through the center of each plot and then along a second transect running perpendicular to that transect. Biological diversity was measured as the species richness presents across the growing season in two central 2500 cm² subplots. Population abundances of individual species within experimental communities (and the overall species composition these add up to) were measured by point-frame in six permanent 2500 cm² subplots. Point-frame measurement involved a series of eight pins dropped vertically at regular intervals into each subplot, with data recorded on the first plant species contacted by each pin as it was lowered into the

vegetation, the plant species the pin landed on, and the plant species touching the pin at a marked height of 5 cm above the ground surface (3 hits per pin × 8 pins per subplot × 6 subplots, for 144 hits per plot). Sampling protocols were specifically designed to leave little potential for cross-interference among sampled variables or years (Suttle et al. 2007, Suttle et al., in review). This sampling regime was in place throughout the experiment, and was the only work undertaken in the 18 plots described in this chapter.

Plant production and species richness data were analyzed with repeated-measures ANOVA. Population abundances and species composition data are fundamentally multivariate (abundance of species *a* in each plot, abundance of species *b* in each plot, and so on across all species present), and best expressed as patterns of similarity and dissimilarity among plots though multivariate ordination. Ordinations were generated from Nonmetric Multidimensional Scaling (NMS or NMDS); these provide a quick view into compositional differences among communities by rendering variation in abundances of all species onto two axes. The ordinations cannot be read for information about any particular species, but instead show overall levels of similarity and dissimilarity in composition of the plots across all their species.

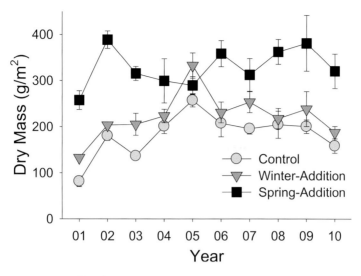

FIGURE 9.3: Effect of watering treatment on primary production. Data represent treatment means plus or minus standard errors for the cumulative biomass accrued across all plant species each year.

These data were collected for only 15 of the 18 total plots under study, or five replicates per treatment.

The most conspicuous feature of these data, regardless of which type of response is considered, is the difference in effect between the winter-addition treatment (an experimentally intensified rainy season) and the spring-addition treatment (an experimentally extended rainy season) (Suttle et al. 2007, Suttle et al., in review). Adding supplemental rainfall during the winter rainy season did not cause significant deviation from the control in any measured variable or year: Winter-addition plots mirrored control plots in plant production (Figures 9.3 and 9.4), plant diversity (Figure 9.5), plant composition (Figure 9.6), and species abundances (Figure 9.7) through every year of the study. Consequently, the focus of this chapter is on the spring-addition plots, which showed large differences from control plots in every measured variable, although those differences took a different form for each response variable considered.

Spring-addition plots showed large and fairly consistent increases in primary production relative to the other two treatments (Figure 9.3). In the first year of the experiment, this difference was largely due to a single group of plants: Nitrogen-fixing forbs (Figure 9.4a). Other plant species showed roughly similar levels of production across all three treatments in the first year of the experiment. Production of nitrogen-fixing forbs in spring-addition plots remained high relative to the other two treatments over most of the following decade, but was no longer the only biomass difference among treatments after the first year. Annual grass production rose sharply in spring-addition plots relative to the other two treatments in the second and third years of the manipulation, and remained at higher levels in the spring-addition treatment over most of the years that followed (Figure 9.4b). Biomass of other forb species (i.e., those not associated with nitrogen-fixing bacteria) in the spring-addition treatment, on the other hand, fell below levels seen in the other two treatments in the third year of precipitation additions and for several years afterward (Figure 9.4c).

For biological diversity, experimental extension of the rainy season caused significant

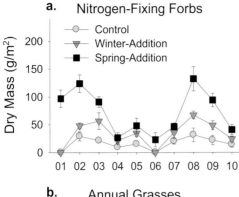

a. Nitrogen-Fixing Forbs

Legend:
- ○ Control
- ▽ Winter-Addition
- ■ Spring-Addition

Dry Mass (g/m²) vs Year (01–10)

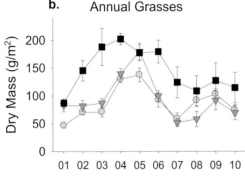

b. Annual Grasses

Dry Mass (g/m²) vs Year (01–10)

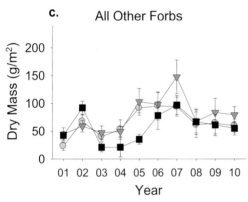

c. All Other Forbs

Dry Mass (g/m²) vs Year (01–10)

Year

FIGURE 9.4: Effects of watering treatment on primary production by different plant groups. Data represent treatment means plus or minus standard errors for the biomass of vascular plants across each year's growing season.

increases relative to ambient levels in the grassland in the first two years of the manipulation, but this effect disappeared during the third year of rainfall addition, and had reversed by the fourth (Figure 9.5). In 2004, four years into the manipulation, species richness in spring-addition plots was starkly lower than in the other two

treatments, and stayed there in 2005. In 2006, six years into the manipulation, one of the wettest months of March on record in California was followed by much higher than normal temperatures over the first few weeks of May; this was accompanied by declines in species richness in control and winter-addition plots near to levels seen in spring-addition plots. Species richness climbed again in control and winter-addition plots following that year, but stayed at significantly lower levels in the spring-addition treatment through the duration of the experiment.

Species composition changed first in a consistent direction across spring-addition plots (as nitrogen-fixing forbs became much more abundant relative to other species) and then in different directions in each different spring-addition plot (as a different species rose to dominate in each). So from responses that were initially all in a similar direction relative to the control (Figure 9.6, 2002), spring-addition plots became steadily more dissimilar from each other as the manipulation continued (Figure 9.6, 2010—note that spring-addition plots represent virtually the entire extent of variability measured across the experiment).

Looking at changes in the abundances of individual species within these communities, we can understand how spring-addition plots have grown so dissimilar from each other over time. Certain individual species have shown wildly different responses to experimental extension of the rainy season from one plot to the next (Figure 9.7). As a result, each replicate plot within the spring-addition treatment is dominated by a single species, but it is a different species in every one. The native perennial bunchgrass *D. californica* exploded in one community subjected to extended rainfall each year but remained at ambient levels in each of the other four (Figure 9.7a). Likewise, the exotic annual grass *B. diandrus*, the annual forb *Geranium dissectum*, and the moss species *Homalothecium pinnatifidum* each rose to very high levels of abundance in a single replicate only. The pattern is less distinct for the nitrogen-fixing legume species *Lotus micranthus*, which has

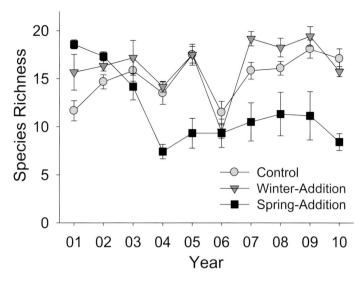

FIGURE 9.5: Effects of watering treatment on biological diversity. Data represent treatment means plus or minus standard errors for the species richness of vascular plants across each year's growing season.

shown generally positive responses to experimentally extended rainfall across all replicate plots (Figures 9.7b–9.7e).

The simplicity of the rainfall additions, together with the detailed understanding of California grassland systems that has emerged out of the long history of ecological research in these systems, helped us develop *a priori* expectations of what simple physiology-driven responses would look like in the different response variables (e.g., Pitt and Heady 1978, Jackson and Roy 1986, Hobbs and Mooney 1995). Allowing actual responses to play out in the full complexity of a natural setting over a decade, we can identify where deviations from physiological expectations arise, and potentially identify what factors emerge to drive these deviations.

The fact that complex dynamics emerged from only one of the two watering treatments, with the other causing no meaningful changes at all, emphasizes that these ecological complexities are of secondary concern to the overlying details of climate change itself. A key takehome point here is that timing matters. Much

of our dialogue on climate change focuses on changes in average annual values and extreme events, but here we see how the same annual change can have very different biological effects depending on the specific seasonal timing over which it unfolds. This illuminates a critical challenge in climate change science and adaptation—uncertainty over regional precipitation changes generally and in the annual timing of those changes specifically. Case in point, in the time since the rainfall experiment described here was established, predictions for precipitation change in California have changed dramatically, with many leading models, including more recent projections from the Hadley Centre, now forecasting decreasing annual rainfall over much of California (e.g., Wilkinson et al. 2002, Hayhoe et al. 2004, Cayan et al. 2006). Unless and until climate models achieve considerable further reductions in uncertainties over precipitation and timing, managers should consider a range of best- and worst-case scenarios and solutions with respect to how these variables will change in their systems.

In terms of understanding the processes underlying responses observed in this study system, the lack of any effect of additional

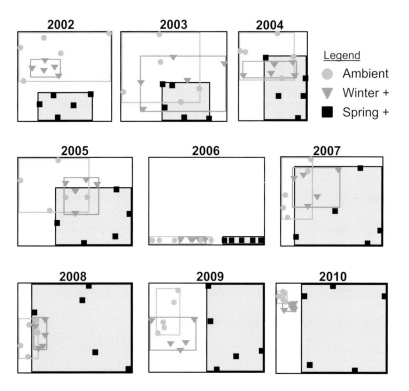

FIGURE 9.6: Treatment-level variability in the species composition of experimental plots. Data reflect patterns of similarity and dissimilarity in relative abundances across all species, as expressed by Nonmetric Multidimensional Scaling (NMS). The compositional similarity and dissimilarity are expressed as distances along two axes, so that the closer two points on a given figure, the more similar those communities. Rectangles are drawn as a visual guide to the total amount of variation along each axis accounted for by the plots of a single treatment.

winter rainfall can be understood on strictly physiological terms for all three sets of response variables: Primary production, biological diversity, and population abundances. There is relatively little overlap between periods of warm temperatures conducive to plant growth and periods of high moisture availability in northern California. Roughly 95% of annual precipitation typically falls between November and April (Major 1988), when temperatures and light levels are low and most plants are able only to use small amounts of water (Evans and Young 1989). Increases in water during this time, therefore, proved to be largely superfluous. Extending the rainy season via water addition through the spring and into summer produced many more dramatic changes in the grassland.

The fact that each response variable behaved differently under experimentally extended spring rainfall points to a different set of processes driving changes at each ecological level considered. Ecosystem functions typically involve a large number of species, making them relatively robust to changes in the number and identity of particular species. Thus primary production—the function studied most closely in experimental plots—changed according to processes we understand well (i.e., physiological tolerances and preferences, which can be tested directly or gleaned from time series data or distribution records). Plant production more than tripled in the first year and doubled in the second year compared to the ambient levels in the grassland (Figure 9.2). Summer is typically a period of high water stress in northern Cali-

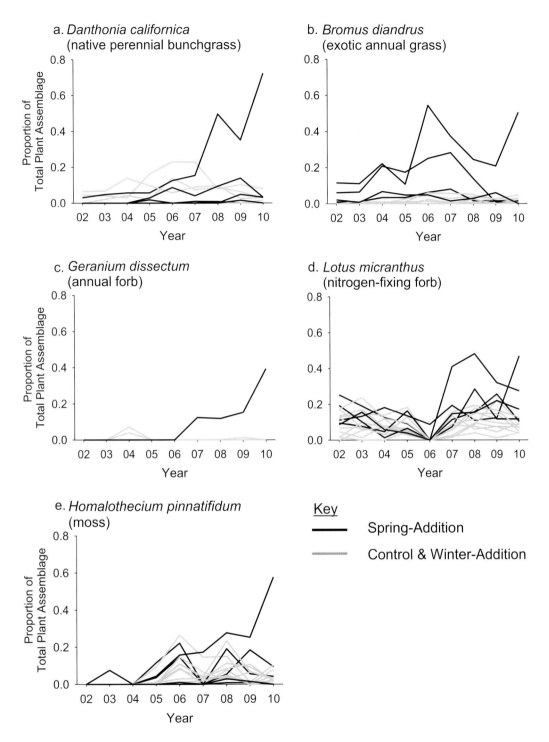

FIGURE 9.7: Effects of watering treatment on individual species abundances. Data represent plot-level population abundances for each of five individual species in five replicate plots of each treatment. Spring-addition plots are represented by black lines; for visual simplicity and to highlight the divergent trajectories evident in spring-addition plots, winter-addition and control plots are represented by lines in the same shade of gray.

fornia grasslands; plants responded to allevia-tion of this stress with increased growth and, in certain species, extended longevity (Suttle et al. 2007). These effects were apparent immedi-ately and remained relatively consistent throughout the duration of the experiment.

Diversity measures are a little more nar-rowly focused than overall production, in that the number of species becomes important, if not their identity. Accordingly, diversity changes in experimental plots have been more complex to model, requiring that we account for a set of species interactions in addition to basic physio-logical responses to climate to understand the reversal in the effect of extended rainfall through time (Suttle et al. 2007). Dividing overall plant production into three broadly defined functional groups (i.e., nitrogen-fixing forbs, annual grasses, and non-fixing forbs) sheds light on the nature of these interactions. The strongest initial response to the extended rainfall season came from nitrogen-fixing forbs (Figure 9.4a). Exotic annual grasses showed no significant response to the first year of the spring-addition treatment, but after the prolif-eration of nitrogen-fixing forbs that year, annual grass production rose dramatically (Figure 9.4b). Exotic grasses are generally the first plants to germinate each year and among the earliest to complete their life cycle and senesce. This early phenology limits direct responses to extended spring rainfall (Pitt and Heady 1978, Jackson and Roy 1986), but allows these plants to benefit in the subsequent growing season as abundant N-fixer litter decomposes and pro-vides increased nitrogen (Bentley and Green 1954, Bentley et al. 1958, Jones et al. 1990). As this process was repeated over several years, the accumulated annual grass litter apparently sup-pressed germination and regrowth of leafy forbs (Figure 9.4c), driving sharp declines in plant species richness (Figure 9.5).

Forb species after forb species disappeared from spring-addition plots until these plots were made up of only a few species of annual grasses and the nitrogen-fixing forb *L. micranthus* (which, with its spindly growth form and nar-row leaves, was able to penetrate the grass litter and remain a productive constituent of spring-addition plots). These dynamics, too, however, turned out to be short-lived, for within all but one of these annual grass-dominated communi-ties, another species took hold and ultimately rose to dominance. The most surprising finding from this research, however, is that the "win-ner" was a different species in every plot. Ten years into the altered precipitation regimes, a different species now dominates each replicate spring-addition plot (Figure 9.7) despite the fact that each of these species is present in every rep-licate community (Suttle et al., in review).

From initially well-mixed communities of roughly similar composition, spring-addition plots have grown increasingly dissimilar from each other over a decade of extended seasonal rainfall (Figure 9.6). Thus a simple manipula-tion of one environmental variable, imposed identically over discrete plots of well-mixed grassland in a single meadow, has given rise to communities that are conspicuously different in their makeup from replicate to replicate. Population trajectories for individual species, very much at the center of aspirations in predic-tive ecology, here show a degree of context dependency that prevents our being able to explain or model them based on experimental data. Many species in the experiment have failed to show any kind of trajectory at all, responding in different directions from repli-cate to replicate under a single experimental treatment. The extent of idiosyncrasy seen in how populations of individual species changed in response to rainfall addition suggests a high degree of uncertainty around predictions of local population changes, as these appear to be governed by factors that vary over fine spatial scales.

The scale of study plots (70 m²) and there-fore of the context dependency seen in popula-tion trajectories make it important to consider what changes might look like if rainfall were added at the scale of whole grasslands. The single-species dominance we see at the scale of experimental plots would likely manifest as increased spatial heterogeneity in vegetation composition at the scale of the meadow as a whole over the decadal scale considered here, but may well lead to dominance by some smaller subset of these species as the new climate regime took hold over longer time periods. Extrapolating from plot-scale to whole-treatment differences, species richness remains lower under the extended-rainfall treatment; even with a different species dominating each replicate plot, there are so few other species able to persist amid those levels of dominance that collective richness across plots is still lower than that measured in winter-addition and control plots. And richness differences notwithstanding, whether we view results at the scale of experimental plots or at the scale of the whole experiment, complexity increases as we narrow from our focus down toward individual species.

SOLUTIONS, ADAPTATION, AND LESSONS

Uncertainty and scale deserve serious consideration in plans for preemptive management around climate change (Wiens and Bachelet 2009). We cannot simply delay action while we wait for more precise forecasts, because in some cases the problems will grow worse and solutions more difficult and in some cases that hoped for precision will remain elusive. As climate scientist Stephen Schneider often said, "This is not a world where you can let the perfect crowd out the good." Where uncertainty in how some species or ecosystem will respond to future climate change arises not from a lack of appropriate empirical data, but from the fundamental nature of the underlying processes, we can act now, even if it means reframing certain priorities. There are many courses of remedial action with effectiveness not bound to any specific vision of the future (e.g., reduce emissions, restore degraded landscapes, expand protected areas, and improve connectedness among them) (Heller and Zavaleta 2008). These are likely to be most effective if undertaken in complement with each other. The step-

wise changes in response complexity from ecosystem to community to population levels in the experiment suggest that, at local scales, changes in individual species will show greater variation and uncertainty than changes in overall diversity or ecosystem function. When management targets are more narrowly ecologically focused, the most robust management actions will be broader and more precautionary, allowing for greater uncertainty in target responses.

Many studies and management efforts across grassland types in California have demonstrated that reversing biological invasions once established and restoring native species once lost is difficult, whether in mesic coastal prairie or arid interior grassland (Corbin et al. 2004). Although the restoration component of resource management and conservation is critical, site-scale efforts (e.g., 1–1000 ha) will be insufficient to achieve protection of the full array of grassland types and constituent communities. Restoration must be coordinated with and complement landscape-scale conservation actions such as establishment of new protected areas, particularly in light of ongoing grassland habitat loss and conversion in the state. Put another way, protecting large areas of grasslands that can be flexibly managed for conservation and compatible uses is a prerequisite for adaptive management or restoration in an uncertain climate future. Protected areas must be large enough and adequately linked so that species with a variety of life histories, from wide-ranging mammals and breeding birds to host-specific pollinating insects and edaphic-specialist plants, can maintain viable populations and potentially migrate to new areas when portions of current ranges become unsuitable.

Because grasslands and savanna cover millions of hectares in California, a complete inventory and biological assessment of sites is impractical. However, landscape-scale ecological research can help to identify and prioritize areas for protection where diverse communities are more likely to persist into the future (Klausmeyer and Shaw 2009, Klausmeyer et al. 2011). For example, recent research along the Central

Coast of California suggests that grassland plant species richness in that ecoregion is correlated with elevation, steep north-facing slopes, and coarse soils, with higher cover of native species on less fertile soils (high C:N ratio, low phosphorous) (Gea-Izquierdo et al. 2007, Gennet, unpublished data). Grassland areas that are current and, to the best of our knowledge, likely future hotspots of species diversity can be identified and prioritized for protection using readily available digital GIS datasets (in this case, digital elevation models, geology, and soil layers).

Surprises are in store, however, so another practical step may be to broaden acceptable ecological outcomes and the scales by which they are defined. Scaling outward from a species-level focus to functional groups, phenological groups, or ecosystem processes, we can expect less idiosyncratic responses to climatic change (Zavaleta et al. 2003, Bai et al. 2004, Hobbs et al. 2007). Scaling outward geographically from individual sites to regional and broader scales, we can expect less idiosyncrasy in responses by individual species (Wiens 1989, Pearson et al. 2004). The goal of maintaining a given mix of species at a given site may be less practical and less sustainable over the long term than that of maintaining native plant cover and functional and phenological group diversity [e.g., managing to retain deeply rooted perennial plants to help control star thistle (*Centaurea solstitialis*)]. The goal of protecting a given species or mix of species, on the other hand, will be more practically accomplished through a regional perspective that views individual sites as part of a broader network. For any particular species or site, however, an expectation of surprises, a monitoring effort sufficient for early detection of the same, and a responsiveness and adaptability built into conservation and management plans are all commensurate with the challenge we face.

Changes in California's climate could have profound consequences for grassland structure and species composition, and some of these will unfold in highly complex ways. With impacts varying by species and life stage, tied strongly to seasonal timing, and propagating widely along interaction networks, species- and site-level predictions of ecological response may carry high levels of uncertainty, and management strategies built upon them should allow for this. The most practical path forward requires acknowledging uncertainties and asking "what then?" These are discussions in which all stakeholders should be engaging. We will not be able to preempt every harmful impact of climate change, but acting along these lines, to borrow a phrase from USGS scientist Nate Stephenson, we should be able to usher in some intended consequences among all the unintended ones.

LITERATURE CITED

Bai, Y., X. Han, J. Wu, Z. Chen, and L. Li. 2004. Ecosystem stability and compensatory effects in the Inner Mongolia grassland. *Nature* 431:181–183.

Bartolome, J. W., S. E. Klukkert, and W. J. Barry. 1986. Opal phytolyths as evidence for displacement of native California grassland. *Madroño* 33:217–222.

Bentley, J. R. and L. R. Green. 1954. Stimulation of native clovers through application of sulfur on California foothill range. *Journal of Range Management* 7:25–30.

Bentley, J. R., L. R. Green, and K. A. Wagnon. 1958. Herbage production and grazing capacity on annual-plant range pastures fertilized with sulfur. *Journal of Range Management* 11:133–140.

Berhe, A. A., K. B. Suttle, S. D. Burton, and J. F. Banfield. 2012. Contingency in the direction and mechanics of soil organic matter responses to increased rainfall. *Plant and Soil* 358:371–383. doi: 10.1007/s11104-012-1156-0.

Burcham, L. T. 1957. *California Range Land: An Historico-Ecological Study of the Range Resource of California*. Division of Forestry, Department of Natural Resources, State of California, Sacramento, CA.

Burcham, L. T. 1961. Cattle and range forage in California 1770-1880. *Agricultural History* 35:140–149.

Cayan, D., A. L. Luers, M. Hanemann, G. Franco, and B. Croes. 2006. *Scenarios of Climate Change in California: An Overview*. California Climate Change Center, Sacramento, CA.

Corbin, J. D. and C. M. D'Antonio. 2003. Competition between native perennial and exotic annual grasses: Implications for a historical species invasion. *Ecology* 85:1273–1283.

Corbin, J. D., C. M. D'Antonio, and S. J. Bainbridge. 2004. Tipping the balance in the restoration of native plants: Experimental approaches to changing the exotic: Native ratio in California grassland. In M. Gordon and L. Bartol (eds), *Experimental Approaches to Conservation Biology.* University of California Press-Cooperative Extension, Berkeley, CA. 154–179.

Cruz-Martinez, K., K. B. Suttle, E. L. Brodie, M. E. Power, G. L. Andersen, and J. F. Banfield. 2009. Despite strong seasonal responses, soil microbial consortia are more resilient to long-term changes in rainfall than overlying grassland. *The ISME Journal* 1:1–9.

D'Antonio, C. M., C. Malmstrom, S. A. Reynolds, and J. Gerlach. 2007. Ecology of invasive non-native species in California grassland. In M. R. Stromberg, J. D. Corbin, and C. M. D'Antonio (eds), *California Grasslands: Ecology and Management.* University of California Press, Berkeley, CA. 67–85.

Evans, R. A. and J. A. Young. 1989. Characterization and analysis of abiotic factors and their influences on vegetation. In L. F. Huennecke and H. Mooney (eds.), *Grassland Structure and Function: California Annual Grassland.* Kluwer Academic Publishers, Dordrecht. 13–28.

Gea-Izquierdo, G., S. Gennet, and J. W. Bartolome. 2007. Assessing plant-nutrient relationships in highly invaded Californian grasslands using non-normal probability distributions. *Applied Vegetation Science* 10:343–350.

Grebmeier, J. M., J. E. Overland, S. E. Moore, E. V. Farley, E. C. Carmak, L. W. Cooper, K. E. Frey, J. H. Helle, F. A. McLaughlin, and S. L. McNutt. 2006. A major ecosystem shift in the northern Bering Sea. *Science* 311: 1461–1464.

Hamilton, J. G. 1997. Changing perceptions of pre-European grasslands in California. *Madroño* 44:311–333.

Hawkes, C. V., S. K. Kivlin, J. Rocca, V. Huguet, M. Thomsen, and K. B. Suttle. 2010. Fungal community responses to precipitation. *Global Change Biology* 17:1637–1645.

Hayes, G. and K. D. Holl. 2003. Cattle grazing impacts on annual forbs and vegetation composition of mesic grasslands in California. *Conservation Biology* 17:1694–1702.

Hayhoe, K., D. Cayan, C. B. Field, P. C. Frumhoff, E. P. Maurer, N. L. Millers, S. C. Moser, S. H. Schneider, K. C. Cahill, E. E. Cleland et al. 2004. Emissions pathways, climate change, and impacts on California. *Proceedings of the National Academy of Science of the United States of America* 101:12422–12427.

Hektner, M. M. and T. C. Foin. 1977. Vegetation analysis of a northern California Coastal prairie: Sea Ranch, Sonoma County, California. *Madroño* 21:83–103.

Heller, N. E. and E. S. Zavaleta. 2008. Biodiversity management in the face of climate change: A review of 22 years of recommendations. *Biological Conservation* 142:14–32.

Hobbs, R. J. and H. A. Mooney. 1995. Spatial and temporal variability in California annual grassland – results from a long-term study. *Journal of Vegetation Science* 6:43–56.

Hobbs, R. J., S. Yates, and H. A. Mooney. 2007. Long-term data reveal complex dynamics in grassland in relation to climate and disturbance. *Ecological Monographs* 77:545–568.

Huenneke, L. F. 1989. Distribution and regional patterns of California grasslands. In Huenneke, L. F. and Mooney, H. A. (eds), *Grassland Structure and Function: California Annual Grassland.* Kluwer Academic Publishers, Dordrecht. 1–12.

Jackson, L. E. and J. Roy. 1986. Growth patterns of mediterranean annual and perennial grasses under simulated rainfall regimes of southern France and California. *Acta Ecologica / Ecologica Plantarum* 7:191–212.

Jones, M. B., M. W. Demment, C. E. Vaughn, G. P. Deo, M. R. Dally, and D. M. Center. 1990. Effects of phosphorus and sulfur fertilization on subclover-grass pasture production as measured by lamb grain. *Journal of Production Agriculture* 3:534–539.

Klausmeyer, K. R. and M. R. Shaw. 2009. Climate change, habitat loss, protected areas and the climate adaptation potential of species in Mediterranean ecosystems worldwide. *PLOS ONE* 4:e6392.

Klausmeyer, K. R., M. R. Shaw, J. B. MacKenzie, and D. R. Cameron. 2011. Landscape-scale indicators of biodiversity's vulnerability to climate change. *Ecosphere* 2:art88.

Mack, R. N. 1989. Temperate grasslands vulnerable to plant invasion: Characteristics and consequences. In J. A. Drake, H. A. Mooney, F. di Castri, R. H. Groves, F. J. Kruger, M. Rejmanek, and M. Williamson (eds), *Biological Invasions: A Global Perspective.* John Wiley and Sons, New York, NY. 155–179.

Major, J. 1988. California climate in relation to vegetation. In M. Barbour and J. Major (eds),

Terrestrial Vegetation of California. California Native Plant Society, Sacramento, CA. 11–74.

Murphy, A. H. 1970. Predicted forage yield based on fall precipitation in California annual grasslands. *Journal of Range Management* 23:363–365.

National Assessment Synthesis Team. 2000. *Climate Change Impacts on the United States: The Potential Consequences of Climate Variability and Change*. US Global Change Research Program, Washington, DC.

Pearson, R. G., T. P. Dawson, and C. Liu. 2004. Modelling species distributions in Britain: A hierarchical integration of climate and land-cover data. *Ecography* 27:285–298.

Pitt, M. D. and H. F. Heady. 1978. Responses of annual vegetation to temperature and rainfall patterns in northern California. *Ecology* 59:336–350.

Post, E. 2013. *Ecology of Climate Change: The Importance of Biotic Interactions*. Princeton University Press, Princeton, NJ, USA.

Stromberg, M. R. and J. R. Griffin. 1996. Long-term patterns in coastal California grasslands in relation to cultivation, gophers, and grazing. *Ecological Applications* 6:1189–1211.

Stromberg, M. R., J. D. Corbin, and C. M. D'Antonio. 2007. *California Grasslands: Ecology and Management*. University of California Press, Berkeley, CA.

Suttle, K. B. and M. A. Thomsen. 2008. Climate change and grassland restoration: Lessons from a rainfall manipulation in California. *Madroño* 54:225–233.

Suttle, K. B., M. A. Thomsen, and M. E. Power. 2007. Species interactions reverse grassland responses to changing climate. *Science* 315:640–642.

Suttle, K. B., M. A. Thomsen, M. Sullivan, D. T. Gerber, and M. E. Power. In review. Limits on predictability emerge in grassland responses to decade-scale climate forcing.

Thomsen, M., C. D'Antonio, K. B. Suttle, and W. P. Sousa. 2006. Ecological resistance seed density, and their interactions determine patterns of invasion in a California coastal grassland. *Ecology Letters* 9:160–170.

Walther, G. R. 2010. Community and ecosystem responses to recent climate change. *Philosophical Transactions of the Royal Society, B* 365:2019–2024.

Wiens, J. A. 1989. Spatial scaling in ecology. *Functional Ecology* 3:385–397.

Wiens, J. A. and D. Bachelet. 2009. Matching the multiple scales of conservation with the multiple scales of climate change. *Conservation Biology* 24:51–62.

Wilkinson, R., K. Clarke, M. Goodchild, J. Reichman, and J. Dozier. 2002. *The Potential Consequences of Climate Variability and Change for California: The California Regional Assessment*. U.S. Global Change Research Program, Washington, DC.

Zavaleta, E. S., M. R. Shaw, N. R. Chiariello, B. D. Thomas, E. E. Cleland, C. B. Field, and H. A. Mooney. 2003. Grassland responses to three years of elevated temperature, CO_2, precipitation, and N deposition. *Ecological Monographs* 73:585–604.

Manager Comments

K. Blake Suttle
in conversation with
Andrea Craig

Suttle: Have you observed changes in the systems that you manage that seem linked to changes in climate? As climate continues to change, what ecological changes do you expect to see?

Craig: Ranchers and other land managers in our area are observing a long string of abnormal weather years, with the timing of precipitation being particularly unusual and unpredictable. Cattle-production operations and wildlife are forced to follow the timing of precipitation. As climate continues to change, I would expect to see a shift in forage and drinking water availability, which would shift the rate and timing of stocking, and migratory patterns and life cycles of wildlife. I would expect to see change in vegetation composition and availability (timing and abundance), leading to a cascade of changes related to soil properties and the organisms most suited to persist and evolve in this landscape.

Suttle: How do you see this information fitting in with the work that you do to conserve grassland systems in California? What kinds of decisions could this type of research be used to inform?

Craig: This information fits in with my rangeland work by informing the decisions involved in determining lease terms, and by informing strategic planning decisions for conservation of the targets that brought the Nature Conservancy to this region. Abnormal weather patterns force us to adapt, and lease and easement terms need to be written to accommodate uncertainty. Climate change research informs my preparation of functional agreements with ranchers and other land managers.

Suttle: Is understanding climate change responses an area where stronger or continued partnership with academic researchers is highly valued? Can you give an example of an ecological study that would be of particular interest to you, and that would have the potential to influence management actions?

Craig: Continued partnership with academic researchers is valued, especially when studies support anecdotal evidence seen on the ground. Examples include: (1) Are precipitation events occurring later in the growing season? (2) Will weather patterns support some summer grazing, or should we expect deer to stay longer at high-elevation habitats? (3) Do we have more grass and less forb forage? (4) Have black bears stopped hibernating, thus pressuring their food supply and other big game? Land managers can certainly make more informed decisions when informed by scientific research. I feel this especially holds true for those managing expansive and complex landscapes that may cover multiple watersheds and have a variation of overlapping ecological systems. The complexity and interconnectivity of these systems is too diverse to manage without a strong research partnership.

Suttle: When you think about your ability to change management approaches or shift conservation priorities in this system, what do you see as the main constraints? Are there particular areas of study (ecological science, climate science, policy, data management, etc.) that could help overcome this constraint?

Craig: The main constraints for me to change a management approach are the leases and conservation easements that are not currently accommodating extreme intra-annual variations seen on the ground. Policies within public agencies, land trusts, and other leasing entities should be modernized to ensure a certain level of flexibility in the terms applied to all new leases and conservation easements, and how these fit into a greater management plan. The degree to which lease and easement terms could accommodate unpredictable change can only be determined with the support of ecological and climate science studies examining weather trends, forage productivity, and changes in biodiversity across the landscape.

Suttle: What do you see as the main challenge to linking information on potential future climatic

(continued)

(continued)

conditions to your work protecting grassland systems into the future?

Craig: The main challenge in linking potential future climate change information to my conservation work is regular and open communication among involved parties: Landowners, ranchers, lessees, field-based staff, senior managers, and others. Monitoring and decision-making frameworks will need to be defined. By sharing information effectively, and being in tune with what is happening on the land, greater flexibility can be built into the whole system.

Suttle: To what extent is monitoring an important management objective at the places where you work? Are monitoring plans set up in a way that captures responses to variation in the seasonality of rainfall? Has this chapter led you to consider any changes in objectives or methods for monitoring (e.g., less species-specific)?

Craig: Monitoring is very important and structured, performed on an annual basis on all properties managed with a conservation easement in this project area. Monitoring plans are not currently set up in a way to capture variation in rainfall seasonality. This study has led me to consider adding to existing monitoring objectives, such as measuring vegetation change outside of the effect of grazing (current quantitative comparisons are for residual dry matter in grazed areas only). This however raises a concern for added staff time and data management responsibilities.

Suttle: Have changes in climate led to changes in your approach to management? Have your tools for addressing the challenges of exotic plants changed or become easier or harder to use (i.e., prescribed fire, use of grazing or herbicides)? Have costs of management changed?

Craig: Weighing in climate change certainly has changed my approach to management and responding to the non-static nature of our environment. During the relatively short time frame in which I've been managing invasive species it has become more challenging to prioritize goals while using limited resources. Prescribed fire is increas-

ingly restricted, and herbicide use, though seemingly cost-effective, also has its limitations. The need for additional monitoring and research has increased, and with an eye on the importance of adequate stewardship, management costs have without a doubt risen.

The Angelo Reserve study results confirm that the biggest rangeland-management challenge the California landowners and cattle operators currently face—uncertainty—is already being exacerbated by climate change. Many ranchers and land managers are already looking for solutions. One landowner grazing under easement in the Lassen Foothills of northeastern California recently asked me, "When is everyone going to stop talking about how weird the weather is and just start planning for it?" Managers at all levels know the question is not whether the future will be increasingly unpredictable; it's what they and policy makers can and should do about it.

Freshwater springs, vernal pools, and swales dot the Lassen Foothills, enabling wildlife and cattle to find water throughout much of the year. However, as in much of California, thin soils limit plant cover to annual grasses and forbs with abbreviated and highly opportunistic life cycles. This vegetation may be adaptable to climate change, and is in turn a highly limiting factor for ranching operations. The most successful managers are those that can be as adaptable and responsive to conditions within a given year as is the vegetation on which the livestock rely.

Here is one solution that would improve intraannual flexibility for managers at the individual property or landscape scale: Legally binding land ownership and management agreements such as leases and conservation easements could be written to better accommodate the increasingly unpredictable nature of California's grazing season. Specifically, season dates defined in leases for turning out and removing cattle should be based on field observations of resource conditions in the field, including weather patterns, water availability, and productivity, rather than predetermined dates. This would enable better utilization of forage, supporting the cattle industry, and would not unduly impact grasslands. Policies within public agencies, land trusts, and other leasing entities should be modernized to ensure these types of flexible terms

(continued)

are applied for all new leases and conservation easements.

Making these decisions in real time will require ranchers and field-based staff to communicate regularly and openly with lessees and agency staff, and for those managers to trust the knowledge and good intentions of those with their eyes on the ground. In some cases this will be a shift. Monitoring and decision-making frameworks for these quick-turnaround decisions will need to be defined. By sharing information and being in tune with what is happening on the land, greater flexibility can be built into the whole system.

Some cattle operators are currently working enough land to juggle abnormal weather patterns, without undue hardship. Operators with less acreage do not have as much flexibility. On a recent mid-April day, at least one local rancher was branding in a steady spring rain, working with a team of cowboys to keep irons hot despite mud- and moisture-soaked coats. We have been catching up with an average year's precipitation the last few weeks which has resulted in, for the second year in a row, ranchers needing to gather cattle out of the valley while conditions are finally favorable and the high country is blanketed in the season's deepest snowpack yet. Consistently abnormal weather patterns associated with climate change may put smaller operators out of business sooner than later. Conservation of this landscape is dependent on keeping the patchwork of ranchers ranching, large scale and small scale alike. Scientific study of climate change will continue to provide valuable information that land managers can integrate into decision-making frameworks. But solutions must also be welcome from those who know each patch of ground best, and must be found quickly. The success of managing for uncertainty will hinge upon the close communication we are able to maintain with our neighbors and other stakeholders in this rural community.

CHAPTER 10

Species Invasions

LINKING CHANGES IN PLANT COMPOSITION TO CHANGES IN CLIMATE

Laura Koteen

Abstract. California's grasslands have been dramatically altered by the invasion of nonnative annual grasses from Mediterranean Europe. Along the coast, this invasion has displaced much of the region's once dominant native perennial grasses. This research investigates the effects of this displacement in community composition on carbon cycling and storage at two sites in coastal northern California. The broader goal is to understand how community changes, brought on by species invasion, have contributed to global climate change through shifting the balance of carbon storage from the soil and plant tissues to the atmosphere. Our coastal research sites had native vegetation growing adjacent to locations where nonnative grasses had invaded. We tracked the processes that affect inputs and outputs of carbon to and from the soil to understand the causes of soil carbon loss following nonnative grass invasion. We found that nonnative grass invasion has resulted in the transfer of an added 30–60 Mg of carbon per hectare from the soil to the atmosphere since the invasion of annual grasses began in the 1700s, becoming widespread by the 1860s. Restoration in select areas may over time be able to reverse the impacts on soil carbon storage described in this chapter. Restoring native grasslands and preventing further large-scale invasion events could contribute to climate change mitigation by reducing carbon lost from California's natural areas.

Key Points

- Changes in land use and the invasion of nonnative plants can reduce the ability of ecosystems to store carbon.
- In California's grasslands, the displacement of native perennial grasses by nonnative annual grasses causes a loss of soil carbon.
- Land cover change is both a cause and effect of climate change because climate is the outcome of interactions between the land surface and the atmosphere.
- Native grassland restoration will likely increase carbon storage if established protocols are followed, if the protocols are maintained until perennial grasses become well established and if the climatic zones of the sites selected support relict stands of native grasses.

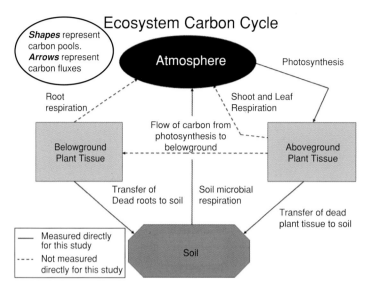

Ecosystem Carbon Cycle

Shapes represent carbon pools.
Arrows represent carbon fluxes

Atmosphere

Photosynthesis

Root respiration

Shoot and Leaf Respiration

Flow of carbon from photosynthesis to belowground

Belowground Plant Tissue

Aboveground Plant Tissue

Transfer of Dead roots to soil

Soil microbial respiration

Transfer of dead plant tissue to soil

— Measured directly for this study
---- Not measured directly for this study

Soil

FIGURE 10.1: An illustration of the study design assessing carbon flux among the primary reservoirs of carbon in grassland ecosystems: the atmosphere, plant biomass, and soil.

INTRODUCTION

Since the onset of European settlement, California's ecosystems have been radically altered. Among the most dramatic of changes is the widespread invasion of nonnative Eurasian grasses into the state's grasslands. Beginning in the 18th century, this invasion has caused the near extirpation of California's native perennial bunch grasses across millions of acres of grassland habitat. Indeed, the displacement of native-grassland vegetation in California is one of the most complete and extensively documented land-cover changes worldwide (Seabloom et al. 2003b). Yet, it is but one example of a phenomenon that is much more widespread.

Because they are so common and a primary cause of biodiversity loss, land-use and land-cover change have long been a focus of ecological research. What may be less appreciated is the effect of land-cover change on climate, and more specifically, on the balance of carbon storage between the terrestrial environment and the atmosphere. Historically, land-use and land-cover change account for one quarter to one-half of all terrestrial-carbon losses to the atmosphere, and thus rival fossil fuel combustion as a cause

of global climate change (House et al. 2005). Biological invasion is among the causes of land-cover change, and a growing number of investigations have now linked species invasion directly to carbon cycle changes (Christian and Wilson 1999, Ehrenfeld 2003, Liao et al. 2008). For this research, we sought to characterize how the invasion of nonnative grasses has altered carbon cycling and storage in California's grasslands.

Land-use and land-cover change can affect global climate change by interrupting the natural processes of carbon exchange that are regulated by plant growth, soil microbial activity, and climate. They affect ecosystem carbon pool sizes and flux rates that can shift the balance of carbon between the atmosphere and storage in both vegetation and soils, which in turn can alter the global greenhouse effect. Carbon pools are the reservoirs in which carbon is stored and include plant tissues, the soil, and the atmosphere. Fluxes describe the rates of exchange of carbon among ecosystem pools. The flux of carbon from the atmosphere to plants occurs via photosynthesis (Figure 10.1). Carbon flux from plant storage to the soil occurs when plant tissues decay and are incorporated into the soil complex by soil microbes and fauna, the process by which plant tissue becomes a part of the soil carbon pool. Carbon flux from the soil to the atmosphere occurs primarily through microbial consumption and respiration of soil organic matter in the form of CO_2. As this study documents, even a seemingly subtle change in species composition (i.e., from one grass type to another) can affect climate by altering ecosystem carbon pool sizes and / or flux rates.

Here we compare carbon pools and fluxes at two locations where both native and nonnative grass types are found in similar environmental settings. Because most grassland carbon is stored in soil (Schlesinger 1997), our first goal was to determine if differences in soil carbon storage exist between native and nonnative grass types. Our second goal was to understand the specific aspects of carbon cycling and storage that differ between native and nonnative grasslands, which might explain soil carbon differences. We began with three hypotheses: (1) Differences in soil carbon result from differences in the amount of root and aboveground carbon that enters the soil due to turnover of plant tissues each year. (2) Differences in soil carbon result from differences in the chemical composition of plant tissues, affecting rates of soil organic matter accumulation and decomposition. (3) Differences in soil carbon result from differences in soil climate beneath the different grassland types. Soil climate and the tissue chemistry of senesced plant matter control the size, activity, and composition of the soil microbial community. These microbial community properties in turn affect flux rates of soil carbon to the atmosphere. Differences in annual plant growth as well as the flux size of plant tissue that enters the soil each year address the first hypothesis.

We predicted that nonnative grass invasion would cause a drop in ecosystem carbon storage due to a shift from perennial plants (natives) to annual ones (nonnative). Each type of plant is associated with a different suite of traits that allows survival through long summer drought. Being perennial, California's native grasses

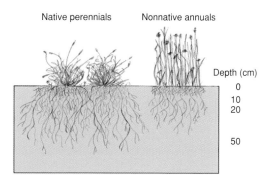

FIGURE 10.2: California's native perennial and nonnative annual grasses. Grass morphology at the time of peak biomass in spring in the native perennial bunch grass community (left) and nonnative annual grass community (right).

maintain a year-round connection with the soil and atmosphere. Deep roots that exploit the full volume of soil for water, a dense aboveground structure that inhibits soil evaporation, and high root production enable their survival through seasonal water scarcity (Figure 10.2, left). Attributes that promote soil carbon accumulation also fit into a strategy of water conservation, as soil organic matter is capable of storing more water than mineral soil (Hudson 1994). In contrast, the nonnative grasses are annuals. They complete their life cycle before the onset of summer drought; they grow from seed each year when autumn rains begin and senesce in April or May. Aboveground, annual grasses are relatively sparse, allowing radiation to penetrate to the soil surface and causing intense drying of the upper soil profile (Figure 10.2, right). Soil desiccation near the surface also results from the annual grass root system, which is concentrated in the top 20 cm.

METHODS

Site Description

We established research plots in two grassland sites: The Tennessee Valley site, which resides in the headlands of the Golden Gate National Recreation Area (elevation 220 m), and the Bolinas Lagoon Preserve site, located on a

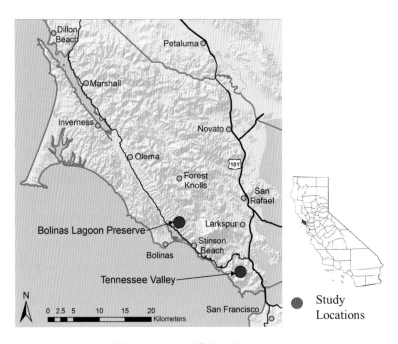

FIGURE 10.3: A map of the study area and field site locations.

private preserve outside Bolinas, California (elevation 168 m; Figure 10.3). Both sites are within a kilometer of the coast and experience a climate with distinct wet and dry seasons. Mean annual precipitation is approximately 900 mm, with high interannual variability (±300 mm standard deviation). The rainy season typically occurs from late October through April, but can extend into May or June. Summers are warm and dry, but moderated by coastal fog. Vegetation at both sites was predominantly grassland interspersed with patches of shrubs. We set up research plots at Tennessee Valley on northeast-facing slopes that ranged from flat to 2%. Slopes at the Bolinas Lagoon Preserve were south-facing and ranged from flat to 5%. The latter site was generally drier than the Tennessee Valley site due the extended direct solar exposure this grassland receives. Soils at both sites are composed of a well-drained sandy loam on bedrock derived from sandstone and shale (Soil Conservation Service 1985).

At each site in the spring of 2003, we set up 2 × 2 m research plots in relatively pure patches of native perennial and nonnative annual grass communities: 10 plots at the Bolinas Lagoon Preserve site, equally divided between native and nonnative grass communities, and 15 plots at the Tennessee Valley site, with five in each of the two native grass communities and five in the nonnative community. Patches were chosen for their similarity in soil properties, land-use history, slope, and aspect. In this study design, native plots represent preinvasion conditions. The relative carbon-storage status in nonnative plots represents the cumulative carbon cycle changes that had accrued since the time of annual grass invasion. The dominant nonnative grass species at Tennessee Valley were slender wild oat, *Avena barbata*, Italian rye grass, *Lolium multiflorum*, rattlesnake grass, *Briza maxima*, and small and rattail fescue, *Vulpia spp*. The dominant native perennial grasses at this site were Hall's bentgrass, *Agrostis halli*, a rhizomatous grass, with continuous aboveground cover, and red fescue, *Festuca rubra*, a bunch grass. In general, bunch grasses are more typical of native perennial grasses in California. At the Bolinas Lagoon Preserve, the five nonnative plots are dominated by the annual grass, false brome, *Brachypodium distachyon*, but also contain common wild oats, *Avena fatua*, and rattlesnake grass, *B. maxima*. The five native bunch grass plots at this site were dominated by a mix of purple needlegrass, *Nassella pulchra*, and California brome, *Bromus carinatus*, or by blue wild rye, *Elymus glaucus*.

Measurements of Soil Carbon Pool Size

At each site, we measured the standing pools of soil carbon in 10 cm intervals by extracting seven soil cores from the surface to 50 cm depth per plot). From each of these cores, we determined the percentage of carbon in each depth interval and then multiplied by the mass of soil in the core to determine the total mass of carbon for each depth.

HYPOTHESIS 1: DIFFERENCES IN BIOMASS AMOUNTS (ANNUAL PRODUCTIVITY) We harvested grasses in 0.25 m² areas in the late spring of 2004 and 2005, and estimated the total annual productivity as equal to total standing aboveground biomass at the time of peak growth (Scurlock et al. 2002, Corbin and D'Antonio 2004, Lauenroth et al. 2006). To determine the belowground flux, we extracted 3.5 cm diameter soil cores from 0–10, 10–20 and 20–50 cm depth in January and April 2004, for nonnatives, and January and June 2004, for natives. Samples were collected at the time of minimum and peak biomass for each grass type (Corbin and D'Antonio 2004, Corbin and D'Antonio 2010). Subsequently, roots were separated from soil to obtain root-density estimates per soil volume. The annual belowground flux was calculated as the difference between root density at peak and minimum biomass plus an estimate of the roots that had decomposed in the intervening time period.

For the years of our study, annual precipitation was below average for the water year 2003–2004 (average of 89% and 83% for Tennessee Valley and the Bolinas Lagoon Preserve, respectively), and higher than average for the water years 2004–2005 (117% and 131%) and 2005–2006 (162% and 144%). We did not measure aboveground productivity directly in 2006 at either site, or root productivity in 2005 or 2006. Instead, we estimated these carbon pools based on reports from the literature that found a positive relationship between aboveground and belowground productivity and precipitation (Jackson and Roy 1986, Dukes et al. 2005, Chou et al. 2008). We also observed an increase in aboveground grass productivity after higher rainfall in 2005.

HYPOTHESIS 2: DIFFERENCES IN PLANT TISSUE CHEMICAL COMPOSITION (LITTER QUALITY) To estimate differences in the chemical composition of plant tissue between native and nonnative grass types, we performed leaf- and root-litter decomposition experiments in each plot (Koteen et al. 2011). Known amounts of senesced plant tissue were sewn into mesh bags and the rate of mass loss tracked over time (Wieder and Lang 1982). Differences in the rate of mass loss served as an indication of differences in tissue chemical composition between grass types, as tissues that are less palatable to soil microbes take longer to decompose. We initiated leaf decomposition experiments in 2003–2004 and 2004–2005, and tracked root-litter decomposition in 2005–2006. To bolster these findings, we also conducted an analysis of secondary compounds in leaf and root samples according to methods outlined in the study of McClaugherty et al. (1985) and Ryan et al. (1990).

HYPOTHESIS 3: DIFFERENCES IN SOIL CLIMATE IN DIFFERENT GRASS TYPES We measured soil respiration on a monthly basis from January 2005 to June 2006 at five sites within each plot, using the LI-COR 6400 infrared gas analyzer (LI-COR Inc., Lincoln, NE). To isolate the respiration of carbon by soil microbes from respiration by roots, we removed the aboveground vegetation from the soil collars where soil respiration measurements were performed, and kept them free of vegetation over the period of observation. Nonetheless, root respiration undoubtedly accounts for some unknown fraction of the amounts we report, as respiration from neighboring roots could still enter the soil column beneath each collar (Pumpanen et al. 2003).

To understand the role of soil climate at different soil depths in promoting or inhibiting soil respiration, we measured soil moisture in 5 cm intervals to 40 cm depth using a soil-moisture-profile probe (Delta-T Devices, Cambridge, UK) regularly over the course of the study. We also measured soil temperature at the surface and at 5, 15 and 35 cm depth, hourly, using HOBO data loggers (Onset Computer Corporation, Bourne, MA) from 2003 to 2006.

To compare the total soil respiration between grass types on an annual basis, we modeled soil respiration for the water years from 2003 to 2006 based on an analysis of the correlation between soil climate variables and measured soil respiration. Because we had nearly continuous measurements of soil moisture and temperature, we used these correlations to compute hourly soil respiration for each grass type, and to produce annual sums. To quantify the level of uncertainty in our modeled annual soil respiration values, we began with the error estimates from the correlation between soil respiration with soil moisture and temperature for each grass type. We next developed estimates for the probability distribution of our model parameters using the open source software program, WinBUGS, and then computed error estimates for each hour of the year via a Markov chain Monte Carlo simulation in Matlab (Verbeeck et al. 2006).

Differences between mean biomass values, flux rates, and litter-decay constants between grass types were determined using ANOVA, with five plots per grass community, for five grass communities across two sites.

As predicted, we found consistently higher pools of carbon in soils of the native perennial grass community relative to the nonnative. At the Bolinas Lagoon site, we found a drop in soil carbon equivalent to 30 Mg per hectare. At the Tennessee Valley site, drops in soil carbon amounted to a loss of 60 Mg per hectare. This difference was most notable at soil depths below 30 cm, but was also evident near the top of the soil profile, at both sites (Table 10.1 and Figures 10.4a and 10.4b). Therefore, we concluded that a net transfer of carbon from the soil to the atmosphere followed nonnative grass invasion. We found support for two out of three hypotheses that explain how this transfer occurred.

Reasons for Differences in Soil Carbon Storage

HYPOTHESIS 1: DIFFERENCES IN BIOMASS AMOUNTS (PLANT PRODUCTIVITY) We found support for our first hypothesis. In 2004, we found higher productivity in the native perennial community relative to nonnative annuals in both aboveground and belowground tissues (Table 10.1). Moreover, the native bunch grasses in our study produced significantly greater fine-root biomass at all soil depths than the nonnative annuals (Table 10.1 and Figures 10.4c and 10.4d). The rhizomatous grass, A. halli, also produced significantly greater root biomass below 10 cm depth. However, root production was significantly lower than the native bunch grasses. High annual rainfall and an extended growing season in 2005 corresponded with a much higher total production in annual and perennial grass types than in 2004. Belowground productivity at both sites in 2005 was significantly higher in all native grasses than in nonnative grasses at all soil depths. Aboveground productivity the same year was not significantly different between grass types at Tennessee Valley, but nonnative aboveground productivity was significantly higher in the nonnative grass type at the Bolinas Lagoon Preserve. In 2006, we assumed that both aboveground and belowground productivity were similar to 2005, given that it was also a high rainfall year.

HYPOTHESIS 2: DIFFERENCES IN TISSUE CHEMICAL COMPOSITION (LITTER QUALITY) We found only small differences in litter quality across the sites, with one exception. From aboveground litter and root decomposition

TABLE 10.1
Productivity, respiration, and soil carbon at the Tennessee Valley (TV) and Bolinas
Lagoon Preserve (BLP) sites in units of kg m^{-2}
Native and nonnative grass communities for the water years 2003–2004, 2004–2005, and 2005–2006

		2003–2004				
Site	Species type	Shoot productivity (carbon) kg m^{-2}	Root productivity (carbon) kg m^{-2}	Total productivity (carbon) kg m^{-2}	Soil respiration kg m^{-2}	
TV	Agrostis (native)	0.33 ± 0.11 [a]	0.21 ± 0.07 [a]	0.54 ± 0.13 [a]	0.42 ± 0.16 [a]	
	Festuca (native)	0.35 ± 0.16 [a]	0.50 ± 0.04 [b]	0.85 ± 0.17 [b]	0.67 ± 0.20 [b]	
	Nonnative	0.12 ± 0.04 [b]	0.14 ± 0.02 [a]	0.26 ± 0.05 [c]	0.44 ± 0.12 [a]	
BLP	Native	0.26 ± 0.15 [a]	0.37 ± 0.05 [a]	0.63 ± 0.16 [a]	*	
	Nonnative	0.17 ± 0.04 [a]	0.20 ± 0.03 [b]	0.36 ± 0.05 [b]	*	

		2004–2005				
Site	Species type	Shoot productivity (carbon) kg m^{-2}	Root productivity (carbon) kg m^{-2}	Total productivity (carbon) kg m^{-2}	Soil respiration kg m^{-2}	Total soil carbon*
TV	Agrostis (native)	0.42 ± 0.08 [a]	0.41 ± 0.14 [a]	0.83 ± 0.16 [a]	0.72 ± 0.18 [a]	17.6 ± 0.4 [a]
	Festuca (native)	0.37 ± 0.13 [a]	1.01 ± 0.08 [b]	1.38 ± 0.16 [b]	0.78 ± 0.26 [a]	16.9 ± 0.4 [a]
	Nonnative	0.22 ± 0.04 [b]	0.14 ± 0.02 [c]	0.35 ± 0.04 [c]	0.77 ± 0.14 [a]	14.9 ± 0.4 [b]
BLP	Native	0.26 ± 0.10 [a]	0.74 ± 0.09 [a]	1.00 ± 0.14 [a]	0.99 ± 0.13 [a]	14.7 ± 0.4 [a]
	Nonnative	0.28 ± 0.07 [b]	0.20 ± 0.03 [b]	0.48 ± 0.07 [b]	0.68 ± 0.15 [b]	9.5 ± 0.2 [b]

		2005–2006				
Site	Species type	Shoot productivity (carbon) kg m^{-2}	Root productivity (carbon) kg m^{-2}	Total productivity (carbon) kg.m^{-2}	Soil respiration kg m^{-2}	
TV	Agrostis (native)	0.42 ± 0.06 [a]	0.41 ± 0.34 [a]	0.83 ± 0.35 [a]	0.74 ± 0.17 [a]	
	Festuca (native)	0.37 ± 0.10 [a]	1.01 ± 0.47 [b]	1.38 ± 0.48 [b]	0.79 ± 0.18 [ab]	
	Nonnative	0.22 ± 0.03 [b]	0.14 ± 0.12 [c]	0.35 ± 0.12 [c]	0.84 ± 0.20 [b]	
BLP	Native	0.26 ± 0.08 [a]	0.74 ± 0.09 [a]	1.00 ± 0.12 [a]	0.76 ± 0.13 [a]	
	Nonnative	0.28 ± 0.05 [b]	0.20 ± 0.03 [b]	0.48 ± 0.06 [b]	0.64 ± 0.14 [a]	

NOTE: Error values represent ±1 standard deviation, n = 5. ANOVA was used to detect differences between mean values. Different letters indicate significant differences at p < 0.05 among measurement categories, and only apply within individual sites.

*The soil carbon values reported here are from a single sampling assay completed in 2005, $n = 7$ cores per plot. Similar values were found in a 2003 sampling. Letters indicate significant differences among grass communities.

experiments, we found significant differences in litter quality between native and nonnative grass types only for the roots at the Tennessee Valley site. At this site, we found the roots of the native perennial bunch grass, *F. rubra*, decomposed significantly more slowly than those of the other *Agrostis* and nonnative grass communities, with the first-order decomposition constant ($k = 0.28$ for *F. rubra*, $k = 0.52$ for *A. halli*, the native rhizomatous grass, and $k = 0.52$ for nonnative grass type, $p = 0.001$). Here, the constant values indicate a rate of mass loss over time, with a higher value indicating a faster rate of decomposition. The analysis of secondary compounds in leaf and root litter generally corroborated the findings of the decomposition assays.

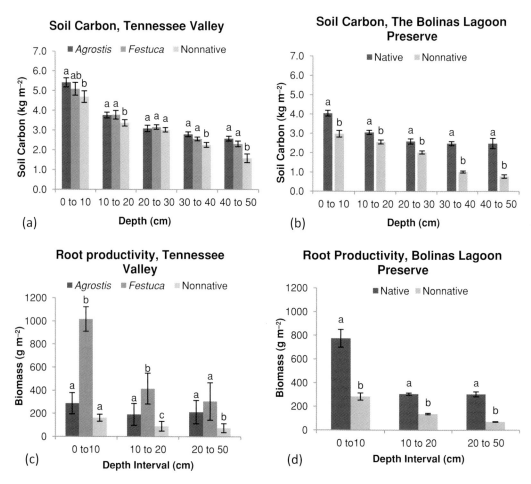

FIGURE 10.4: Soil carbon storage and root productivity as a function of soil depth. Soil carbon storage in native perennial and nonnative annual grass communities at (a) Tennessee Valley and (b) the Bolinas Lagoon Preserve. Annual root productivity for all grass types averaged between the dry year 2003–2004 and the wet year 2004–2005 at (c) Tennessee Valley and (d) the Bolinas Lagoon Preserve. At the Tennessee Valley site, *Festuca* is a native perennial bunch grass and *Agrostis* is a native perennial rhizomatous grass. Error bars represent ±1 standard error.

HYPOTHESIS 3: DIFFERENCES IN AMOUNTS OF SOIL RESPIRATION Differences in soil respiration rates explain differences in soil carbon storage at both of these sites because net carbon storage is the difference between annual carbon inputs into the soil (annual productivity) and outputs from the soil (soil respiration) (Figure 10.1). At both the Tennessee Valley and Bolinas sites, we found no consistent pattern in soil respiration differences between native and nonnative grass communities that held across all years (Table 10.1). These results would seem to disprove the hypothesis that soil respiration contributes to the drop in soil carbon storage in

nonnative grasslands. In each site comparison, for each individual year, however, soil carbon fluxes into the soil due to plant growth, and senescence exceeds or equals carbon fluxes out of the soil from soil respiration in the native grass types, and the reverse is true for the nonnative grass types (Figure 10.5).

Soil respiration rates were positively correlated with (1) the timing of organic plant fluxes into the soil, (2) the soil carbon pool size, (3) the product of soil temperature and moisture, and (4) soil moisture as the dominant factor (i.e., soil fluxes from the soil were highest during the months that the decomposing plant tis-

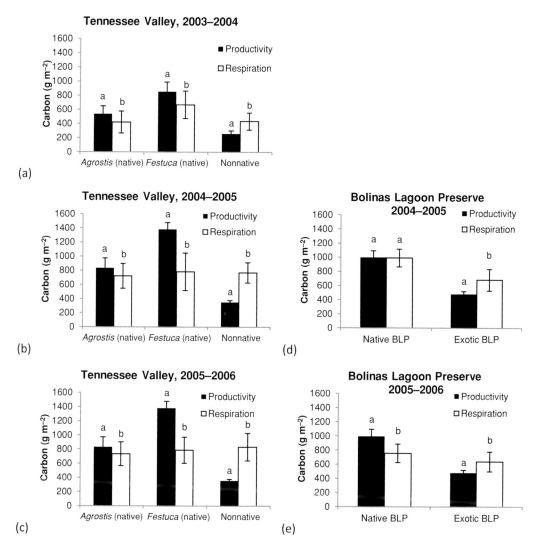

FIGURE 10.5: Annual productivity and soil respiration for native perennial and nonnative annual grasses, Tennessee Valley and the Bolinas Lagoon Preserve. Annual production of biomass carbon and annual soil respiration for the water years extending from October through September for 2003–2004 (a), 2004–2005 (b), and 2005–2006 (c), for the Tennessee Valley and Bolinas Lagoon Preserve Field sites (d and e). *Agrostis* is the native rhizomatous grass and *Festuca* is the native bunch grass. Error bars represent ±1 standard deviation. Different letters represent significant differences at $p < 0.05$ between annual productivity and soil respiration within individual grass communities.

sues were actively entering the soil) (Koteen et al. 2011). We also found that soil moisture differs with depth along the soil profile among grass types, and with respect to the location of soil carbon, which is highest at the top of the soil profile and declines with soil depth (Figures 10.4a and 10.4b). At both sites, soil moisture was higher at the top of the soil profile in native grasses relative to nonnative grasses and

lower at the bottom of the soil profile (Figures 10.6 and 10.7). Soil temperature was consistently negatively correlated with soil moisture (data not shown).

We further examined the possibility that differences in soil properties (i.e., soil texture, soil pH, soil rock content) might explain differences in soil carbon storage. In general, we found small differences in these properties

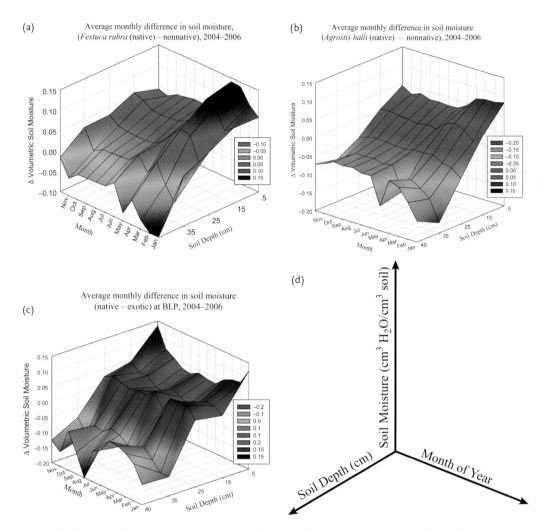

FIGURE 10.6: The difference between grass types in volumetric soil moisture along the soil profile. (a) Differences between the *Festuca* (native bunch grass) community and the nonnative community at Tennessee Valley (b) between the *Agrostis* (native rhizomatous grass) community and the nonnative community at Tennessee Valley, and (c) the native bunch grasses and nonnative communities at the Bolinas Lagoon Preserve, and (d) a diagram of axes defining the variables represented on each axis.

among soils of different grass types, with the exception of soil texture. Small but significant differences in soil texture may have contributed to, but do not explain, the differences we found in soil carbon storage.

DISCUSSION

Our findings indicate that the invasion of California's grasslands by nonnative grasses from Mediterranean Europe has caused a substantial drop in soil carbon storage equaling from 30 to

60 Mg per hectare at the sites examined. We attribute the drop in soil carbon to the difference in the cumulative annual net carbon flux between native and nonnative grass types since the time of annual grass invasion.

The loss of soil carbon appears to stem from key differences in annual and perennial plant traits that evolved in response to periods of seasonal water scarcity. Native perennials possess deep roots, and a dense aboveground structure that inhibits soil evaporation. Nonnative annuals, in contrast, have shallow roots, a short life

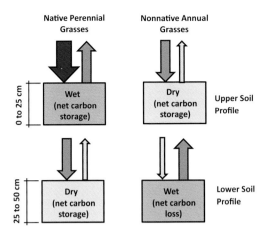

FIGURE 10.7: Contrasting carbon cycling dynamics in coastal native perennial and nonnative annual grass communities within individual research sites. Diagram of theorized annual carbon inputs and outputs to and from the soil. The inputs and outputs (indicated by arrows) are relative to each other in the same depth profile and indicate average conditions over the course of the annual cycle. Soil depth divisions between the upper and lower soil profiles are approximate and vary by species. The size of the soil respiration term is highly correlated with the soil moisture content. See Discussion text for a detailed explanation of the contrasting plant strategies that produce differences in soil moisture content and biomass inputs.

cycle, and a sparse plant canopy. In this case, the same strategies employed to survive the summer drought also explain the differences we found between grass types in carbon cycling and storage.

In addition to differences in plant traits, the difference in growing-season length also helps explain the loss of soil carbon following nonnative grass invasion through its effect on the balance between Net Primary Production (NPP) and soil respiration in each grass type, especially in years of high rainfall (2004–2006). Whereas both productivity and soil respiration in Tennessee Valley vary positively with water availability in native perennial grasses (Figure 10.5), other studies have shown soil respiration varies more than NPP as a function of water availability, in soils dominated by nonnative annuals (Ma et al. 2007, Zhang et al. 2010). The timing of annual grass senescence is, at least in part, internally set by the need for reproduction to occur every year

regardless of environmental conditions, and senescence closely follows flowering in this grass type (Jackson and Roy 1986, Jackson et al. 1988). Therefore, productivity is capped in the annual grass type by constraints on growing-season length, whereas soil respiration is not. Our findings indicate that this imbalance in soil fluxes into and out of soil accounts for the loss in soil carbon in the annual grass type. The timing of summer dormancy in native perennial grasses varies to a lesser degree than in annual grasses (Laude 1953). And the higher NPP in years of high rainfall in most cases supersedes the loss of soil carbon due to higher soil respiration, leading to net soil carbon accumulation where native perennial grasses are found.

The differences in plant strategies with regard to water use also help explain the pattern of soil carbon differences between grass types in the upper and lower soil profiles. Near the soil surface, soil carbon storage is greater in the perennial grass community. However, the difference is small given the large differences in total productivity between native and nonnative grass communities, and significant in most, but not all, comparisons (Figures 10.4a and 10.4b). This outcome is consistent with the interpretation that both biomass inputs and soil respiration are high in the upper soil profile of the perennial grass type and relatively low in the nonnative grass type (Figure 10.7). We attribute differences in soil-respiration rates to differences in soil moisture in upper soil layers between grass types. We found that the dense aboveground cover and deeper root system of perennial grasses lead to a relatively even drawdown of soil moisture along the depth of the soil profile. In contrast, the relatively sparse aboveground cover of nonnative grasses produces a soil moisture profile that is dry for much of the year near the soil surface, suppressing soil respiration. Deeper in the soil, the carbon pool sizes diverge between grass types with significantly greater carbon storage present where native perennial grasses are found. At these depths, perennial grasses have both a greater biomass flux into soil and lower respiratory losses

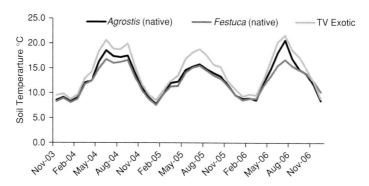

Soil temperature, Tennessee Valley, 2004–2006

FIGURE 10.8: Soil temperature near the top of the soil profile for native and nonnative grass communities at the Tennessee Valley site.

(Figures 10.4c, 10.4d, and 10.7). In contrast, very low inputs from shallow-rooted annual grasses, coupled with high soil respiration rates associated with untapped soil moisture beneath the rooting zone, have caused a loss in soil carbon at these depths.

These findings are largely consistent with other studies investigating nonnative grass invasions into grasslands. Generally, similar research has found that invading species possess higher aboveground and lower belowground productivity, greater seed production, lower belowground allocation, and a shallower rooting depth in comparison to the native grasses they displace (Christian and Wilson 1999, Wilsey and Polley 2006, Adair and Burke 2010). These findings differ in that we also found consistently higher aboveground productivity in the native grasses at the Tennessee Valley site.

In addition to promoting higher carbon storage, we found that native perennial grasses also influence the microclimate in ways that may increase their resilience to continued increases in regional temperature. Specifically, we found that the denser structure and more complete cover of perennial grasses lead to a moister, cooler soil surface environment when compared to soil surfaces under nonnative grasses, particularly at the Tennessee Valley site (Figure 10.8). Thus, perennial grasses may promote the persistence of associated plant and animal species that are close to the upper limits of species-specific temperature or drought tolerances. And where natives remain in small patches in largely annual dominated grasslands, they may serve as refugia to associated species during periods in which climatic conditions in nonnative patches are unfavorable.

Lastly, we note that the changes we document in the grasslands of northern coastal California are similar to changes that have occurred elsewhere in the state and in many temperate grasslands worldwide where deep-rooted vegetation has been replaced by shallow-rooted crop species, such as cereal crops. Therefore, in documenting the change in soil carbon storage within coastal California's grasslands, we are focusing on a phenomenon that may be much more widespread (Houghton 1999).

CONCLUSION

With concerns that global climate change will reconfigure our land surface with ecosystems of novel character and uncertain function, much ecological research is devoted to understanding how climate change alters the composition of terrestrial ecosystems and instigates carbon feedbacks. Although most changes in community composition in California's grasslands are not the result of a changing climate, they are an excellent example of the types of changes climate change may bring about, and therefore warrant further study. Moreover, because these vegetation shifts occurred long ago, these systems are advanced in their progression toward a new equilibrium with respect to resident species and carbon cycle changes. As such, they offer insights into ecosystem-level processes and changes that may not yet be apparent in ecosystems currently in a state of transition. Further, California's altered grasslands may be representative of ecosystem-level changes that may increase in the future. They are composed largely of invasive species and support lower species diversity (Stromberg et al. 2001, Coleman and Levine 2007). As the climate changes, many species will experience heightened physiological stress, disease, and disturbance (Koteen 2002, Roy et al. 2004, Evans et al. 2008), causing land areas to open up (Westerling and Bryant 2006). By definition, invasive species are less hampered by competitive pressures than native species and may therefore have an advantage in colonizing areas of recent disturbance.

LESSONS FOR GRASSLAND RESTORATION FROM EXPERIMENTAL AND OBSERVATIONAL STUDIES

The primary solution to soil carbon loss with nonnative grass invasion is restoration of native perennial grass communities (Box 10.1). An informed approach would consider the soil types, future climate, and species for which restoration efforts might have the maximum effect in increasing soil carbon storage. Although the changes documented here represent a modest contribution to atmospheric CO_2, native perennial grasses do store significantly more carbon than nonnative annuals, and their loss represents the near disappearance of a unique and iconic ecosystem that once populated large areas of California. Therefore, the restoration of native perennial grasses to some of the locations they once occupied would simultaneously sequester carbon and restore native biodiversity. However, if one's goal is to simply seques-

---

BOX 10.1 · Native Grass Restoration as a Carbon Offset

If we assume the site-averaged difference in soil carbon storage between native and nonnative grasslands from this study (45 Mg ha⁻¹) can be applied more generally to grasslands in California, the potential opens up for grassland restoration activities to gain financial support through state or federal carbon offset programs. According to the 2007 EPA estimates, a vehicle of average fuel efficiency emits approximately 5 metric tons of CO_2 per year (EPA 2009). Therefore, in carbon currency, we estimate that the restoration of 500 ha of grassland to native species would be equivalent to removing more than 4000 cars from the road for a year. In practice, however, the total gain in grassland soil carbon storage would accrue over many years. Therefore, in this context, and other biological settings, the actual funds to support such efforts would require established rules for reimbursement over time, and a carbon market interested in supporting projects that support both climate and conservation goals.

---

ter the maximum amount of carbon without regard for benefits to biodiversity, prevented emissions, reforestation, or afforestation would be superior options (Hudiburg et al. 2009, Battles et al. 2014). Other land-based options for increasing soil carbon storage in California, where nonnative annual grasslands are likely to persist, include applying organic amendments to soil grazed pastures and other grasslands (DeLonge and Ryals 2013, Ryals et al. 2014).

When designing restoration protocols for California's native grasses, several considerations should be instituted in order to ensure restoration success. Restoration efforts should focus on locations where reconversion to exotic annual grasses is least likely, should draw from seed sources local to the area where restoration is taking place, and should focus on species native to the region. Because remnant native perennial grasslands are found primarily along California's coast, and especially along the northern coast, presumably restoration efforts would be most successful in these regions. Perennial-grass persistence along the coast is thought to be aided by the existence of moisture inputs from the summer fog cycle (Corbin et al. 2005). In light of climatic changes, regions that possess landscape features that mitigate projected temperature increases and enhanced drought (i.e., northerly aspect) or which are projected to receive summer precipitation or an extended rainy season may also be good locations for restoration efforts. Any climatic changes that increase soil water availability after nonnative annual grass senescence would uniquely benefit natives. Regions other than those considered strictly coastal may be good candidates for native grass restoration as well. Beyond geographic location, the absence of prior cultivation is strongly associated with the persistence of native perennial grasses (Stromberg and Griffin 1996).

A second measure that will increase grassland restoration success is the maintenance of management protocols until perennial grasses become well established (Stromberg et al. 2007). Methods that seek to overcome the relative seed limitation of perennial grasses during establishment, as well as space and light limitations, may aid in producing self-sustaining native grass communities (DiVittorio et al. 2007). Several recent experiments have found that multiple species of native perennial grasses were superior competitors to exotic annuals, or achieved a state of persistent coexistence with nonnative annuals under a range of climatic conditions and treatments when steps were taken to overcome barriers to establishment (Seabloom et al. 2003, Corbin and D'Antonio 2004, Suttle and Thomsen 2007).

In a number of earlier studies, nonnative annual grasses proved to be strong competitors against the native grass, *N. pulchra*, in a range of experiments in California's Central Valley (Dyer and Rice 1997, Dyer and Rice 1999). Overall, it appears unlikely that annual grasses are superior competitors to many well-established native perennial grasses, and that at least in some locations, the reverse may be true.

Overall, the results of this comparative study indicate that a native perennial grass restoration effort would yield multiple benefits for the state of California. Native perennial grasses are found to positively impact global climate relative to annual grasses by sequestering more carbon in the soil. They also support more diverse ecosystems than nonnative annuals, and may provide associated taxa with microclimatic conditions more favorable to their persistence during hotter and drier portions of the year. Therefore, efforts that seek to restore native grasses would serve both climate and biodiversity goals. In general, the expectation that perennial grass communities function in ways that promote climate change objectives relative to nonnative annuals are in good agreement with this study's findings and with an understanding of their fundamental differences in their perennial or annual life cycles.

LITERATURE CITED

Adair, E. C. and I. C. Burke. 2010. Plant phenology and life span influence soil pool dynamics: *Bromus tectorum* invasion of perennial C_3–C_4 grass communities. *Plant and Soil* 335(1–2):255–269.

Battles, J. J., P. Gonzalez, T. Robards, B. M. Collins, and D. S. Saah. 2014. *California Forest and Rangeland Greenhouse Gas Inventory Development*, CA Air Resources Board Agreement 10-778. California Air Resources Board, Sacramento, CA.

Chou, W. W., W. L. Silver, R. D. Jackson, A. W. Thompson, and B. Allen-Diaz. 2008. The sensitivity of annual grassland carbon cycling to the quantity and timing of rainfall. *Global Change Biology* 14(6):1382–1394.

Christian, J. M. and S. D. Wilson. 1999. Long-term ecosystem impacts of an introduced grass in the northern Great Plains. *Ecology* 80(7):2397–2407.

Coleman, H. M. and J. M. Levine. 2007. Mechanisms underlying the impacts of exotic annual grasses in a coastal California meadow. *Biological Invasions* 9(1):65–71.

Corbin, J. D. and C. M. D'Antonio. 2004. Competition between native perennial and exotic annual grasses: Implications for an historical invasion. *Ecology* 85(5):1273–1283.

Corbin, J. D. and C. M. D'Antonio. 2010. Not novel, just better: Competition between native and non-native plants in California grasslands that share species traits. *Plant Ecology* 209(1):71–81.

Corbin, J. D., M. A. Thomsen, T. E. Dawson, and C. M. D'Antonio. 2005. Summer water use by California coastal prairie grasses: Fog, drought, and community composition. *Oecologia* 145(4):511–521.

DeLonge, M. S., R. Ryals, and W. L. Silver. 2013. A lifecycle model to evaluate carbon sequestration potential and greenhouse gas dynamics of managed grasslands. *Ecosystems* 16:962–979.

DiVittorio, C. T., J. D. Corbin, and C. M. D'Antonio. 2007. Spatial and temporal patterns of seed dispersal: An important determinant of grassland invasion. *Ecological Applications* 17(2):311–316.

Dukes, J. S., N. R. Chiariello, E. E. Cleland, L. A Moore, M. R. Shaw, S. Thayer, T. Tobeck, H. A. Mooney, and C. B. Field. 2005. Responses of grassland production to single and multiple global environmental changes. *PLOS Biology* 3(10):1829–1837.

Dyer, A. R. and K. J. Rice. 1997. Intraspecific and diffuse competition: The response of *Nassella pulchra* in a California grassland. *Ecological Applications* 7(2):484–492.

Dyer, A. R. and K. J. Rice. 1999. Effects of competition on resource availability and growth of a California bunchgrass. *Ecology* 80(8):2697–2710.

Ehrenfeld, J. G. 2003. Effects of exotic plant invasions on soil nutrient cycling processes. *Ecosystems* 6(6):503–523.

EPA. 2009. *Inventory of U.S. Greenhouse Gas Emissions and Sinks: 1990-2007*. US EPA, Washington, DC.

Evans, N., A. Baierl, M. A. Semenov, P. Gladders, and B. D. L. Fitt. 2008. Range and severity of a plant disease increased by global warming. *Journal of the Royal Society Interface* 5(22):525–531.

Houghton, R. A. 1999. The annual net flux of carbon to the atmosphere from changes in land use 1850-1990. *Tellus Series B: Chemical and Physical Meteorology* 51(2):298–313.

House, J., V. Brovkin, R. Betts, R. Costanza, M. A. Silva Dias, E. Holland, C. Le Quéré, N. K. Phat, U. Riebesell, M. Scholes et al. 2006. Climate and air quality. In P. Kabat and S. Nishioka (eds), *Millennium Ecosystem Assessment 2006 – Current State and Trends. Findings of the Condition and Trends Working Group (Ecosystems and Human Well-Being)* 1. Island Press, Washington, DC. 355–390.

Hudiburg, T., B. Law, D. P. Turner, J. Campbell, D. Donato, and M. Duane. 2009. Carbon dynamics of Oregon and Northern California forests and potential land-based carbon storage. *Ecological Applications* 19(1):163–180.

Hudson, B. D. 1994. Soil organic matter and available water capacity. *Journal of Soil and Water Conservation* 49(2):189–194.

Jackson, L. E. and J. Roy. 1986. Growth patterns of Mediterranean annual and perennial grasses under simulated rainfall regimes of southern France and California. *Acta Oecologica* 7(2):191–212.

Jackson, L. E., R. B. Strauss, M. K. Firestone, and J. W. Bartolome. 1988. Plant and soil nitrogen dynamics in California annual grassland. *Plant and Soil* 110(1):9–17.

Koteen, L. 2002. Climate change, whitebark pine, and grizzly bears in the Greater Yellowstone Ecosystem. In S. H. Schneider and T. L. Root (eds), *Wildlife Responses to Climate Change: North American Case Studies*. Island Press, Washington, DC. 343–414.

Koteen, L. E., D. D. Baldocchi, and J. Harte. 2011. Invasion of non-native grasses causes a drop in soil carbon storage in California grasslands. *Environmental Research Letters* 6(4):044001.

Laude, H. M. 1953. The nature of summer dormancy in perennial grasses. *Botanical Gazette* 114(3):284–292.

Lauenroth, W. K., A. A. Wade, M. A. Williamson, B. E. Rose, S. Kumar, and D. P. Cariveau. 2006. Uncertainty in calculations of net primary production for grasslands. *Ecosystems* 9(5):843–851.

Liao, C. Z., R. H. Peng, Y. Luo, X. H. Zhou, X. W. Wu, C. M. Fang, J. K. Chen, and B. Li. 2008. Altered ecosystem carbon and nitrogen cycles by plant invasion: A meta-analysis. *New Phytologist* 177(3):706–714.

Ma, S., D. D. Baldocchi, L. Xu, and T. Hehn. 2007. Inter-annual variability in carbon dioxide exchange of an oak / grass savanna and open grassland in California. *Agricultural and Forest Meteorology* 147(3–4):157–171.

McClaugherty, C. A., J. Pastor, J. D. Aber, and J. M. Melillo. 1985. Forest litter decomposition in relation to soil nitrogen dynamics and litter quality. *Ecology* 66(1):266–275.

Pumpanen, J., H. Ilvesniemi, and P. Hari. 2003. A process-based model for predicting soil carbon dioxide efflux and concentration. *Soil Science Society of America Journal* 67(2):402–413.

Roy, B. A., S. Gusewell, and J. Harte. 2004. Response of plant pathogens and herbivores to a warming experiment. *Ecology* 85(9):2570–2581.

Ryals, R., M. Kaiser, M. S. Torn, A. A. Berheb, and W. L. Silvera. 2014. Impacts of organic matter amendments on carbon and nitrogen dynamics in grassland soils. *Soil Biology and Biochemistry* 68:52–61.

Ryan, M.G., J.M. Melillo, and A. Ricca. 1990. A comparison of methods for determining proximate carbon fractions of forest litter. *Canadian Journal of Forest Research* 20(2):166–171.

Scurlock, J.M.O., K. Johnson, and R.J. Olson. 2002. Estimating net primary productivity from grassland biomass dynamics measurements. *Global Change Biology* 8(8):736–753.

Schlesinger, W.H. 1997. *Biogeochemistry*, 2nd ed. Academic Press, San Diego, CA. 588 pp.

Seabloom, E.W., E.T. Borer, V.L. Boucher, R.S. Burton, K.L. Cottingham, L. Goldwasser, W.K. Gram, B.E. Kendall, and F. Micheli. 2003a. Competition, seed limitation, disturbance, and reestablishment of California native annual forbs. *Ecological Applications* 13(3):575–592.

Seabloom, E.W., W.S. Harpole, O.J. Reichman, and D. Tilman. 2003b. Invasion, competitive dominance, and resource use by exotic and native California grassland species. *Proceedings of the National Academy of Sciences of the United States of America* 100(23):13384–13389.

Soil Conservation Service. 1985. *Soil Survey of Marin County, California*. USDA, Sacramento, CA. 229 pp.

Stromberg, M.R. and J.R. Griffin. 1996. Long-term patterns in coastal California grasslands in relation to cultivation, gophers, and grazing. *Ecological Applications* 6(4):1189–1211.

Stromberg, M.R., P. Kephart, and V. Yadon. 2001. Composition, invasibility, and diversity in coastal California grasslands. *Madrono* 48(4):236–252.

Stromberg, M.R., C.M. D'Antonio, T.P. Young, J. Wirka, and P.R. Kephart. 2007. California grassland restoration. In M.R. Stromberg, J.D. Corbin, and C.M. D'Antonio (eds), *California Grasslands: Ecology and Management*. University of California Press, Berkeley, CA. 254–280.

Suttle, K.B. and M.A. Thomsen. 2007. Climate change and grassland restoration in California: Lessons from six years of rainfall manipulation in a north coast grassland. *Madrono* 54(3):225–233.

Verbeeck, H., R. Samson, F. Verdonck, and R. Lemeur. 2006. Parameter sensitivity and uncertainty of the forest carbon flux model FORUG: A Monte Carlo analysis. *Tree Physiology* 26(6):807–817.

Westerling, A. and B. Bryant. 2006. *Climate Change and Wildfire in and Around California: Fire Modeling and Loss Modeling*. California Climate Change Center, Sacramento, CA. 28 pp.

Wieder, R.K. and G.E. Lang. 1982. A critique of the analytical methods used in examining decomposition data obtained from litter bags. *Ecology* 63:1636–1642.

Wilsey, B.J. and H.W. Polley. 2006. Aboveground productivity and root-shoot allocation differ between native and introduced grass species. *Oecologia* 150(2):300–309.

Zhang, L., B.K. Wylie, L. Ji, T.G. Gilmanov, and L.L. Tieszen. 2010. Climate-driven interannual variability in net ecosystem exchange in the Northern Great Plains grasslands. *Rangeland Ecology & Management* 63(1):40–50.

Manager Comments

Laura Koteen
in conversation with
Mark Stromberg

Koteen: In my work, I identify restoration of native grasslands as an important component of responding to climate change in California. Can you comment on what you see as the main impediments to restoration of native grasslands? Ecological? Social? Economic?

Stromberg: Ecological impediments include climatic factors. If one is using seed, native grasses have the greatest chance of successful establishment, and are most competitive in a good year, which means a warm, wet year: a good warm, wet fall to get them well germinated and with sufficient growth, and a good wet spring to solidify their establishment. However, these years are kind of rare. Their chances of successful establishment and survival are also best along the coast regions, where the fog belt is active in summer, because it is wetter and cooler. It is unclear how climate change may affect restoration efforts vis-à-vis the establishment phase: whether it will increase the frequency of warm, wet years that enhance restoration success, or whether it will alter rainfall patterns in ways that hinder native grass establishment. It is also possible that climatic conditions will shift so that exotic grasses and / or other community actors are favored. For example, exotic perennial grasses have already greatly increased their presence in grass communities along the coast, and brush encroachment of both native and nonnative shrubs has increased in recent decades.

The biggest social impediment is the loss of a connection with nature. People don't have a good knowledge of California's natural history. They don't know that perennial bunch grasses were once the norm in many places (e.g., Berkeley) and that our grasslands are composed of largely nonnative grasses. They are not able to identify native from nonnative grasses by sight. People do notice that the flowers have been lost over time. However, they are not aware that spring flower shows are gone because the native grass community that supports them is gone, as well as the native processes, such as grazing or fire that maintain native grassland by reducing the pressure of invasive species.

From an economic perspective, grass restoration is expensive, especially over large areas. To keep it affordable, grass restoration should be confined to those areas where grasses are self-sustaining once well established (i.e., the coastal terraces, interior areas with deep soil, often subirrigated).

Koteen: In addition to possible carbon storage / climate benefits, what additional benefits do you see to native grass restoration?

Stromberg: I see many benefits to native grassland restoration, including connecting people back to nature and to California's natural history if the restoration builds in an educational component. I also see the potential for better water quality through reduced erosion, and wildlife benefits. In learning of the carbon / climate benefits of restoring grasslands from this research, I see the potential for using native grassland restoration to obtain funding through carbon offsets. There is a lot of root tissue stored below ground in native grasses. In dry years, a lot of that root carbon will become incorporated into the soil.

Koteen: Are there particular locations you would prioritize in terms of promoting native grassland restoration?

Stromberg: Different areas are important for different reasons. I would include the fog zone all along the undeveloped coastline of California, and in the near coastal region in protected natural areas, where there is a marine influence and inputs of moisture from fog. Not only does the climate support grasses in these locations, if the plant communities within the coastal terraces are restored, but they also have the potential to become really spectacularly diverse native areas. Of course, it is also unclear if climate change may impact the summer fog cycle in California, which provides the moisture inputs which allow native grass communities to flourish. The native coastal terrace grassland communities of California are among the top 1% of diverse grassland communi-

(continued)

ties worldwide. Good restoration could occur from San Diego County where protected areas remain and north from Santa Barbara up through Mendocino County.

I would also prioritize the San Francisco Bay Area, because there, one-third of the coastal areas are in protected lands, generally open spaces supported by the local communities. There is also a knowledge base and a cultural base that values preserving these open spaces and restoring native vegetation. Moreover, it's an urban area. In addition to having ecological value, grasses there have an educational value. If they are in the Bay Area, people will see them and learn about them, and may then appreciate that they are a rich part of California's natural heritage. Similarly, rooftop restorations, such as the California Academy of Science, are also good areas to prioritize, again because of the educational value of having them in such high impact places. Restoration efforts on rooftops and in heavily frequented areas also provide opportunities for people to learn about the carbon storage benefits associated with native grass restoration. The restoration of Crissy Field, near the Golden Gate Bridge, is another good example of an area where lots of people see and experience native grasses and are thus educated about them.

Agricultural areas can also be good candidates for restoration, such as areas of the Central Valley where the soil remains damp. This includes low-level swampy areas, locations of natural springs as well as areas that receive subsurface flows from irrigation runoff. At the farm scale, good places include strips of grassland in between crops. If placed between crops, such hedgerows could provide important habitat for native pollinators, and serve as pollination sources for crops. Roadsides, ditches, and slopes are also good locations for native grasses. The deep roots of native grasses hold the soil together and reduce erosion and reduce or eliminate annual road and ditch grading. Particularly in light of possible increases in storm intensity, which may accompany global climate change, the value of erosion reduction will increase.

Koteen: Coming back to carbon, to what extent do your management objectives incorporate carbon storage? Do you see opportunities to highlight the value of these functions and their dependence on management actions that promote native species with state agencies or funding sources?

Stromberg: The University of California Natural Reserve System, where I work, has not previously highlighted these functions, and the natural reserve system has primarily research and education functions. However, I do see the value in promoting carbon storage through native grassland restoration and have seen landowners in Arizona and California tout these benefits in other restoration efforts. There is also a project in the Dakotas for native grassland preservation supported by a grant obtained from Ducks Unlimited and conservation partners based on carbon storage potential of Midwestern native grasslands. In addition, I have been involved with efforts that explicitly recognize the carbon storage potential of oak woodlands and which attempt to educate private landowners about the potential revenue for management activities to increase carbon storage in oak woodlands through the Climate Action Reserve, a state program. Such revenues could potentially be available for grassland restoration as well.

Koteen: Has this chapter led you to consider any changes in objectives or methods for monitoring at the University of California Reserves?

Stromberg: I think it would be useful to set up long-term monitoring of carbon accumulation in soils that have been restored to native grassland. UC Reserves can be good locations for such monitoring and can be used to highlight ecosystem services provided by plant communities.

Koteen: Are there other ways you can point to that we should be thinking about climate change in reference to grassland management than the considerations suggested by this study?

Stromberg: I think carbon credits could be used for purchasing some of the areas of native grassland that remain and preserving them. I think any environmental impact review that is considering restoration of native grassland should have enhanced carbon storage as part of the standard text.

PART III

Perspectives for Framing Biological Impacts of Rapid Climate Change

CHAPTER 11

Evolutionary Conservation under Climate Change

Jason P. Sexton and Alden B. Griffith

Abstract. Faced with rapid climatic change, we can lower extinction probabilities and maximize the potential for species to track rapid climate change by facilitating evolutionary adaptation. Evolutionary-management alternatives range from intensive interventions in extreme cases (e.g., rescuing a species by breeding it with another species) to more subtle alternatives (e.g., conserving land to act as gene-flow corridors among populations). Resource managers may improve conservation results by applying evolutionary theory in their assessments of climate change risk, taking steps to maximize the adaptive potential of species of concern, and intervening when necessary, followed by careful monitoring of intervention outcomes. We discuss a variety of "evolutionary risk assessment" approaches, including identifying barriers to biological adaptation and evaluating the adaptive potential of populations and species, and make recommendations for operating within an evolutionarily minded framework in conservation biology.

Key Points

- Evolutionary adaptation is an important way living things cope with new or changing environments and is one means by which some species may avoid extinction under rapid climate change.
- Genetic variation among individuals is the means or currency by which evolutionary adaptation occurs and it can be measured, but it can also be in limited supply.
- It is critical that resource managers and conservationists maximize the adaptive capacity of species by incorporating evolutionarily minded approaches into their plans and actions.
- There is a wide variety of actions to help maximize evolutionary potential, including estimating genetic variation within populations, protecting gene-flow corridors, including assisted breeding or assisted-immigration actions, identifying and removing human-made gene-flow barriers, and limiting actions that reduce population size and genetic variation.

How humans can positively influence evolutionary processes to maintain diverse ecosystems is a relatively new scientific and management topic, yet it is a vital one. Evolution has produced the tremendous biodiversity that we now see rapidly disappearing through human-caused global change. Although humans cannot control evolution on a grand scale, we can take actions to protect and promote the evolutionary processes that create and maintain biodiversity; that is to say, we can manage to maximize adaptive genetic variation, which is the backbone of biological adaptation, and ultimately, biodiversity.

We know that species are already responding to the effects of human-caused climate change in several ways: Changing phenologies, shifting geographic ranges, and changing abundances (Root et al. 2003, Parmesan 2006, Anderegg and Root, this volume). Many of these shifts will be necessary for species to adjust to a rapidly changing climate, but may not always be fully realized due to constraints imposed by habitat availability and biological limits to migration and dispersal (Geber and Dawson 1993, Sexton et al. 2009). For those species that cannot move to suitable habitat, and have reached their limits of environmental tolerance, adaptation via evolutionary processes (see Rice and Emery 2003), or extinction, is the only option in the absence of human intervention (Ackerly 2003).

Rapid evolution to changing conditions, including human-induced conditions, is not a new concept (Hendry et al. 2010). In the 19th century, Charles Darwin observed rapid evolution in moths in response to industrial air pollution. Agriculturalists use breeding and genetic technologies to increase yields under various growth environments by changing the traits of domesticated plants and animals. For example, modern-day corn is the product of human-induced evolution from thousands of years of domestication, with rapid evolution occurring during the last half-century through intensive breeding and hybridization research (Moeller and Wang 2008). The theoretical and quantitative underpinnings of evolutionary processes are strong and often well understood (more so than many important ecological processes) due to the rich history of genetics research and recent advances (e.g., genomics). Nevertheless, many questions remain about which naturally occurring species can adapt quickly enough to the environmental changes imposed by climate change (Gienapp et al. 2008).

In general, a species tolerates multiple environments within its geographic range through population-based local adaptation, broad environmental tolerance, or both (Bradshaw 1965). Thus, if a species has an overall negative response to increasing temperatures, for example, the fate of that species may lie within the adaptive capacity of individual populations that are more tolerant of increased temperatures than other populations. If these populations are well connected, gene flow can assist populations in adapting to the new temperature regime throughout the species range (Geber and Dawson 1993, Aitken and Whitlock 2013).

The idea that evolutionary and ecological events can occur on similar timescales has gradually gained traction (Hendry et al. 2011, Sgrò et al. 2011). Adaptations to climate on a yearly basis have been observed through long-term studies (Grant and Grant 2002), and the past few decades have revealed a slew of studies illustrating rapid responses (Kinnison and Hairston 2007). Additionally, many cases of rapid evolution are observed from studying the evolutionary ecology of invasive species (Box 11.1), and these cases illustrate key factors that enable rapid evolution during climate change, namely high levels of adaptive genetic variation (Skelly et al. 2007). However, although much can and has been learned from the adaptive potential of invasive species (and the role of humans in influencing adaptive potentials), there are many unknowns about how to move forward to protect native species from climate change and how to effectively and ethically

BOX 11.1 · Rapid Evolution and Adaptive Lessons from Invasive Species

Invasive species give us clues as to how management can maximize evolutionary responses in native species (e.g. Griffith et al. 1989). The key is genetic variation, which can be enhanced through maintaining large populations, maintaining adaptive gene flow among populations, and perhaps through interpopulation (and interspecies) breeding to produce new, adaptive genetic combinations (Box Figure 11.1.1). Rapidly changing environments are creating new opportunities for biological invasions (including diseases) (Harvell et al. 2002). Whereas some biological invasions have colonized multiple environments through broad phenotypic plasticity (e.g., Williams et al. 1995), others have benefited from population-based differences, multiple introductions, and rapid evolution. Nowhere is the adaptive potential of biological invasions more apparent than when we attempt to control or eradicate them (Carroll 2011). Many organisms have evolved resistance to chemical control (e.g., herbicides and vaccines), and resistance often evolves in less than 20 years (Palumbi 2001).

Genetic variation available for rapid evolution has been documented in many invasive species (Reznick and Ghalambor 2001, Maron et al. 2004). For example, Sexton et al. (2002) found potentially adaptive differences in root investment and plasticity between invaded northern and southern populations of saltcedar (*Tamarix* spp.) in the United States, and there appears to be an adaptive cline in wood tissue resistance to frost damage (Friedman et al. 2008). Saltcedar has had multiple introductions, and its high genetic variation has been enhanced through hybridization between at least two species in its invaded range (Gaskin and Schaal 2002). Leger and Rice (2007) found that

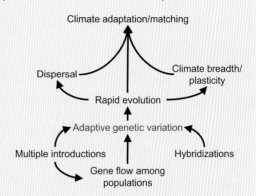

BOX FIGURE 11.1.1: Illustration of factors that can increase adaptive genetic variation, which can then increase climate adaptation through rapid evolution, increased dispersal ability, and increased climate tolerance (breadth or plasticity).

invasive populations of California poppies (*Eschscholzia californica*) in Chile have evolved trait differences along climate gradients in a similar fashion to native California populations since their introduction 150 years ago. The fly, *Drosophila subobscura*, has evolved a strikingly similar cline in wing size within 20 years in introduced populations in North America—but in different parts of the wing—to clines found in the Old World (Huey et al. 2000). Additionally, several invasions have experienced *increased* genetic variability within their introduced ranges (likely due to multiple introductions and genetic reshuffling), which may have strong implications for adaptive potential (Dlugosch and Parker 2008). A worst-case evolutionary scenario under climate change would involve decreasing adaptive potentials in native species and increasing them in biological invasions.

maximize adaptive potential. Resource managers can facilitate adaptive processes (Ashley et al. 2003, Sgrò et al. 2011), indirectly or *extrinsically* (see Klausmeyer and Shaw 2009), by maximizing the potential for species to adapt (e.g., by protecting or providing restored habitats that are likely to be viable under future climate condition) or, directly, through efforts to improve the *intrinsic* adaptive potential of spe-

cies, whether by intentional breeding or by movement of individuals to increase genetic variation to respond to natural selection. In both cases, species can and should be viewed as potential sets of populations that can respond differently to different climatic conditions, and not singly, as entities that either fit or do not fit within particular climate envelopes (see Holmstrom et al. 2010, for an example of the consequences of not viewing species as a set of distinct populations). Ultimately, it is the fitness of individuals in particular environments that will affect population growth rates and the success of a given species across its range (Kinnison and Hairston 2007). Evolutionarily minded management can increase the likelihood that populations are able to evolve and adapt to rapidly changing environments, and can minimize the risk of unintended consequences (e.g., Holmstrom et al. 2010).

In this chapter, we review the basic information necessary for resource managers and conservation biologists to apply evolutionary theory and principles to rapid climate change and other human-induced disturbances. We focus here on single-species or single-taxon conservation responses, but evolutionary management can and should be used in community and ecosystem contexts as well (see Mace and Purvis 2008). Evolutionary management that forestalls extinctions under climate change and habitat degradation is a new field of research combining conservation genetics and evolutionary biology. We review the following: (1) Constraints on and barriers to evolutionary adaptation to climate change, (2) how to evaluate the adaptive potential of populations and species of concern, (3) possible actions to maximize adaptation, and (4) management implications and tools. Climate change involves complex changes to a variety of environmental variables (e.g., temperature, precipitation, atmospheric carbon dioxide concentrations). For simplicity, we will discuss adaptive examples in the context of warming climates, but it is important to recognize that changes in response to different climate variables in different ecosystems will have diverse evolutionary consequences.

Constraints to Evolutionary Adaptations to Climate Change

Demographic constraints to evolution arise mostly from small population sizes among other risk factors (Holt 2003) (see Box 11.2).

Small populations tend to be more limited in their genetic variation and thus have fewer genetic resources upon which natural selection may act (e.g., Kelly et al. 2011). For small populations to persist in the face of climate change, they may need to either be moved to appropriate climates or be rescued by other populations through the process of migration or gene flow (Sexton et al. 2011, Aitken and whitlock 2013).

The adaptive potential to respond to climate change may be present within the species as a whole, but it may be distributed across many populations that vary in isolation from one another. Ironically, this isolation among populations may have historically allowed the genetic variation in climate-related traits to sort and build over the species' history. For example, a species with populations distributed across a wide range of environments may have adaptations to cold tolerance at one end of its geographic range, and adaptations to heat tolerance at the other. If individuals from different environments were to breed, the species as a whole might utilize a greater amount of their adaptive potential in response to climate. Yet there is also the danger of increasing the genetic load (i.e., maladaptation) when mixing populations (Aitken and Whitlock 2013). Therefore, increasing adaptive potential by using populations better adapted to predicted future environments (discussed below) is ideal. Thinking holistically in this vein about how species' populations are distributed along climatic gradients can help in designing adaptive, climate change strategies.

Even when the genetic variation required to adapt rapidly is present within a population, the arrangement of that variation may preclude rapid evolution. The trait combinations we see in individuals (e.g., delayed flowering and slow leaf development) can often be considered as "packages" that have proven successful and been selected for over time in certain environments. Owing to genetic correlations among traits, it may be very difficult to produce individuals with the most favorable combinations of trait values in time to track climate evolution-

BOX 11.2 · Assessing Risk Factors to Climate Change Adaptation

Many factors can reduce or bring "risk" to the process of climate adaptation. Box Table 11.2.1 lists some population- and species-level factors that can forestall climate adaptation and uses California populations of Chinook salmon (*Oncorhynchus tshawytscha*) as an example. Presence of a risk factor (represented by "X") is determined from an assessment of *average* population estimates based on biological opinion or from available data. As a species, Chinook salmon may have few evolutionary risk factors and may be able to adapt to climate change with monitoring and intervention. On the other hand, California populations of Chinook salmon are a small subset of Chinook populations and occur at the southern range boundary of the species range. Rapid adaptation in California pop-

ulations may not be possible without genetic rescue or other management actions. We note that genetic rescue is a complex issue (Tallmon et al. 2004) and is not equivalent to hatchery supplementation, which may actually decrease genetic diversity and reproductive potential of supplemented populations (Dylan 2008).

Such coarse tools can be used to begin a dialogue with multiple stakeholders and can lead to more sophisticated models predicting population responses. This cursory assessment assumes that the species must adapt in place and cannot migrate to suitable habitats in future climates. Such tools can be catered to specific systems to incorporate a wide range of factors, including ecological, geographic, and human impacts.

BOX TABLE 11.2.1
Risk factors for the Chinook salmon (Oncorhynchus tshawytscha) *at both the population and species level.*

	Whole range	California populations	Management action
Population-level risk factors			
Small populations or habitat fragmentation		X	Promote breeding program while maintaining or increasing genetic diversity; create migration corridors
Coadapted genes	unknown	unknown	
Isolated populations		X	Assess populations for unique ecological traits
Low within-population genetic variation		X	Introduce individuals from nearby populations; stop hatchery supplementation
Species-level risk factors			
Low reproductive turnover			
Inbreeding			
Poor dispersal			
Small climate range		X	Assess performance of individuals in predicted future environments
Low reproductive rate	X	X	
Tally of risk factors	1	5	

arily (Davis et al. 2005). Etterson and Shaw (2001) found that populations of the prairie annual plant *Chamaecrista fasciculata* had the genetic variation to produce individuals capable of tracking potential future climates, but that correlations among important traits could forestall this process. For example, a positive correlation between leaf number and leaf thickness suggests that natural selection cannot easily respond to higher temperatures, which favor individuals with both more and thinner leaves.

Mating system (asexual, selfing, mixed mating, etc.) and life history (short-lived vs. long-lived) may affect a population's adaptive potential over short timescales. The type of mating system is generally fixed within a species, but it can also vary so that some populations differ from one another. When individuals only rarely interbreed, it can take longer for adaptations to arise because recombination is occurring less widely, and thus producing novel combinations at a slower rate. Hence, less outcrossing populations or species may have slower evolutionary rates of response to selection and may require higher levels of human intervention (e.g., assisted breeding) to adapt. Similarly, organisms that reproduce less often and in fewer numbers may have fewer opportunities to produce large numbers of genetic variants that may be successful in novel or rapidly changing environments.

Evolution may also be constrained if behavioral responses become maladaptive due to a decoupling of perceived cues and actual environmental conditions / fitness. Climate change and other human-induced environmental impacts are likely increasing the prevalence of such "evolutionary traps," which may limit or alter the evolutionary potential of certain species (Robertson et al. 2013).

The variety of constraints to evolutionary processes does not necessarily represent constraints for management actions (Lankau et al. 2011). On the contrary, recognizing those constraints may often be the impetus for action. Knowledge of how adaptive potential is con-strained increases the chance for successful management and reduces the risk of unintended consequences. The examples of evolutionary constraints discussed above—small population sizes, distribution of genetic variation, trait correlations, mating system, life history, behavioral responses—all inform our ability to incorporate evolutionary processes in management and serve as guides for applied research.

Evaluating Future Responses and Adaptive Potential

The vulnerability to, and hence consequences of, climate change will differ among species. Given life history information, one can assess the feasibility of migration to, and the likely quality of, potentially appropriate habitats. For example, as climatic conditions change over time, the most favorable climate for a population or species could occur within human-developed areas, which otherwise could be inhospitable to a species. When these preferred conditions are projected to occur in future locations that are too far away for individuals to migrate to (or too degraded to be habitable), the potential for climate tolerance and adaptation in current locations should be estimated. This process begins with identifying which traits and trait values will be most critical as the climate changes, and then determining adaptive potentials of these traits in various populations.

The adaptive potential to respond to climate change must be evaluated on two levels: Population-level factors (e.g., demographic and genetic constraints discussed above, connectedness, etc.) and species-level factors (e.g., life history characteristics such as mating system and generation time). These are not mutually exclusive. Many factors vary by species *and* population (e.g., morphological traits, type of mating system) and we suggest that taxa be evaluated from both of the following viewpoints: What is the potential of a species to adapt to climate change given its populations' characteristics?

What is the potential of a population to evolve rapidly given its species-level life history?

However, how does one actually evaluate the adaptive potential of a species or population? Adaptation is the product of natural selection acting on heritable genetic variation. Therefore, in order to fully evaluate adaptive potential, we must determine which traits are important under climate change, measure the amount of genetic variation distributed within and among populations, and estimate the strength of selection due to climate change (how much of an advantage to survival and reproduction certain trait values convey) while recognizing the constraints and barriers discussed above. Although this can be time-consuming, we hope to demonstrate that even a partial or basic knowledge of a species' or population's adaptive potential can be very informative from a management standpoint. Additionally, information from each of the above steps can help guide or necessitate investigation of other steps.

Identifying Climate-Related Trait Variation

Variation in climate-related traits is often best explored across environmental gradients. Ecotypes, and adaptations to associated climates, have long been demonstrated in plants (Turesson 1925, Clausen et al. 1940) and animals (Mayr 1947), and the process of ecotypic differentiation can have major effects on life history evolution (Rezende et al. 2004, Bradshaw and Holzapfel 2007). Climate-related trait variation across environmental gradients is often queued by photoperiod differences related to latitude (Bradshaw and Holzapfel 2007). For this reason, gene flow from warmer latitudes may not realize immediate benefits in rapidly warming climates if the newly derived population is maladapted to the location's photoperiod (Bradshaw and Holzapfel 2006, Visser 2008). Therefore, one should strike a balance between combining climate-related genes in new, adaptive ways and minimizing a maladaptive genetic load.

Flexibility is a big advantage in rapidly changing climates, and individuals that can successfully inhabit a wide range of habitats will be favored. For simplicity, we refer to individuals possessing broad environmental tolerance or high phenotypic plasticity (Bradshaw 1965, Ghalambor et al. 2007) as "generalists." Generalists are favored in environments that shift from year to year (Levins 1968). Thus, some populations of the same species may have more generalist characteristics than others, and there can be a clear genetic basis to these differences (Via et al. 1995, Sultan 2001). In fact, generalist performance is not fixed, but can evolve as a trait itself, and genetic variation for plasticity among important climate traits can and should be evaluated when possible (Via et al. 1995), especially with respect to its role in climate change responses (Hendry et al. 2008). Richards et al. (2006) discuss phenotypic plasticity in an invasion biology context and suggest ways to evaluate its role in colonizing new environments. A *relatively* quick assay for generalist genotypes is to raise individuals in two environments that approach both ends of a species' range for a particular environmental variable (e.g., temperature, precipitation). Generalist genotypes will be those that perform relatively well in both environments.

Similar to the advantages conferred by a generalist genotype, an organism's dispersal ability, which can involve many heritable traits, can be of high importance in variable or shifting environments (Johnson and Gaines 1990). Darling et al. (2008) observed greater dispersal ability—via larger-winged fruit—in peripheral populations of coastal dune sand verbena (*Abronia umbellata*) in California and Oregon. They posit that since range margins can shift back and forth from year to year, heightened dispersal is favored in these populations. The legs of cane toad (*Bufo marinus*) individuals at the invasive front of Queensland populations were found to be longer, giving them greater dispersal ability (Phillips et al. 2006). Dispersal ability can also be important in the context of disturbances such as fire and dieback from disease, which are anticipated to increase due to climate change.

Estimating Adaptive Potential

There are two main ways to assess the genetic variation of a population relevant to climate change. One approach is to quantify and monitor the amount of overall genetic variation within or among populations using variation in DNA, without respect to any particular trait (this is often referred to as "neutral genetic variation") (see Schwartz et al. 2007). This approach has a long history in conservation biology and roughly represents a bet-hedging strategy. In essence, it is assumed that the more overall genetic variability within a population, the more likely the population will avoid problems associated with detrimental genes and inbreeding depression, and the more likely it will be able to respond to natural selection. Additionally, if some populations are known to contain high levels of overall genetic diversity, they may represent good candidates for sources of genetic variation in managed breeding programs. Grivet et al. (2008) present a good example of this approach, where molecular variation is coupled with spatial data to identify potentially important regions (i.e., highly diverse areas with little protection) of the species range of California valley oak. Table 11.1 lists case studies where molecular data have been used to infer population processes that can help inform conservation decisions. Genetic-marker data may also be used to estimate the connectedness and conservation status of natural populations in the face of global change (Schwartz et al. 2007, Lawrence et al. 2008). Although neutral genetic divergence (based on molecular techniques) does tend to positively correlate with variation in actual phenotypic traits, it is also important to note that for many species, the relationship is weak or absent and thus may grant little predictive power (McKay and Latta 2002, Leinonen et al. 2008). Nevertheless, genetic diversity of molecular markers is often used to assess the importance of populations for economic value (crop plants, etc.), and Jump et al. (2009) argue that this "option value" should also be applied for its potential conservation value in wild species.

A second approach, and a more direct one, is to examine genetically based variability in important climate-related phenotypic traits (i.e., physical traits like body size, time to development, and leaf shape). In order to maximize adaptive potential in the successful management of populations, there must be sufficient genetic variation in traits of interest. Controlled conditions (e.g., common garden experiments in plants and laboratory trials in animals), in which individuals from different locations are grown together within similar conditions, have long been used to minimize environmentally caused variation in phenotypes to expose underlying heritable variation. [See Kellermann et al. 2009 for an example of estimating genetic variation in climate-related traits among many species of fruit flies (*Drosophila* spp.).] Genetic variation can also be measured in the field for organisms not suited to labs or gardens (see Conner and Hartl 2004). If these experiments are conducted in more than one environment, they can also help determine the degree of phenotypic plasticity and reveal generalist genotypes. This combined approach assesses both genetic variation and the potential need for management in the first place; that is to say, perhaps high plasticity will be sufficient to tolerate environmental change for some species. Nevertheless, a recent experiment revealed population responses to climate change are lagging in the mouse-ear cress, *Arabidopsis thaliana* (Wilczek et al. 2014).

Beyond estimates of genetic variability, it is important to understand how this variability responds to selection. There are two main techniques for estimating evolutionary responses (or response rate) of populations to predicted future environments: (1) *Retrospective temporal analyses,* where changes in populations are correlated with observed changes in the environment, and (2) *selection experiments,* where experimental populations are exposed to changing or target conditions for several generations and the resulting changes to the population are measured (see Box 1 in Reusch and Wood 2007). Recent concern over the evolutionary consequences of cli-

TABLE 11.1

Case studies where molecular marker data have been used to infer population processes relevant to climate change

California Tiger Salamander (*Ambystoma californiense*)	Desert Bighorn Sheep (*Ovis canadensis nelsoni*)	Joshua Tree (*Yucca brevifolia*)	Shasta Crayfish (*Pacifastacus fortis*)
		Background	
• Three distinct populations in California protected under the U.S. Endangered Species Act. • Climate change may alter the magnitude and timing of vernal pools, which are critical for breeding.	• Listed in 1998 as an endangered species under the U.S. Endangered Species Act. • The sheep are restricted to montane habitat, which is surrounded by lower-elevation desert. Warming temperatures reduces the size of these "sky islands."	• This desert tree species has been present in southern California since the late Pleistocene.[9] • Range change projection model run with input of CO_2 doubling predicts major range contractions, including the extirpation of Joshua Tree National Park populations.[10]	• Listed in 1988 as an endangered species under the U.S. Endangered Species Act. • Dependent upon lava cobble and boulder substrate and cool, freshwater springs.[13] • Threats from the invasive signal crayfish (*Pacifastacus leniusculus*), other nonnative aquatic species, and habitat disturbance.[13]
		Genetics	
• Neutral genetic analysis shows strong genetic differentiation across space.[1] • Endangered and isolated subpopulations are particularly genetically differentiated.[1] • Substantial genetic variation also found at smaller scales among breeding ponds.[2]	• Ewes form home-range groups that are demographically and genetically distinct.[4,5] • Lower-elevation populations have lower-genetic diversity, whereas high-elevation regions act as "reservoirs of genetic diversity."[6]	• Species range is divided among morphologically distinct populations, apparently driven by pollinator coevolution.[11] • Geographic structure among regions (east, west, and north) revealed by neutral molecular markers.[12]	• Genetic analysis revealed significant structure between geographic locations that are likely magnified due to hydrologic impoundments.[14] • Greatest genetic diversity found where Shasta and signal crayfish have been sympatric for relatively long periods.[14]
		Migration and gene flow	
• Grassland and chaparral provide dispersal corridors, while oak woodland creates a costly barrier to movement, and thus gene flow.[2]	• Subpopulations may be part of a larger metapopulation, but local extinctions have greatly outnumbered natural recolonizations.[7]	• Gene flow appears to be fairly unrestricted among populations despite the above patterns.[12]	• Large stretches of unsuitable habitat and water diversions interrupt the potential for gene flow between most areas of occupied habitat.[14-15]

(Continued)

TABLE 11.1

(Continued)

California Tiger Salamander (*Ambystoma californiense*)	Desert Bighorn Sheep (*Ovis canadensis nelsoni*)	Joshua Tree (*Yucca brevifolia*)	Shasta Crayfish (*Pacifastacus fortis*)
Constraints and consequences			
• Hybridization with the invasive barred tiger salamander increases immediate fitness, but reduces native genetic diversity and may limit adaptive potential during periods of selection.[3]	• Road networks and other anthropogenic barriers have been shown to greatly reduce gene flow among subpopulations. This has already caused substantial drops in genetic diversity within only a few decades.[8]	• Seeds lack rapid dispersal mechanisms, thus colonization of new areas may be very slow. • Pollen-mediated gene flow is determined strictly by specialized mutualists (yucca moths).[12]	• Long generation time, low fecundity, competition, and habitat degradation are threatening the survival of the species.[15] • Removal of invasive crayfish and habitat restoration is ongoing and helping to conserve remaining populations.[15]

[1]Shaffer et al. 2004; [2]Wang et al. 2009; [3]Fitzpatrick and Shaffer 2007; [4]Rubin et al. 1998; [5]Boyce et al. 1999; [6]Epps et al. 2006; [7]Epps et al. 2004; [8]Epps et al. 2005; [9]Holmgren and Betancourt 2008; [10]Cole et al. 2011 ; [11]Godsoe et al. 2008; [12]Smith et al. 2008; [13]USFWS 1998; [14]Petersen et al., unpublished data; [15]Spring Rivers 2009.

mate change has driven many such studies. For example, evolved differences in reproductive timing were detected in field mustard (*Brassica rapa*) grown from seeds collected before and after a five-year drought in southern California (Franks et al. 2007). Similarly, recent adaptive genetic shifts in the timing of reproduction and development in response to rapid climate change have been detected in animals, including insects, birds, and mammals (Bradshaw and Holzapfel 2006). To maximize predictive power of population-response estimates, the future environment should be estimated and many individuals should be sampled, observed, or manipulated for several generations. Nevertheless, much can also be learned from single-generation observations of natural selection in rapidly changing environments. A simple understanding of a species' life history (e.g., generation times, self-fertilization rates) or the results from studies on analogous organisms may inform the likelihood of a response to selection.

Assessing gene flow potential across climatic gradients is useful for understanding adaptive potential, especially if there are populations assumed to be favored in future climates. Historical and contemporary gene flow can be estimated through the use of molecular genetic markers (see examples in Excoffier and Heckel 2006). Alternatively, estimates of gene flow potential can be made with basic observations of seasonal timing of reproduction coupled with life history information. For example, populations that are well connected spatially, overlap in reproductive timing, and can disperse easily have potentially high gene flow.

Utilizing Evolutionary Potential—Goals and Potential Actions

Adaptive potential often means the availability of adaptive genetic variation within a species. In the worst-case scenario, we are forced to attempt to manipulate evolutionary outcomes to save a species. In the best-case scenario, we are preserving evolutionary processes and allowing for adaptive change to occur on its own (Latta 2008). In this section, we progress from evolutionarily minded practices that vary between low and high levels of intervention. These measures vary from regional, ecosystem-level responses to population- and individual-level actions. To maximize chances for success, researchers should initiate a working relationship with resource managers, and managers should seek input from researchers (Richardson et al. 2009).

Prioritizing Populations

Management plans focused on preserving evolutionary processes would target large, genetically diverse populations as adaptive sources, but would also consider habitat quality and availability (Lande 1988). These may be useful in safeguarding unique populations at risk of going extinct, such as populations that have been historically isolated. Populations showing evidence of having unique properties (either morphological or from molecular markers) may require prioritization and special protections (Lesica and Allendorf 1995, Crandall et al. 2000). These should also include portions of the species range that best represent future climates, such as relatively warm areas (lower latitudes and elevations). Warm-adapted populations may have useful genetic attributes for adapting to future climates, but they may also be the first to be extirpated (as they may already be living near their physiological limits) depending on the ability of individuals to migrate (Figure 11.1; but see Crimmins et al. 2011 for how distributions along gradients may shift counterintuitively in response to climate change). Thus, it may be important to conserve seeds, eggs, or individuals from ecologically extreme areas of the range to ensure their availability (see Havens et al. 2006 for seed-banking examples). Maintaining natural habitat linkages between neighboring populations could also be prioritized, especially across ecological and geophysical gradients and transitions that support and have promoted diversification (Anderson and Ferree 2010). If contemporary and historical genetic and ecological data are available, populations can be defined in a more

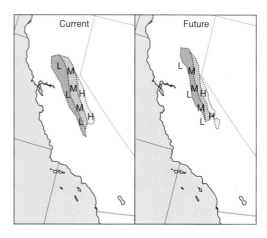

FIGURE 11.1: Hypothetical example of how climate zones with locally adapted genotypes may shift in the future. In the current scenario, there are three climate belts with three adapted genotypes (occupying differing elevations of the Sierra Nevada mountains), such that individuals with the "L" (low-elevation) genotype are favored in the warmest climate zone and individuals with the "H" (high-elevation) genotype are favored in the coolest zone. Without migration, genotypes and climate zones are no longer matched in the future scenario. L can withstand the warmest temperatures and could lend useful genes to "M" (middle-elevation), but many of the L populations might go extinct because they are beyond their physiological limits in future scenarios. New areas are potentially suitable for the species in future climates, but colonization to these areas may be slower than rapid climate change permits. Some new climate zones are effectively islands that cannot be easily reached through natural migration.

holistic context (see Crandall et al. 2000). Overall, it is likely that trade-offs between preserving unique biological units (e.g., populations) and maximizing climate adaptive potential will exist. See Carroll (2011) for a perspective on incorporating eco-evolutionary management in rapidly changing ecosystems.

Assisting Gene Flow and Migration

During climate shifts, selection pressures will result in local differences in individual survival and reproduction, but these same pressures can also cause a replacement of local individuals with migrant ones (Davis et al. 2005). This process is not all-or-nothing. Complete replacement of local individuals by immigrating ones is one

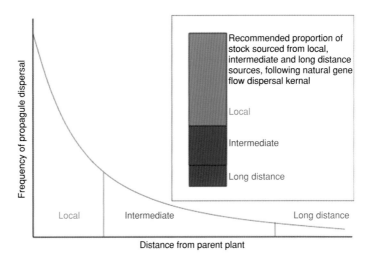

FIGURE 11.2: Potential stocking or seeding scheme to maximize adaptive potential at a given population under environmental or habitat change. The stocking proportions are intended to mimic the natural gene flow proportions a well-connected population might encounter and could be used to restore a population, to bolster depauperate populations, or for managed translocation outside the current range. Source: Figure taken from Sgrò et al. (2011).

possible outcome, while the alternative is a mixing of genotypes. The latter case may be more favorable since climate, while important, is not the only driver of biological distributions. Local individuals may perform poorly in a rapidly shifting climate, but they may possess special adaptations key to their existence in a particular place. For example, individuals may be locally adapted to a particular soil type or predation regime. Immigrants may have a favored climate tolerance, but may be unable to perform well with other novel conditions. Hence, new combinations—produced through sexual reproduction—may allow for successful climate matching while retaining local advantages (adaptations) (Kinnison and Hairston 2007). Sgrò et al. (2011) present an example for how one might stock or seed individuals from varying distances with the goal of maximizing adaptive potential under climate change (Figure 11.2).

Given the highly modified nature of many modern landscapes, management actions that promote dispersal of individuals will often be necessary to help species respond to environmental stress. These actions can be indirect / passive (e.g., removing barriers to move-ment, creating corridors) or direct / active (e.g., managed relocation). Using corridors to encourage movement is a widely accepted practice in conservation biology, but it is undervalued for its potential to facilitate rapid adaptation within species by increasing gene flow (although see Loss et al. 2011, Groves et al. 2012). The theory is the same: Both individuals and their genes need to be able to freely move to preferred sites. In the context of climate change and evolutionary adaptation, this means maximizing connectivity among populations and across climatic gradients. In this vein, *evolutionary processes* are being conserved, not just individuals or populations (Crandall et al. 2000, Latta 2008).

When it is clear that there is no gene flow among populations and population declines can be attributed to shifting climates, managed relocation may be considered (see Richardson et al. 2009). We are unaware of programs that have yet used managed relocation to address climate change concerns, but there are examples of matings between populations being used for conservation purposes. Tallmon et al. (2004) highlight several examples where genetic rescue using a small number of immigrants has proven successful in the conservation of small populations (e.g., prairie chicken, Florida panther). Sexton et al. (2011) showed that experimental gene flow between isolated populations occurring along the same warm, low-elevation range limit of the California Sierra Nevada endemic plant, *Mimulus laciniatus*, increased fitness of individuals at the range limit. In another example, individual tidepool copepod, *Tigriopus californicus*, populations have limited potential for adaptation to heat stress (Kelly et al. 2011); however, experimental crosses between heat-tolerant populations further increased tolerance to heat stress (Pereira et al. 2014). Thus, gene flow among populations occupying similarly stressful environments may provide benefits in rapidly changing environments while simultaneously reducing genetic load. Highly endangered populations of the mountain pygmy possum (*Burramys parvus*) in alpine Australia (Mitrovski et al. 2008) have recently produced interbred individuals from introductions from distant mountains, which may improve the target population's adaptive capacity (A.R. Weeks, pers. comm.). Important questions remain: How many individuals are necessary to assist adaptive evolution and where should they come from? What is the balance between the benefits of novel gene introductions and the cost of altering local adaptive genetic combinations? As discussed above, source populations should consist of individuals with traits conferring an advantage in a new climate regime. Nevertheless, as the rate of evolution depends on many factors (e.g., strength of selection, genetic diversity of locals plus migrants), it is difficult to know how many migrants will be required. When managed gene flow or relocation seems appropriate, we suggest using initially low numbers of individuals followed by long-term monitoring.

Population Restoration

If local populations are extirpated due to climate change, there is little chance that restoring populations using original, local individuals (if they are somehow available from seed storage or captive populations, etc.) will be successful. Climate-matching (as close as possible) candidate populations with the geographic area to be restored is a sensible strategy, as well as introducing a mix of individuals from varying ecological or "adaptive" distances to ensure adequate levels of genetic variation (see Sgrò et al. 2011). Montalvo and Ellstrand (2000) found that greater fitness of the plant *Lotus scoparius* in restored southern Californian coastal sage scrub populations was most highly correlated with greater genetic and environmental similarity to source populations. However, molecular markers may not be reliable for estimating the suitability of populations for restoration. Knapp and Rice (1998) found that the source environment of populations was the best predictor of successful matching of phenotypic trait values with restoration sites. Managers should seek as many sources of information as possible to inform restoration transplant decisions, but there is no substitute for performance-based information.

Captive Breeding and Species Hybridization

Interventions such as captive breeding can be used to save species or populations, but should only be employed when there is a reasonable chance for success and when there is no other alternative (e.g., captive breeding in the California condor, a species that had reached a population low of 22 individuals). A few important general points are worth considering. Microevolution (genetic changes within populations) is always occurring. Hence, populations will respond evolutionarily (and possibly behaviorally) to the captive environment, often to the detriment of reintroduced wild populations. Frankham (2008) describes how this process can happen and advises that population-based, captivity-program replicates be used and that individuals be kept in captivity for as few generations as possible to avoid the accumulation of harmful genes (deleterious alleles).

Species hybridization is a means of acquiring new adaptations (see Box 11.1 on rapid evolution in invasive species) and is a natural mechanism for speciation in certain groups (Rieseberg 2006), yet it has not to our knowledge been employed as a tool for evolutionary adaptation to climate change, but see Tallmon et al. (2004). While we do not advocate this as a mainstream conservation strategy, it is worth noting that hybridization could be a preferred, last-ditch alternative to the extinction of endemic species that cannot survive rapid climate change (see Fitzpatrick and Shaffer 2007). Hybridization with closely related, climate-tolerant species might be possible for certain species; however, as is true for managed relocation, philosophical and ethical factors must certainly be considered (Richardson et al. 2009).

Incorporating New Management Perspectives and Tools

For which cases do we need to make evolutionary management a priority in response to climate change? Until recently, management actions that explicitly consider evolutionary processes have typically been reserved for extreme cases, such as captive breeding programs or the introduction of individuals from distant populations (Tallmon et al. 2004). For such cases, genetic diversity and/or structure have become critically impaired due to anthropogenic influences, and any attempt to conserve or restore healthy populations depends on our knowledge of evolutionary biology. In the case of anthropogenic climate change, we can anticipate widespread changes to selection pressures and evolutionary processes, and we have the chance to incorporate evolutionary concerns at an early management stage. As a first step, surveys of the genetic diversity of populations across broad ranges would help prioritize the need for action and provide baseline data from which to draw in the future. While this "snapshot" approach has limitations (e.g., neutral genetic diversity not necessarily

corresponding to diversity of climate-related traits), it provides a coarse view of which species and populations may be limited in their evolutionary response. This type of dataset, combined with knowledge of life histories and population sizes, could be very powerful from a management standpoint. Furthermore, resources for incorporating such evolutionary concepts in policy and management are becoming more widely available (Hendry et al. 2010).

The magnitude and duration of human-caused climate change depends largely on our current and future behaviors (greenhouse gas emissions, land use decisions, etc.), which can only be estimated as scenarios. Hence, we do not know how long or large of a climate shift to manage for. Millar et al. (2007) have suggested that management strategies targeting the ecological effects of climate change can be categorized as either *resistance* (managing systems to avoid change), *resilience* (managing systems for an eventual return to previous states), or *response* (managing systems to accommodate change). For many species, maximizing adaptive potential may improve the efficacy of all three strategies. From a management standpoint, evolution is an extremely powerful and pervasive tool by which populations cope with environmental change. However, the process of natural selection is entirely reactive and never proactive. We have the unique position of affecting evolutionary processes in order to anticipate future selection regimes. A rapid evolutionary response requires strong selection pressure and enough genetic variability to confer advantageous (or at least tolerant) phenotypes—climate change will provide the former and humans should manage for the latter.

One central question that remains is *how much do we need to know in order to incorporate evolutionary processes into climate change management?* The degree to which we can positively influence evolutionary responses is a new scientific and management frontier, but there is growing recognition for the need to incorporate evolutionary understanding into management plans designed to accommodate change (Hoffmann and Sgrò 2011, Groves et al. 2012, Stein et al. 2013). At the very least, we should incorporate knowledge of evolutionary processes into decision-making, and at most directly influence evolutionary trajectories for positive outcomes. Most importantly, well-intended management actions could fail or have long-lasting negative evolutionary consequences if evolutionary concerns are not taken into account (Ashley et al. 2003, Mace and Purvis 2008). Through climate change, humans are imposing a tremendous force in altering natural selection pressures, and the composition of the earth's biotas will likely shift dramatically if current trends continue. In trying to preserve biodiversity, we must be sure not to lose sight of the evolutionary processes that created it in the first place.

ACKNOWLEDGMENTS

We thank David Ackerly, Jessica Blois, Kim Hall, Mark Herzog, Rebecca Quinones, Kevin Rice, Terry Root, and Blake Suttle for helpful comments. Brad Shaffer, Jessica Petersen, and Ian Wang generously provided information about the population genetics of their respective study systems. J.P.S. was supported by the NSF Responding to Rapid Environmental Change (REACH) Integrative Graduate Education and Research Traineeship (IGERT) at UC Davis (NSF-DGE 0801430) and NSF award no. DEB 1003009 during the writing of this chapter.

LITERATURE CITED

Ackerly, D. D. 2003. Community assembly, niche conservatism, and adaptive evolution in changing environments. *International Journal of Plant Sciences* 164:S165–S184.

Aitken, S. N. and M. C. Whitlock. 2013. Assisted gene flow to facilitate local adaptation to climate change. *Annual Review of Ecology and Systematics* 44:367–388.

Anderson, M. G. and C. E. Ferree. 2010. Conserving the stage: Climate change and the geophysical underpinnings of species diversity. *PLOS ONE* 5:e11554.

Ashley, M. V., M. F. Willson, O. R. W. Pergams, D. J. O'Dowd, S. M. Gende, and J. S. Brown. 2003. Evolutionarily enlightened management. *Biological Conservation* 111:115–123.

Boyce, W. M., R. R. Ramey, T. C. Rodwell, E. S. Rubin, and R. S. Singer. 1999. Population subdivision among desert bighorn sheep (*Ovis canadensis*) ewes revealed by mitochondrial DNA analysis. *Molecular Ecology* 8: 99–106.

Bradshaw, A. D. 1965. Evolutionary significance of phenotypic plasticity in plants. *Advances in Genetics* 13:115–155.

Bradshaw, W. E. and C. M. Holzapfel. 2006. Evolutionary response to rapid climate change. *Science* 312:1477–1488.

Bradshaw, W. E. and C. M. Holzapfel. 2007. Genetic response to rapid climate change: It's seasonal timing that matters. *Molecular Ecology* 17:157–166.

Carroll, S. P. 2011. Conciliation biology: The eco-evolutionary management of permanently invaded biotic systems. *Evolutionary Applications* 4:184–199.

Clausen, J., D. D. Keck, and W. M. Hiesey. 1940. *Experimental Studies on the Nature of species. I. Effect of Varied Environments on western North American Plants.* Carnegie Institution of Washington Publication 520, Washington, DC.

Cole, K. L., K. Ironside, J. Eischeid, G. Garfin, P. B. Duffy, and C. Toney. 2011. Past and ongoing shifts in Joshua tree distribution support future modeled range contraction. *Ecological Applications* 21:137–149.

Conner, J. K. and D. L. Hartl. 2004. *A Primer of Ecological Genetics.* Sinauer Associates, Sunderland, MA.

Crandall, K., O. Bininda-Emonds, G. Mace, and R. Wayne. 2000. Considering evolutionary processes in conservation biology. *Trends in Ecology and Evolution* 15:290–295.

Crimmins, S. M., S. Z. Dobrowski, J. A. Greenberg, J. T. Abatzoglou, and A. R. Mynsberge. 2011. Changes in climatic water balance drive downhill shifts in plant species' optimum elevations. *Science* 331:324–327.

Darling, E., K. E. Samis, and C. G. Eckert. 2008. Increased seed dispersal potential towards geographic range limits in a Pacific coast dune plant. *New Phytologist* 178:424–435.

Davis, M. B., R. G. Shaw, and J. R. Etterson. 2005. Evolutionary responses to changing climate. *Ecology* 86:1704–1714.

Dlugosch, K. M. and I. M. Parker. 2008. Founding events in species invasions: genetic variation, adaptive evolution, and the role of multiple introductions. *Molecular Ecology* 17:431–449.

Dylan, J. F. 2008. How well can captive breeding programs conserve biodiversity? A review of salmonids. *Evolutionary Applications* 1:535–586.

Epps, C. W., D. R. McCullough, J. D. Wehausen, V. C. Bleich, and J. L. Rechel. 2004. Effects of climate change on population persistence of desert-dwelling mountain sheep in California. *Conservation Biology* 18:102–113.

Epps, C. W., P. Palsboll, J. D. Wehausen, G. K. Roderick, and D. R. McCullough. 2006. Elevation and connectivity define genetic refugia for mountain sheep as climate warms. *Molecular Ecology* 15:4295–4302.

Epps, C. W., P. J. Palsboll, J. D. Wehausen, G. K. Roderick, R. R. Ramey, and D. R. McCullough. 2005. Highways block gene flow and cause a rapid decline in genetic diversity of desert bighorn sheep. *Ecology Letters* 8:1029–1038.

Etterson, J. R. and R. G. Shaw. 2001. Constraint to adaptive evolution in response to global warming. *Science* 294:151–154.

Excoffier, L. and G. Heckel. 2006. Computer programs for population genetics data analysis: A survival guide. *Nature Reviews Genetics* 7:745–758.

Fitzpatrick, B. M. and H. B. Shaffer. 2007. Hybrid vigor between native and introduced salamanders raises new challenges for conservation. *Proceedings of the National Academy of Sciences of the United States of America* 104:15793–15798.

Frankham, R. 2008. Genetic adaptation to captivity in species conservation programs. *Molecular Ecology* 17:325–333.

Franks, S. J., S. Sim, and A. E. Weis. 2007. Rapid evolution of flowering time by an annual plant in response to a climate fluctuation. *Proceedings of the National Academy of Sciences of the United States of America* 104:1278–1282.

Friedman, J. M., J. E. Roelle, J. F. Gaskin, A. E. Pepper, and J. R. Manhart. 2008. Latitudinal variation in cold hardiness in introduced *Tamarix* and native *Populus*. *Evolutionary Applications* 1:598–607.

Gaskin, J. F. and B. A. Schaal. 2002. Hybrid *Tamarix* widespread in U.S. invasion and undetected in native Asian range. *Proceedings of the National Academy of Sciences of the United States of America* 99:11256–11259.

Geber, M. A. and T. E. Dawson. 1993. Evolutionary responses of plants to global change. In P. M. Kareiva, J. G. Kingsolver, and R. B. Huey (eds), *Biotic Interactions and Global Change*. Sinauer, Sunderland, MA. 179–197.

Ghalambor, C. K., J. K. McKay, S. P. Carroll, and D. N. Reznick. 2007. Adaptive versus non-adaptive phenotypic plasticity and the potential for contemporary adaptation in new environments. *Functional Ecology* 21:394–407.

Gienapp, P., C. Teplitsky, J. S. Alho, J. A. Mills, and J. Merila. 2008. Climate change and evolution: Disentangling environmental and genetic responses. *Molecular Ecology* 17:167–178.

Godsoe, W., J. B. Yoder, C. I. Smith, and O. Pellmyr. 2008. Coevolution and divergence in the Joshua tree / yucca moth mutualism. *American Naturalist* 171:816–823.

Grant, P. R. and B. R. Grant. 2002. Unpredictable evolution in a 30-year study of Darwin's finches. *Science* 296:707–711.

Grivet, D., V. L. Sork, R. D. Westfall, and F. W. Davis. 2008. Conserving the evolutionary potential of California valley oak (*Quercus lobata* Née): A multivariate genetic approach to conservation planning. *Molecular Ecology* 17:139–156.

Griffith, B., J. M. Scott, J. W. Carpenter, and C. Reed. 1989. Translocation as a species conservation tool: status and strategy. *Science* 245:477–480.

Groves, C. R., E. T. Game, M. G. Anderson, M. Cross, C. Enquist, Z. Ferdana, E. Girvetz, A. Gondor, K. R. Hall, J. Higgins et al. 2012. Incorporating climate change into systematic conservation planning. *Biodiversity and Conservation* 21:1651–1671.

Harvell, C. D., C. E. Mitchell, J. R. Ward, S. Altizer, A. P. Dobson, R. S. Ostfeld, and M. D. Samuel. 2002. Climate warming and disease risks for terrestrial and marine biota. *Science* 296:2158–2162.

Havens, K., P. Vitt, M. Maunder, E. O. Guerrant Jr., and K. Dixon. 2006. Ex situ plant conservation and beyond. *Bioscience* 56:525–531.

Hendry, A. P., T. J. Farrugia, and M. T. Kinnison. 2008. Human influences on rates of phenotypic change in wild animal populations. *Molecular Ecology* 17:20–29.

Hendry, A. P., M. T. Kinnison, M. Heino, T. Day, T. B. Smith, G. Fitt, C. T. Bergstrom, J. Oakeshott, P. S. Jørgensen, M. P. Zalucki et al. 2011. Evolutionary principles and their practical application. *Evolutionary Applications* 4:159–183.

Hendry, A. P., L. G. Lohmann, E. Conti, J. Cracraft, K. A. Crandall, D. P. Faith, C. Haeuser, C. A. Joly, K. Kogure, A. Larigauderie et al. 2010. Evolutionary biology in biodiversity science, conservation, and policy: A call to action. *Evolution* 64:1517–1528.

Hoffmann, A. A. and C. M. Sgrò. 2011. Climate change and evolutionary adaptation. *Nature* 470:479–485.

Holmgren, C. A. and J. L. Betancourt. 2008. *A Long-Term Vegetation History of the Mojave-*

Colorado Desert Ecotone at Joshua Tree National Park. Final Report Prepared for Joshua Tree National Park Association.

Holmstrom, R. M., J. R. Etterson, and D. J. Schimpf. 2010. Dune restoration introduces genetically distinct American beachgrass, *Ammophila breviligulata*, into a threatened local population. *Restoration Ecology* 18:426–437.

Holt, R. D. 2003. On the evolutionary ecology of species' ranges. *Evolutionary Ecology Research* 5:159–178.

Huey, R. B., G. W. Gilchrist, M. L. Carlson, D. Berrigan, and L. Serra. 2000. Rapid evolution of a geographic cline in size in an introduced fly. *Science* 287:308–309.

Johnson, M. L. and M. S. Gaines. 1990. Evolution of dispersal: Theoretical models and empirical tests using birds and mammals. *Annual Reviews in Ecology and Systematics* 21:449–480.

Jump, A. S., R. Marchant, and J. Peñuelas. 2009. Environmental change and the option value of genetic diversity. *Trends in Plant Science* 14: 51–58.

Kelly, M. W., E. Sanford, and R. K. Grosberg. 2011. Limited potential for adaptation to climate change in a broadly distributed marine crustacean. *Proceedings of the Royal Society B: Biological Sciences* 279:349–356. doi: 10.1098 / rspb.2011 .0542.

Kellermann, V., B. van Heerwaarden, C. M. Sgro, and A. A. Hoffmann. 2009. Fundamental evolutionary limits in ecological traits drive *Drosophila* species distributions. *Science* 325:1244–1246.

Kinnison, M. and N. Hairston. 2007. Eco-evolutionary conservation biology: Contemporary evolution and the dynamics of persistence. *Functional Ecology* 21:444–454.

Klausmeyer, K. R. and M. R. Shaw. 2009. Climate change, habitat loss, protected areas and the climate adaptation potential of species in mediterranean ecosystems worldwide. *PLOS ONE* 4:e6392.

Knapp, E. E. and K. J. Rice. 1998. Comparison of isozymes and quantitative traits for evaluating patterns of genetic variation in purple needle-grass (*Nassella pulchra*). *Conservation Biology* 12:1031–1041.

Lande, R. 1988. Genetics and demography in biological conservation. *Science* 241:1455–1460.

Lankau, R., P. S. Jørgensen, D. J. Harris, and A. Sih. 2011. Incorporating evolutionary principles into environmental management and policy. *Evolutionary Applications* 4:315–325.

Latta, R. G. 2008. Conservation genetics as applied evolution: From genetic pattern to evolutionary process. *Evolutionary Applications* 1:84–94.

Lawrence, H. A., G. A. Taylor, D. E. Crockett, C. D. Millar, and D. M. Lambert. 2008. New genetic approach to detecting individuals of rare and endangered species. *Conservation Biology* 22:1267–1276.

Leinonen, T., R. B. O'Hara, J. M. Cano, and J. Merila. 2008. Comparative studies of quantitative trait and neutral marker divergence: A meta-analysis. *Journal of Evolutionary Biology* 21:1–17.

Lesica, P. and F. W. Allendorf. 1995. When are peripheral populations valuable for conservation? *Conservation Biology* 9:753–760.

Levins, R. 1968. *Evolution in Changing Environments*. Princeton University Press, Princeton, NJ.

Leger, E. A. and K. J. Rice. 2007. Assessing the speed and predictability of local adaptation in invasive California poppies (*Eschscholzia californica*). *Journal of Evolutionary Biology* 20:1090.

Loss, S. R., L. A. Terwilliger, and A. C. Peterson. 2011. Assisted colonization: Integrating conservation strategies in the face of climate change. *Biological Conservation* 144:92–100.

Mace, G. M. and A. Purvis. 2008. Evolutionary biology and practical conservation: Bridging a widening gap. *Molecular Ecology* 17:9–19.

Maron, J. L., V. Montserrat, R. Bommarco, S. Elmendorf, and P. Beardsley. 2004. Rapid evolution of an invasive plant. *Ecological Monographs* 74:261–280.

Mayr, E. 1947. Ecological factors in speciation. *Evolution* 1:263–288.

McKay, J. K. and R. G. Latta. 2002. Adaptive population divergence: Markers, QTL and traits. *Trends in Ecology and Evolution* 17:285–291.

Millar, C. I., N. L. Stephenson, and S. L. Stephens. 2007. Climate change and forests of the future: Managing in the face of uncertainty. *Ecological Applications* 17:2145–2151.

Mitrovski, P., A. A. Hoffmann, D. A. Heinze, and A. R. Weeks. 2008. Rapid loss of genetic variation in an endangered possum. *Biology Letters* 4:134–138.

Moeller, L. and K. Wang. 2008. Engineering with precision: Tools for the new generation of transgenic crops. *Bioscience* 58:391–401.

Montalvo, A. M. and N. C. Ellstrand. 2000. Transplantation of the subshrub *Lotus scoparius*: Testing the home-site advantage hypothesis. *Conservation Biology* 14:1034–1045.

Palumbi, S. R. 2001. Humans as the world's greatest evolutionary force. *Science* 293:1786–1790.

Parmesan, C. 2006. Ecological and evolutionary responses to recent climate change. *Annual Review of Ecology, Evolution, and Systematics* 37:637–669.

Pearce, C. M. and D. G. Smith. 2001. Plains cottonwood's last stand: Can it survive invasion of Russian olive onto the Milk River, Montana floodplain? *Environmental Management* 28:623–637.

Pereira, R. J., F. S. Barreto, and Burton, R. S. 2014. Ecological novelty by hybridization: Experimental evidence for increased thermal tolerance by transgressive segregation in *Tigriopus californicus*. *Evolution* 68:204–215.

Petersen, J. L., M. J. Ellis, and B. P. May. Conservation genetics of California's endangered Shasta crayfish (*Pacifastacus fortis*). Unpublished data.

Phillips, B., G. Brown, J. Webb, and R. Shine. 2006. Invasion and the evolution of speed in toads. *Nature* 439:803–803.

Reusch, T. B. H. and T. E. Wood. 2007. Molecular ecology of global change. *Molecular Ecology* 16:3973–3992.

Rezende, E. L., F. Bozinovic, and T. Garland. 2004. Climatic adaptation and the evolution of basal and maximum rates of metabolism in rodents. *Evolution* 58:1361–1374.

Reznick, D. N. and C. K. Ghalambor. 2001. The population ecology of contemporary adaptations: what empirical studies reveal about the conditions that promote adaptive evolution. *Genetica* 112–113:183–198.

Rice, K. J. and N. C. Emery. 2003. Managing microevolution: Restoration in the face of global change. *Frontiers in Ecology and the Environment* 1:469–478.

Richards, C. L., O. Bossdorf, N. Z. Muth, J. Gurevitch, and M. Pigliucci. 2006. Jack of all trades, master of some? On the role of phenotypic plasticity in plant invasions. *Ecology Letters* 9:981–993.

Richardson, D. M., J. J. Hellmann, J. S. McLachlan, D. F. Sax, M. W. Schwartz, P. Gonzalez, E. J. Brennan, A. Camacho, T. L. Root, O. E. Sala et al. 2009. Multidimensional evaluation of managed relocation. *Proceedings of the National Academy of Sciences of the United States of America* 106:9721–9724.

Rieseberg, L. H. 2006. Hybrid speciation in wild sunflowers. *Annals of the Missouri Botanical Garden* 93:34–48.

Robertson, B. A., J. S. Rehage, and A. Sih. 2013. Ecological novelty and the emergence of evolutionary traps. *Trends in Ecology and Evolution* 28:552–560.

Root, T. L., J. T. Price, K. R. Hall, S. H. Schneider, C. Rosenzweig, and J. A. Pounds. 2003. Fingerprints of global warming on wild animals and plants. *Nature* 421:57–60.

Rubin, E. S., W. M. Boyce, M. C. Jorgensen, S. G. Torres, C. L. Hayes, C. S. O'Brien, and D. A. Jessup. 1998. Distribution and abundance of bighorn sheep in the peninsular ranges, California. *Wildlife Society Bulletin* 26:539–551.

Schwartz, M., G. Luikart, and R. Waples. 2007. Genetic monitoring as a promising tool for conservation and management. *Trends in Ecology and Evolution* 22:25–33.

Sexton, J. P., P. J. Mcintyre, A. L. Angert, and K. J. Rice. 2009. Evolution and ecology of species range limits. *Annual Review of Ecology, Evolution, and Systematics* 40:415–436.

Sexton, J. P., S. Y. Strauss, and K. J. Rice. 2011. Gene flow increases fitness at the warm edge of a species' range. *Proceedings of the National Academy of Sciences of the United States of America* 108:11704–11709. doi: 10.1073 / pnas.1100404108.

Sexton, J. P., J. K. McKay, and A. Sala. 2002. Plasticity and genetic diversity may allow saltcedar to invade cold climates in North America. *Ecological Applications* 12:1652–1660.

Sgrò, C. M., A. J. Lowe, and A. A. Hoffmann. 2011. Building evolutionary resilience for conserving biodiversity under climate change. *Evolutionary Applications* 4:326–337.

Shaffer, H. B., G. B. Pauly, J. C. Oliver, and P. C. Trenham. 2004. The molecular phylogenetics of endangerment: Cryptic variation and historical phylogeography of the California tiger salamander, *Ambystoma californiense*. *Molecular Ecology* 13:3033–3049.

Skelly, D. K., L. N. Joseph, H. P. Possingham, L. K. Freidenburg, T. J. Farrugia, M. T. Kinnison, and A. P. Hendry. 2007. Evolutionary responses to climate change. *Conservation Biology* 21:1353–1355.

Smith, C. I., W. K. W. Godsoe, S. Tank, J. B. Yoder, and O. Pellmyr. 2008. Distinguishing coevolution from covicariance in an obligate pollination mutualism: Asynchronous divergence in Joshua tree and its pollinators. *Evolution* 62:2676–2687.

Spring Rivers. 2009. *Shasta Crayfish Technical Review Committee Summary Report*. Prepared for Pacific Gas and Electric Company, Environmental Services by Spring Rivers Ecological Sciences LLC of Cassel, California. Pacific Gas and Electric Company, San Francisco, CA.

Stein, B. A., A. Staudt, M. S. Cross, N. S. Dubois, C. Enquist, R. Griffis, L. J. Hansen, J. J. Hellmann, J. J. Lawler, E. J. Nelson et al. 2013. Preparing for and managing change: Climate adaptation for biodiversity and ecosystems. *Frontiers in Ecology and the Environment* 11:502–510.

Sultan, S. E. 2001. Phenotypic plasticity and ecological breadth in plants. *American Zoologist* 41:1599.

Tallmon, D. A., G. Luikart, and R. S. Waples. 2004. The alluring simplicity and complex reality of genetic rescue. *Trends in Ecology and Evolution* 19:489–496.

Turesson, G. 1925. The plant species in relation to habitat and climate. *Hereditas* 6:147–236.

USFWS. 1998. *Shasta Crayfish Recovery Plan*. United States Fish and Wildlife Service, Portland, OR, USA.

Via, S., R. Gomulkiewicz, G. De Jong, S. M. Scheiner, C. D. Schlichting, and P. H. Van Tienderen. 1995. Adaptive phenotypic plasticity: Consensus and controversy. *Trends in Ecology and Evolution* 5:212–217.

Visser, M. E. 2008. Keeping up with a warming world; assessing the rate of adaptation to climate change. *Proceedings of the Royal Society B: Biological Sciences* 275:649–659.

Wang, I. J., W. K. Savage, and H. B. Shaffer. 2009. Landscape genetics and least cost path analysis reveal unexpected dispersal routes in the California tiger salamander (*Ambystoma californiense*). *Molecular Ecology* 18:1365–1374.

Wilczek, A. M., M. D. Cooper, T. M. Korves, and J. Schmitt. 2014. Lagging adaptation to warming climate in *Arabidopsis thaliana*. *Proceedings of the National Academy of Sciences of the United States of America* 111:7906–7913.

Williams, D. G., R. N. Mack, and R. A. Black. 1995. Ecophysiology of introduced *Pennisetum setaceum* on Hawaii - the role of phenotypic plasticity. *Ecology* 76:1569–1580.

Manager Comments

Jason P. Sexton and Alden B. Griffith
in conversation with
Rob Klinger

Sexton and Griffith: Rob, how much attention do you see being paid to evolutionary concerns, and what are the constraints to greater consideration?

Klinger: As your chapter points out, the idea that evolutionary considerations need to be integral to management planning as proactive rather than just reactive strategies is not new. But your discussion also underscores some of the potential conflicts and challenges evolutionary arguments will have to effectively address to become significant components of management programs. Ultimately, what will determine the degree to which evolutionary perspectives become major components of management programs will depend on three factors: (1) How well evolutionary considerations dovetail with broader management strategies, especially in relation to the timescales at which most management programs are typically implemented and evaluated; (2) the institutional capacity of agencies to manage climate change-related issues; and (3) the degree to which our perceptions and conservation values change in such an uncertain environmental and management setting.

Sexton and Griffith: Are there principles or mandates within the current framework for management that can help promote greater consideration of evolutionary factors and adaptive potential? Similarly, are current paradigms in conflict with this goal?

Klinger: Key to incorporating evolutionary considerations into climate change management will be the ability to answer the question "In practical terms, what can we do about that?" This takes the discussion from the abstract to the concrete, and it is where your point about focusing on first principles (e.g., maintaining connectivity) becomes extremely important. This would be especially so in highly threatened and fragmented ecosystems with large numbers of endemic and rare species (e.g., vernal pool systems in the Central Valley and coastal scrub in southern California), or for one or a few high-profile species in ecosystems where

changes from climatic shifts are expected to be rapid and pronounced (e.g., fishers and spotted owls in the Sierra Nevada / Cascade range). Some agencies such as the U.S. Fish & Wildlife Service (USFWS) and the California Department of Fish & Game (CDFG) will be focusing on particular species, but it will be critical for these agencies to develop approaches to decide on which species they devote resources to. Though originally proposed in the context of invasive species management, Brooks and Klinger (2009) described a process for integrating prioritization with predictive modeling for a suite of potential target species. Not only could this approach be modified for groups of species threatened by climate shifts, but it would also provide management agencies with a systematic and concrete set of options and be a way of directly incorporating evolutionary principles into a prioritization effort.

Incorporating evolutionary considerations into climate change management programs will be something most managers and scientists have no problem with conceptually or in principle. However, the reality of actually having them as primary components of direct management action is another matter. Over the last 30 years, management of natural resources in the United States has evolved from a focus on particular species or groups of species such as game animals, timber, charismatic, or flagship species, and threatened or endangered species, to a far greater emphasis on ecosystem management conducted within an adaptive management framework. Most resource management agencies, whether governmental or nongovernmental, adopted the ecosystem management philosophy at least a decade ago and proposals to base management on evolutionary considerations could, in many ways, be perceived as a return to the era of single-species management. Regardless of whether it is a charismatic species such as a bighorn sheep or a more obscure species such as a salamander or herbaceous plant, management programs based on individual species are, by definition, very limited in scope and often very expensive. One of the basic ideas behind ecosystem management is that large-scale management actions will benefit many species, including those that are rare. The notion of ecosystem management being based on one or a few flagship rare species is certainly not new (although these tend to be large, charismatic species); nevertheless, the strategy of most agencies is to focus on ecosystem and ecological processes, and the stressors on those processes, and there is little evidence indicating this will change. Evolutionary considerations are only one of many issues that will need to be taken into account in management programs developed around climate change, and it may be difficult for agencies to justify devoting resources to projects unless they have outcomes that can be measured within a decade (at the outside) and have management implications beyond just one or a few species.

Sexton and Griffith: In your experience, how would you describe the current capacity of agencies to consider and incorporate the many important management issues stemming from climate change, including evolutionary considerations?

Klinger: Governmental and nongovernmental resource agencies are faced with an overwhelming number of issues related to climate shifts, but the rate and magnitude of environmental changes have caught most of them unprepared to effectively deal with them. Consequently, the institutional capacity of many of these agencies for managing climate-induced changes is in a nascent state. Most are rapidly trying to develop integrated climate change strategies and programs, but because agency capacity is limited, many potentially useful studies and projects are competing for a limited pool of financial and logistical resources. And of course, this pool of funds depends on congressional support, which we have all observed can wax and wane. So, not only is this pool of funds limited, but it is also unpredictable. Nevertheless, it is apparent that some resources will continue to be devoted to management strategies and programs focused on climatic shifts. For example, the U.S. Forest Service and National Park Service have been developing national climate change programs structured around adaptation and mitigation of change. Each program has pilot projects in California, but neither of them is directly addressing evolutionary issues. It is during the developmental stage of such programs that evolutionary perspectives may have the best chance to become integrated into programmatic goals and strategies. This would increase the likelihood that agencies have the capacity to support studies and programs with a significant evolutionary/genetic focus. However, as noted above, the key to effectively including evolutionary perspectives in climate change programs will be dovetailing these perspectives into existing management strategies and demonstrating concretely how this directly benefits management programs.

Sexton and Griffith: Are there ways in which you see evolutionary perspectives in management naturally dovetailing with future shifts in management priorities, perspectives, and practices as the climate continues to change?

Klinger: There is often an underlying assumption that management is absolutely necessary to try and maintain ecosystems in some familiar configuration, which in the context of climate change has become especially manifest with notions such as managing ecosystems for resistance and resilience, and managed relocation. But climate change, especially rapid change, has made it apparent that the development of novel ecosystems is almost ensured, and even those that keep some semblance of their structure and processes will likely have different geographic boundaries.

Attempting to manage resources for desired conditions based on the past will often be ill-fated, and this is forcing a reevaluation of not just what conditions we find desirable, but also whether it is feasible to attain those conditions. As your chapter implies, there is little sense in managing for resistance to change when the agent of change cannot be altered in any meaningful way, and managing for resilience does not make much sense when ecosystems will not be persisting in an equilibrium or historical state. Given the level of ecological uncertainty that already exists with the direction and extent of climate-associated changes, the added controversy, uncertainty, and expense associated with moving species means that evolutionary, as well as ecological, arguments would have to be extremely compelling to justify such systematic efforts.

What complicates the evolutionary perspective on climate change is that adaptations are individual

(continued)

(continued)

traits that become translated at the population level, but entire ecosystems are undergoing transformations now (for examples in California, see Cayan et al. 2008). It is not unreasonable to think that many species have already begun to adapt to these changes, but identifying which species these are and then translating individual responses into entire communities poses tremendous challenges. Millar et al. (2007) described a framework on how to think in terms of ecosystem responses in the future, but what is needed is a similar framework to help scientists and managers think through the transformations that are happening now. This becomes particularly important when considering calls for fairly rapid and controversial management actions, such as managed relocation (McLachlan et al. 2007). There are abundant examples of unintended consequences from well-intentioned management actions, so a sad irony would be undertaking costly climate response programs that were unnecessary or even counterproductive. At this point, one of the best strategies may be simple patience combined with good monitoring, particularly when thinking in evolutionary terms.

LITERATURE CITED

Brooks, M. L. and R. C. Klinger. 2009. Practical considerations for early detection monitoring of plant invasions. Chapter 2. In Inderjit (ed.), *Management of Invasive Weeds*. Springer, New York, NY. 9–33.

Cayan, D. R., E. P. Maurer, M. D. Dettinger, M. Tyree, and K. Hayhoe. 2008. Climate change scenarios for the California region. *Climatic Change* 87:S21–S42.

McLachlan, J. S., J. J. Hellman, and M. W. Schwartz. 2007. A framework for debate of assisted migration in an era of climate change. *Conservation Biology* 21:297–302.

Millar, C. I., N. L. Stephenson, and S. L. Stephens. 2007. Climate change and forests of the future: Managing in the face of uncertainty. *Ecological Applications* 17:2145–2151.

CHAPTER 12

Fossils Predict Biological Responses to Future Climate Change

Jessica L. Blois and Elizabeth A. Hadly

Abstract. The next few centuries will most likely see rapid and severe climate change. The legacy of climate change will unfold in biological systems over both short and long timescales. While policy often focuses on the effects of climate change over the next decades and century, a longer-term perspective allows better understanding of how species and communities will most likely respond fully to changing climates of the future. The fossil record provides such a long-term perspective, and illustrates a variety of impacts associated with climate change. Moreover, the fossil record can show whether the estimates of current and future climate change and the corresponding biological responses are within the range of historic variability. In addition to playing a role in extinction events, changes in past climates affected species abundances and geographic ranges, population demographics, genetic diversity, and morphological features of individual species such as body size. Over longer time periods, climate affected the macroevolution and species diversity of mammals by influencing rates and magnitudes of immigration, emigration, extinction,

Key Points

- Climate influences mammals and other species in many different ways, with ample evidence in the fossil record supporting effects on abundance, genetics, morphology, geographic distributions, extinction, and speciation.
- The effects of climate change are primarily on population processes, such as abundance, genetic, and morphological change, and on range shifts. These interacting processes can affect larger-scale processes such as speciation and extinction.
- The fossil record highlights the potential for unexpected and significant changes within future mammalian communities— communities of the future are likely to be very different than those on the landscape today.
- One of the primary ways managers can positively influence outcomes in the face of climate change is by increasing the overall resilience of biological systems, but as climate change accelerates beyond the capacity of species to naturally respond and adapt, more active management of populations may be necessary.

and speciation. Because rates and magnitudes of climate change over the next few centuries are mostly predicted to be greater than past climate changes and will interact with other human impacts, a wide variety of biological processes are likely to show greater changes than they have at any time in the past. These processes include geographic range shifts, microevolutionary changes in populations (e.g., loss or gain of adaptive genotypes), and biotic turnover more typical of longer temporal scales. Overall, California is likely to experience increasing abundances of some highly adaptable species, but an overall loss of diversity; the sum of all these processes will likely result in novel communities in California.

INTRODUCTION

Environments of the future are likely to be substantially different from past or present environments due to rapid and large-magnitude climate change (Williams et al. 2007, IPCC 2013). The predicted magnitude and rate of climate change, in the context of other human impacts, has serious implications for the persistence of California's native ecosystems. A factor that complicates society's ability to address these changes is that biological responses to climate change will unfold over both short (10–100 years) and long (1000–1,000,000 years) timescales, much longer than the annual to decadal timescales over which policy decisions are made. While studies of responses to recent (e.g., past 100 years) and ongoing climate change are highly useful (see review by Anderegg and Root, Chapter 3), the fossil record provides additional insight into how the changes we are observing today may translate into longer-term ecological and evolutionary trends.

The fossil record also helps constrain uncertainty about the effects of climate change by providing benchmarks to assess whether current and projected climate change and biological responses are within the range of historical variability. Throughout paleohistory, change

has been the normal state for both the climate (Figure 12.1) and biological systems, though these changes primarily occurred without the added stressors of human impacts such as habitat destruction, hunting, and species introductions. Paleontological data can help determine which species are most vulnerable to climate change, which traits are associated with vulnerability, and which types of biological changes are most likely (Flessa and Jackson 2005, Hadly and Barnosky 2009, Dietl and Flessa 2011).

In this chapter, we highlight the influence of climate on multiple levels of the biological hierarchy, from genes to communities, using the mammalian fossil record (Table 12.1; Blois and Hadly 2009). Mammals are dominant elements of modern ecosystems and anthropogenic climate change has already affected many of them (Beever et al. 2003, Schwartz and Armitage 2004). Ecologically, mammals help structure biological communities since they function as predators, herbivores, seed dispersers, and scavengers (Huntley and Reichman 1994, Dirzo et al. 2014). California contains several important fossil sites to study the influence of climate change on mammals. In particular, fossil deposits from two caves in northern California (Samwell Cave and Potter Creek Cave) contain a rich mammalian assemblage from the late Pleistocene (see Figure 12.1 for timeline and climatic context; Sinclair 1904; Furlong 1906). We recently excavated a new deposit from Samwell Cave that encompasses both Holocene and late Pleistocene mammalian communities (Samwell Cave Popcorn Dome; Blois 2009, Blois et al. 2010), completing a time series of the mammalian community in this region over the past 18,000 years. Fossils from these sites show many changes through time within both individual species and the overall mammalian community.

BIOTIC RESPONSES TO PAST CLIMATE CHANGE

The biotic responses to paleoclimate change reviewed here fall into six categories spanning

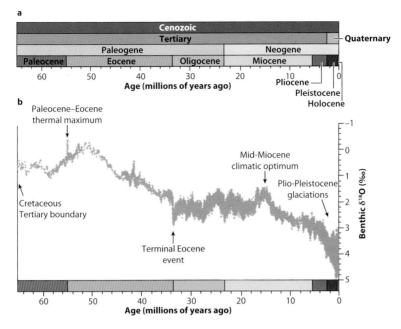

FIGURE 12.1: (a) Cenozoic timeline and (b) climate reconstruction, modified from Zachos et al. 2008. Source: Reprinted from Blois and Hadly (2009).

the biological hierarchy: Abundance change, genetic change, morphologic change, range shifts, speciation, and extinction (Blois and Hadly 2009).

Abundance change

At the most fundamental level, climate may influence population size and/or density (here referred to as "abundance") (Table 12.1). The abundance of a population is ultimately determined by the carrying capacity of the environment (Andrewartha and Birch 1954), which may change as climate changes influence the mammal's preferred habitat (del Monte-Luna et al. 2004). Overall population connectivity and the resilience of a species are influenced by the distribution of the abundance of individuals and populations across a species range (Lundberg et al. 2000). Thus, abundance changes influence other aspects of populations and species, such as genetic diversity, morphology, and geographic ranges.

In northern California, small mammals have shown significant abundance changes over the past 18,000 years (Figure 12.2; Blois et al. 2010). The greatest changes occurred during the Pleistocene–Holocene transition around 11 thousand years ago (Kya) (Figures 12.1 and 12.2). During this time period, northern California transitioned from cold and mesic conditions with low fire frequency to very warm and dry conditions (Daniels et al. 2005). In the Holocene, deer mice (*Peromyscus*) started to dominate the small mammal community, concurrent with decreases in the relative abundance of wood rats (*Neotoma*), ground squirrels (*Spermophilus*), some gophers (*Thomomys*), and many other species (Figure 12.2).

Change in the abundance of two animals in particular stand out, both likely occurring in response to climate changes during the Pleistocene–Holocene transition. First, the mountain beaver (*Aplodontia rufa*) became locally extirpated from the region around the caves (Figure 12.2f). Remains of this animal are found in northern California cave deposits that date to the

TABLE 12.1
Methods to examine relationships between different attributes of fossil data and climate.

Fossil data type	Attribute	Relevance
	Abundance	Changes in abundance can indicate fluctuations in available habitats and resources, local extirpations, and local colonizations.
Fossils from a single site through time	Ancient DNA (aDNA)	aDNA from individuals through time can indicate population structure, genetic diversity, dispersal, and colonization. aDNA can also help identify fossils to species.
	Morphology (e.g., body size, tooth shape and size, limb length and shape)	Changes in morphology can indicate changes in the way species interact with climate and their environment. Over long time periods, changes in morphology can also indicate speciation.
	Community structure	Changes in the community structure through time can indicate shifts in habitat structure or altered interactions between species.
	Geographic range shifts	Shifts in the geographic ranges of species are primarily caused by climate changes altering the suitability of habitats within and around species range edges.
Species lists from multiple sites through time	Diversity	Changes in the species diversity through time can be caused by species losses via emigration, extirpation, or extinction and species gains due to colonization and, at very long timescales, speciation.
	Extinction	Climate change may influence extinction by leading to habitat changes or abundance changes that decrease the resilience of populations and species, for example.
	Speciation	Climate change may cause speciation by isolating populations into separate habitats or by sustained morphologic change.

Last Glacial Maximum (LGM) at 18 Kya, but disappear from the deposits by 13 Kya. A series of warm and dry periods within the late Pleistocene and early and middle Holocene (Daniels et al. 2005) may have caused local extirpation of mountain beaver populations as the landscape became drier. Today, the mountain beaver is found only in cooler, mesic locations in the region, generally at higher elevations. These animals are tied to mesic habitats because their kidneys are very inefficient, requiring that they constantly drink water and consume water-rich plant material (Carraway and Verts 1993). Second, the gopher *Thomomys* cf. *mazama* decreased in abundance and was completely replaced by a different gopher (*Thomomys bottae*) during the Pleistocene–Holocene transition (Figures 12.2b;

Blois et al. 2010). The climate changes at this transition and the environmental niches occupied currently by the different types of gophers suggest that climate influenced the local extirpation of *Thomomys* cf. *mazama* (Blois 2009, Blois et al. 2010) because this gopher is confined to generally higher elevations in the region today.

The outcome of abundance changes within individual species, particularly the decline of formerly common species, was a net loss of local diversity as measured both by a decline in species richness (the number of species present) and a reduction in evenness (similarity in the abundance of different species) (Blois et al. 2010). Today's mammal community is very different from the community present during the Pleistocene. These examples illus-

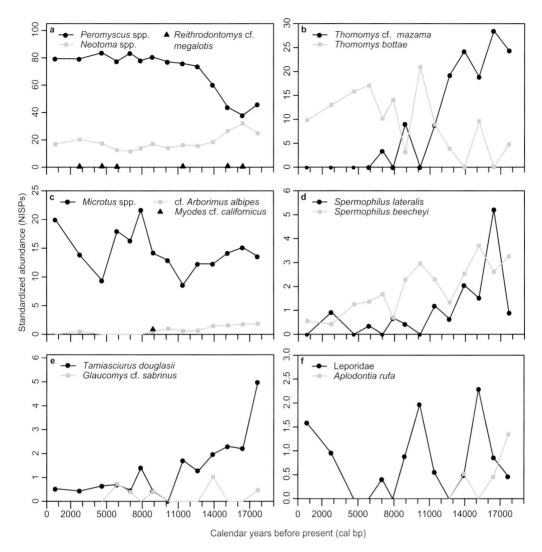

FIGURE 12.2: Abundance of key small mammal taxa in the Samwell Cave Popcorn Dome deposit. Source: Modified from Blois et al. (2010); supplementary information).

trate that biotic change is dynamic and adjustments to population abundance can cause local extirpation, range shifts, and biodiversity loss, up to global extinction of species (Table 12.1).

Genetic change

Genetic change is a result of population-level processes such as recombination, mutation, selection, random genetic drift, and gene flow (Charlesworth et al. 2003), all of which take place in mammals over tens to thousands of years. Most of these processes may be influenced by climate, either directly such as when climate selects for or against key traits or indirectly through abundance or habitat change, which influences gene flow and genetic drift. Ancient DNA (aDNA) provides the opportunity to study the impact of climate change on the evolutionary dynamics within past populations and species (Table 12.1). aDNA is DNA that has survived in the remains of ancient organisms. Under optimal preservation conditions such as in permafrost settings, aDNA from mammals can be as old as 50,000–65,000 years or older (Willerslev et al. 2003). Few examples of aDNA studies exist for California populations of terrestrial mammals (Blois 2009), though increasing attention is being paid to museum specimens that extend DNA samples back to the early 1900s (e.g., Perrine et al. 2007). Our work on ancient genetic data from Samwell Cave confirmed the replacement event among gophers (described above) at the Pleistocene–Holocene transition (Blois et al. 2010). We are also using aDNA to investigate causes of the expansion of *T. bottae* in northern California (Figure 12.2b). In North America and elsewhere, aDNA has been used to identify species that are morphologically indistinguishable from one another in the fossil record and thus help describe past geographic ranges of species (Willerslev et al. 2003, Gilbert et al. 2008). It has also been used to identify all organisms (plant, animal, microbial, etc.) present at a particular time and place from the DNA preserved in ancient sediments (Kuch et al. 2002, Willerslev et al. 2003, Willerslev et al. 2014).

aDNA can also be used to characterize the structure and genetic diversity of fossil animal populations (e.g., Hadly et al. 2004, Hofreiter et al. 2004). For example, aDNA combined with population genetic models has been used to discriminate if populations were connected or isolated during periods of climate change (Hadly et al. 2004, Hofreiter et al. 2004), the probable size of ancient populations (Chan et al. 2006), survival of clades across periods of climate change (Brace et al. 2012, Foote et al. 2013, Fulton et al. 2013), and whether changes in population size are consistent with what is known about the population biology of the species and the climatic history of the region (Hadly et al. 2004, Lorenzen et al. 2011). Genetic change may become increasingly important for species to successfully adapt to anthropogenic climate change (e.g., Jump and Penuelas 2005; Sexton and Griffith, Chapter 11).

Morphologic change

Trends in the average size or shape of mammals are among the most commonly documented responses to climate in the fossil record (Table 12.1; Blois and Hadly 2009). Changes in climate may affect the morphology of individuals directly or indirectly. For example, mammals are generally larger in cold or high latitude climates (Ashton et al. 2000), presumably because larger bodies are more efficient at maintaining high internal body temperatures. Climate may also act directly by changing the amount of time available to forage during favorable weather or by influencing the fasting endurance of individuals (Millar and Hickling 1990, Sinervo et al. 2010).

Climate change also may influence animals indirectly via effects on vegetation and habitat. For example, tooth size, shape, or structure may change due to changing type and quality of food resources (Patton and Brylski 1987). Additionally, as resources become more or less abundant due to climate change, body size should change accordingly (Lomolino 2005). Finally, changes in habitat structure (e.g., transitions from a forested habitat to grassland) may cause changes in limb morphology, such as the increase in body size and lengthening of limbs (facilitating greater running ability) that coincided with the expansion of more open grasslands in North America and elsewhere (MacFadden 1992, Janis 2008).

Often, trends in morphology across space, such as the pattern of mammals being larger at higher latitudes, are used to predict changes through time (e.g., Smith and Betancourt 1998, McGuire 2010). These patterns do not always hold, though, as demonstrated by time-series data from the fossil record. For example, mammals should be smaller today than during the LGM because it is warmer today. However, the body size of California ground squirrels (*Spermophilus beecheyi*) in the region around Samwell and Potter Creek Caves increased between the LGM (cold and dry) and today (warm and relatively mesic) (Blois et al. 2008). Contrary to expectations based on temperature gradients across space, Blois et al. (2008) found that body size variation across both time and space was primarily explained by differences in precipitation. Space–time comparisons such as this are useful for narrowing down which element of climate change will have the greatest impact on particular species and communities.

Range shifts

Fossil data are useful for detecting range shifts because they allow researchers to track species ranges over thousands of years (Table 12.1). These data have contributed to evidence that range shifts are one of the strongest responses a species can have to climate change, and show that while substantial range adjustments can happen within decades (Parmesan and Yohe 2003, Root et al. 2003), in some cases these responses occur over millennia (Graham et al. 1996, Lyons 2003).

While a single site such as the Samwell Cave Popcorn Dome deposit is not sufficient to detail the full geographic range shifts of California mammals since the LGM (Table 12.1), it does provide evidence for local range adjustments through time. For example, neither *A. rufa* nor *Thomomys* cf. *mazama* is found locally in the Samwell Cave area today, but both were there in the past, indicating that the ranges of these animals adjusted to their present-day ranges by moving northward and upward in elevation through time (Blois et al. 2010). Similarly, McGuire (2011) documented the past occurrence of the long-tailed vole (*Microtus longicaudus*) at two late Quaternary sites in the San Francisco Bay Area, substantially removed from their present range limits, indicating range shift or contraction through time. Integration of the fossil deposits from northern California with data from other localities throughout California and North America can increase the power to observe range changes through time (Graham et al. 1996, McGuire and Davis 2013). For example, range shifts in response to Late Pleistocene warming occurred in most North American mammals (Graham et al. 1996, Lyons 2003), with over approximately 30% of the mammals experiencing substantial range shifts between the LGM and the

Holocene (Lyons 2003). The range shifts between the LGM and the Holocene showed strong directionality during periods of significant climate change (Graham et al. 1996, Lyons 2003). Additionally, poleward range limits, which are located in areas that typically experience greater magnitude climate changes than other areas in the range, shifted more than the limits facing the equator, on average, during periods of more significant warming such as the transition from the LGM to the Holocene (Lyons 2003). These findings are consistent with observations that the majority of species showing range shifts over the past century have moved poleward (Parmesan and Yohe 2003, Root et al. 2003, Hickling et al. 2006). Modern studies have also shown strong links with climate, presumably temperature, along the poleward edge of ranges, at least in many birds (Root 1988).

Speciation

Prolonged and unidirectional periods of climate change can result in speciation and / or lineage diversification (Table 12.1). Climate change may contribute to speciation by creating patchy habitats within the geographic range of a species (Vrba 1992, Gavrilets et al. 1998, Barnosky 2001), each supporting an isolated population that may proceed down independent evolutionary trajectories. Prolonged environmental change may also cause morphologic change leading to speciation (Gingerich 1985, Martin 1993). Both types of speciation are common and often simultaneously observed in the fossil record.

In California, several sites have long enough fossil records to investigate speciation, particularly the Barstow Formation in southeastern California. Pagnac (2006) surmised that speciation within the Miocene horse lineage *Scaphohippus* may have been related to climate. Rodents within the Barstow Formation are also a good example of morphologic evolution and speciation (Lindsay 1972), though no thorough investigations of the links with climate have been made using these deposits. Revisiting these deposits may be a fruitful area of future research on the effects of climate on speciation within California mammals.

Extinction

Extinction is the certain outcome for all species; in fact, most of the mammals that have inhabited the earth have already become extinct (Avise et al. 1998, Alroy 2000). The fossil record can be used to identify factors associated with extinction risk (Table 12.1) such as properties of mammals themselves (e.g., body size, degree of specialization, and rarity) as well as other factors such as habitat reduction, disease, and human hunting (e.g., Thomas et al. 2004a, Harrison et al. 2008, Nogués-Bravo et al. 2008), and determine how these factors interact with climate change. Species that are highly specialized to particular climate regimes and / or habitat types will be most vulnerable to extinction (e.g., Menendez et al. 2006), but with even greater amounts of climate change, the same mechanisms may affect less specialized species as well.

Many California fossil localities contain records of extinct animals (e.g., Sinclair 1904, Furlong 1906, Springer et al. 2010). For example, 24% of the mammals at Potter Creek Cave and Samwell Cave went globally extinct at the end of the Pleistocene, all but two of which were >44 kg (these very large animals are termed "megafauna"). A similar picture emerges from fossil deposits in southern California: 43% of the mammals went extinct, with a distinct bias toward larger body sizes (Stock and Harris 1992). The Pleistocene megafaunal extinctions in California and the rest of North America were likely due to the effects of climate change and human hunting, as well as other impacts such as disease (Barnosky et al. 2004, Nogués-Bravo et al. 2008). These same cascading factors are also likely to lead to future extinctions (Thomas et al. 2004b, Rosenzweig et al. 2008, Davidson et al. 2009, Barnosky et al. 2011).

Overall biotic change

In mammals, several episodes of significant, community-wide biological change seen in the past 65 million years occurred during periods when the rate of climate change was high (Figure 12.3) (Barnosky et al. 2003). Given that current and future rates of average global climate change are much higher than anything recorded by the fossil record (Figure 12.3), a large amount of ecological change should be expected in the future, possibly at a level not encountered before in the fossil record.

SYNERGIES AND SOLUTIONS

Overall, the California fossil record highlights the many ways climate may influence mammals and points to the potential for unexpected and significant changes within future mammalian communities. The effects of climate change are on population processes, such as abundance, genetic, and morphological change, and range shifts. These processes interact with one another during times of significant climate change and can affect larger-scale processes such as speciation and extinction. Ultimately, all of these processes combine to affect the overall structure of mammalian diversity. Thus, changes to the population processes can be early signals of the effects of anthropogenic climate change on the biological system (e.g., Ceballos and Ehrlich 2002, Post et al. 2009, Drake and Griffen 2010). Some of these signals are already being seen today (e.g., Parmesan and Yohe 2003, Root et al. 2003, Parmesan 2006, Rosenzweig et al. 2008) and the fossil record offers several concrete lessons to biologists, managers, and policy-makers to address the changes ahead (Hadly and Barnosky 2009).

As this review has demonstrated, species and communities will not remain static in the future. Greater amounts of climate change will likely lead to greater ecological changes (Figure 12.3; Barnosky et al. 2003), including population losses, range shifts, and extinction. Given the dominance of range shifts as a major

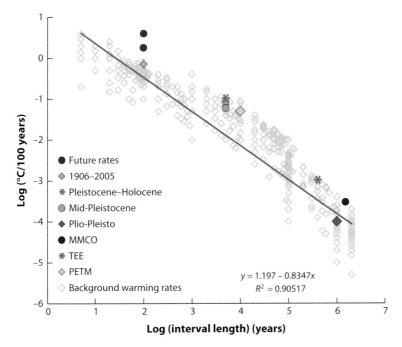

FIGURE 12.3: Per-100-year temperature change rates for different intervals throughout the Cenozoic, on a log–log scale. The best-fit line is shown, as well as the fit of the line (R^2). All rates plotted are rates of warming, except for the Terminal Eocene Event (TEE), which is rate of cooling. The 1906–2005 per-100-year warming rate is estimated at 0.74 ± 0.18 based on the IPCC 2007 assessment; the low and high 95% bounds are plotted. Future rates are based on the IPCC 2007 assessment of "best estimate" of future temperature in 2090–2099 for different emission scenarios. The low estimate is based on the B1 emissions scenario and the high estimate is based on the A1F1 emissions scenario. Past and future 100-year data are from Solomon et al. (2007), TEE data are from Zachos et al. (2001), and Paleocene–Eocene Thermal Maximum (PETM) data are from Zachos et al. (2008). All other data are as in Figure 2 (top) from Barnosky et al. (2003). MMCO: Mid-Miocene Climatic Optimum. Source: Modified from Blois and Hadly (2009).

response to climate change, significant abundance declines within populations, population losses within species, the prevalence of introduced species, and the threat of widespread extinctions, many communities of the future will be very different than those of today (Williams and Jackson 2007, Stralberg et al. 2009).

Proactive management will need to balance needs of current species with the goal of allowing communities to adapt and change over time. For example, if particular species are identified as high priority for conservation, detailed knowledge of the life history of those species and how life history is influenced by climate is very helpful because species are influenced by climate in different ways (e.g., McCain and King 2014) and there are many

possible pathways to extinction (e.g., Davidson et al. 2009). However, we will still lose some species due to climate change. It will be useful to decide ahead of time which species are most important for conservation purposes and what amount of species loss is tolerable. Can key ecosystem functions be maintained, but with a reduced or different set of species?

The effects of climate change have and will continue to interact with other impacts such as invasion of nonnative species, habitat degradation, and habitat loss (Vermeij 1991, D'Antonia and Vitousek 1992). One of the primary ways managers can positively influence outcomes in the face of climate change is by increasing the overall resilience of biological systems. This can be done by preserving habitats, creating or expanding corridors between habitats, and preventing the invasion or establishment of particularly detrimental nonnative species. More active management of populations themselves may also be necessary. For example, if forestalling the loss of a particular species is the goal, the evolutionary timescale may need to be "sped up," perhaps by transplanting individuals to increase genetic diversity or maintain genetic connectivity between populations. Additionally, if natural range shifts become impossible due to the pace of climate change or barriers to dispersal, transplanting populations into new habitats where they can be managed may be necessary (e.g., McLachlan et al. 2007, Richardson et al. 2009).

CONCLUSION

Mammals respond to climate change at all levels of biological organization. Climate affects population size and density, as well as habitat connectivity, which interact with the strength of climate change on individuals themselves to influence genetic and morphologic processes. Climate change also forces species to shift their geographic range or adapt to the new climate conditions if climates become intolerable. Finally, all of these processes interact to influence longer-term processes such as speciation and extinction, and ultimately how diversity across the globe is shaped through time.

All of these changes should be expected in the future; indeed, many of them are already observed in response to climate change over the past century. Given the diverse impacts of climate and humans over the past few centuries, we may already be nearing the sixth mass extinction event in earth's history (Barnosky et al. 2011). The fossil record demonstrates that communities are very resilient to climate change, but also that surprises are likely and many species will respond in unknown ways. Life has persisted on earth for over 5 billion years and will persist for many more years to come. However, the shape of modern-day communities and ecosystems may be vastly different than those of the past, leading to unknown

effects on human societies dependent on those ecosystems. It is up to human society to decide what our global biotic systems will be.

ACKNOWLEDGMENTS

The authors would like to thank C. Millar, J. Sexton, J. Dorman, K. Hall, T. Root, C. Howell, and M. Herzog for comments that greatly improved the manuscript. This research was supported by a BICCCA-PIREA fellowship to JLB and NSF grants EAR-0545648 and EAR-0719429 to EAH.

LITERATURE CITED

Alroy, J. 2000. New methods for quantifying macroevolutionary patterns and processes. *Paleobiology* 26:707–733.

Andrewartha, H.G. and L.C. Birch. 1954. *The Distribution and Abundance of Animals.* University of Chicago Press, Chicago, IL.

Ashton, K.G., M.C. Tracy, and A. de Queiroz. 2000. Is Bergmann's rule valid for mammals? *The American Naturalist* 156:390–415.

Avise, J.C., D. Walker, and G.C. Johns. 1998. Speciation durations and Pleistocene effects on vertebrate phylogeography. *Proceedings of the Royal Society of London, Series B: Biological Sciences* 265:1707–1712.

Barnosky, A.D. 2001. Distinguishing the effects of the Red Queen and Court Jester on Miocene mammal evolution in the northern Rocky Mountains. *Journal of Vertebrate Paleontology* 21:172–185.

Barnosky, A.D., E.A. Hadly, and C.J. Bell. 2003. Mammalian response to global warming on varied temporal scales. *Journal of Mammalogy* 84:354–368.

Barnosky, A.D., P.L. Koch, R.S. Feranec, S.L. Wing, and A.B. Shabel. 2004. Assessing the causes of late Pleistocene extinctions on the continents. *Science* 306:70–75.

Barnosky, A.D., N. Matzke, S. Tomiya, G.O.U. Wogan, B. Swartz, T.B. Quental, C.R. Marshall, J.L. McGuire, E.L. Lindsey, K.C. Maguire et al. 2011. Has the earth's sixth mass extinction already arrived? *Nature* 471:51–57.

Beever, E.A., P.E. Brussard, and J. Berger. 2003. Patterns of apparent extirpation among isolated populations of pikas (*Ochotona princeps*) in the Great Basin. *Journal of Mammalogy* 84:37–54.

Blois, J. L. 2009. *Ecological Responses to Paleoclimatic Change: Insights from Mammalian Populations, Species and Communities.* Stanford University, Stanford, CA.

Blois, J. L. and E. A. Hadly. 2009. Mammalian response to Cenozoic climatic change. *Annual Review of Earth and Planetary Sciences* 37:181–208.

Blois, J. L., R. Feranec, and E. A. Hadly. 2008. Environmental influences on spatial and temporal patterns of body-size variation in California ground squirrels (*Spermophilus beecheyi*). *Journal of Biogeography* 35:602–613.

Blois, J. L., J. L. McGuire, and E. A. Hadly. 2010. Small mammal diversity loss in response to late-Pleistocene climatic change. *Nature* 465:771–774.

Brace, S., E. Palkopoulou, L. Dalén, A. M. Lister, R. Miller, M. Otte, M. Germonpre, S. P. E. Blockley, J. R. Stewart, and I. Barnes. 2012. Serial population extinctions in a small mammal indicate Late Pleistocene ecosystem instability. *Proceedings of the National Academy of Sciences of the United States of America* 109:20532–20536.

Carraway, L. N. and B. J. Verts. 1993. *Aplodontia rufa. Mammalian Species* 431:1–10.

Ceballos, G. and P. Ehrlich. 2002. Mammal population losses and the extinction crisis. *Science* 296:904–907.

Chan, Y. L., C. N. K. Anderson, and E. A. Hadly. 2006. Bayesian estimation of the timing and severity of a population bottleneck from ancient DNA. *PloS Genetics* 2:451–460.

Charlesworth, B., D. Charlesworth, and N. H. Barton. 2003. The effects of genetic and geographic structure on neutral variation. *Annual Review of Ecology, Evolution, and Systematics* 34:99–125.

Daniels, M. L., R. S. Anderson, and C. Whitlock. 2005. Vegetation and fire history since the Late Pleistocene from the Trinity Mountains, northwestern California, USA. *Holocene* 15:1062–1071.

D'Antonia, C. M. and P. M. Vitousek. 1992. Biological invasions by exotic grasses, the grass / fire cycle, and global change. *Annual Review of Ecology and Systematics* 23:63–87.

Davidson, A. D., M. J. Hamilton, A. G. Boyer, J. H. Brown, and G. Ceballos. 2009. Multiple ecological pathways to extinction in mammals. *Proceedings of the National Academy of Sciences of the United States of America* 106:10702–10705.

del Monte-Luna, P., B. W. Brook, M. J. Zetina-Rejon, and V. Cruz-Escalona. 2004. The carrying capacity of ecosystems. *Global Ecology and Biogeography* 13:485–495.

Dietl, G. P. and K. W. Flessa. 2011. Conservation paleobiology: Putting the dead to work. *Trends in Ecology & Evolution* 26:30–37.

Dirzo, R., H. S. Young, M. Galetti, G. Ceballos, N. J. B. Isaac, and B. Collen. 2014. Defaunation in the Anthropocene. *Science* 345:401–406.

Drake, J. M. and B. D. Griffen. 2010. Early warning signals of extinction in deteriorating environments. *Nature* 467:456–459.

Flessa, K. W. and S. T. Jackson. 2005. *The Geological Record of Ecological Dynamics: Understanding the Biotic Effects of Future Environmental Change.* National Academy Press, Washington, DC.

Foote, D., K. Kaschner, S. E. Schultze, C. Garilao, S. Y. W. Ho, K. Post, T. F. G. Higham, C. Stokowska, H. van der Es, C. B. Embling et al. 2013. Ancient DNA reveals that bowhead whale lineages survived Late Pleistocene climate change and habitat shifts. *Nature Communications* 4:1677.

Fulton, T. L., R. W. Norris, and R. W. Graham. 2013. Ancient DNA supports southern survival of Richardson's collared lemming (*Dicrostonyx richardsoni*) during the last glacial maximum. *Molecular Ecology* 22:2540–2548.

Furlong, E. L. 1906. The exploration of Samwel Cave. *American Journal of Science* 172:235–247.

Gavrilets, S., H. Li, and M. D. Vose. 1998. Rapid parapatric speciation on holey adaptive landscapes. *Proceedings of the Royal Society of London, Series B: Biological Sciences* 265:1483–1489.

Gilbert, M. T. P., D. L. Jenkins, A. Gotherstrom, and N. Naveran. 2008. DNA from pre-Clovis human coprolites in Oregon, North America. *Science* 320:786–789.

Gingerich, P. D. 1985. Species in the fossil record: Concepts, trends, and transitions. *Paleobiology* 11:27–41.

Graham, R. W., E. L. Lundelius Jr., M. A. Graham, E. K. Schroeder, R. S. Toomey III, E. Anderson, A. D. Barnosky, J. A. Burns, C. S. Churcher, D. K. Grayson et al. 1996. Spatial response of mammals to Late Quaternary environmental fluctuations. *Science* 272:1601–1606.

Hadly, E. A. and A. D. Barnosky. 2009. Vertebrate fossils and the future of conservation biology. In G. P. Dietl and K. W. Flessa (eds), *Conservation Paleobiology: Using the Past to Manage for the Future. The Paleontological Society Papers.* The Paleontological Society, Boulder, CO. 39–59.

Hadly, E. A., U. Ramakrishnan, Y. L. Chan, M. van Tuinen, K. O'Keefe, P. A. Spaeth, and C. J. Conroy. 2004. Genetic response to climatic change: Insights from ancient DNA and phylochronology. *PLOS Biology* 2:e290.

Harrison, S., J. H. Viers, J. L. Thorne, and J. B. Grace. 2008. Favorable environments and the persistence of naturally rare species. *Conservation Letters* 1:65–74.

Hickling, R., D. B. Roy, J. K. Hill, R. Fox, and C. D. Thomas. 2006. The distributions of a wide range of taxonomic groups are expanding polewards. *Global Change Biology* 12:450–455.

Hofreiter, M., G. Rabeder, V. Jaenicke-Després, G. Withalm, D. Nagel, M. Paunovic, G. Jambresic, and S. Pääbo. 2004. Evidence for reproductive isolation between cave bear populations. *Current Biology* 14:40–43.

Huntly, N. and O. J. Reichman. 1994. Effects of subterranean mammalian herbivores on vegetation. *Journal of Mammalogy* 75:852–859.

Intergovernmental Panel on Climate Change (IPCC). 2013. *Climate Change 2013: The Physical Science Basis. Contribution of Working Group I to the Fifth Assessment Report of the Intergovernmental Panel on Climate Change.* Cambridge University Press, New York.

Janis, C. M. 2008. An evolutionary history of browsing and grazing ungulates. In I. J. Gordon and H. H. T. Prins (eds), *The Ecology of Browsing and Grazing.* Springer, Berlin. 21–45.

Jump, A. and J. Penuelas. 2005. Running to stand still: Adaptation and the response of plants to rapid climate change. *Ecology Letters* 8:1010–1020.

Kuch, M., N. Rohland, J. L. Betancourt, and C. Latorre. 2002. Molecular analysis of a 11 700-year-old rodent midden from the Atacama Desert, Chile. *Molecular Ecology* 11:913–924.

Lindsay, E. H. 1972. *Small Mammal Fossils from the Barstow Formation, California.* University of California Publications in Geological Sciences, Berkeley, CA.

Lomolino, M. V. 2005. Body size evolution in insular vertebrates: Generality of the island rule. *Journal of Biogeography* 32:1683–1699.

Lorenzen, E. D., D. Nógues-Bravo, L. Orlando, J. Weinstock, J. Binladen, K. A. Marske, A. Ugan, M. K. Borregaard, M. T. P. Gilbert, R. Nielsen et al. 2011. Species-specific responses of Late Quaternary megafauna to climate and humans. *Nature* 479:359–364.

Lundberg, P., E. Ranta, J. Ripa, and V. Kaitala. 2000. Population variability in space and time. *Trends in Ecology & Evolution* 15:460–464.

Lyons, S. 2003. A quantitative assessment of the range shifts of Pleistocene mammals. *Journal of Mammalogy* 84:385–402.

MacFadden, B. J. 1992. *Fossil Horses: Systematics, Paleobiology and Evolution of the Family Equidae.* Cambridge University Press, New York.

Martin, R. A. 1993. Patterns of variation and speciation in Quaternary rodents. In R. A. Martin and A. D. Barnosky (eds), *Morphological Change in Quaternary Mammals of North America.* Cambridge University Press, New York. 226–280.

McCain, C. M. and S. R. B. King. 2014. Body size and activity times mediate mammalian responses to climate change. *Global Change Biology* 20:1760–1769.

McGuire, J. 2010. Geometric morphometrics of vole (*Microtus californicus*) dentition as a new paleoclimate proxy: Shape change along geographic and climatic clines. *Quaternary International* 212:198–205.

McGuire, J. L. 2011. Identifying California *Microtus* species using geometric morphometrics documents Quaternary geographic range contractions. *Journal of Mammalogy* 92:1383–1394.

McGuire, J. L. and E. B. Davis. 2013. Using the palaeontological record of *Microtus* to test species distribution models and reveal responses to climate change. *Journal of Biogeography* 40:1490–1500.

McLachlan, J. S., J. J. Hellmann, and M. W. Schwartz. 2007. A framework for debate of assisted migration in an era of climate change. *Conservation Biology* 21:297–302.

Menendez, R, A. G. Megias, J. K. Hill, B. Braschler, S. G. Willis, Y. Collingham, R. Fox, D. B. Roy, and C. D. Thomas. 2006. Species richness changes lag behind climate change. *Proceedings of the Royal Society of London, Series B: Biological Sciences* 273:1465–1470.

Millar, J. S. and G. J. Hickling. 1990. Fasting endurance and the evolution of mammalian body size. *Functional Ecology* 4:5–12.

Nogués-Bravo, D., J. Rodríguez, J. Hortal, P. Batra, and M. B. Araújo. 2008. Climate change, humans, and the extinction of the woolly mammoth. *PLOS Biology* 6:e79.

Pagnac, D. 2006. *Scaphohippus,* a new genus of horse (Mammalia: Equidae) from the Barstow Formation of California. *Journal of Mammalian Evolution* 13:37–61.

Parmesan, C. 2006. Ecological and evolutionary responses to recent climate change. *Annual Review of Ecology, Evolution and Systematics* 37:637–669.

Parmesan, C. and G. Yohe. 2003. A globally coherent fingerprint of climate change impacts across natural systems. *Nature* 421:37–42.

Patton, J. L. and P. V. Brylski. 1987. Pocket gophers in alfalfa fields: Causes and consequences of habitat-related body size variation. *American Naturalist* 130:493–506.

Perrine, J., J. Pollinger, B. Sacks, R. Barrett, and R. Wayne. 2007. Genetic evidence for the persistence of the critically endangered Sierra Nevada red fox in California. *Conservation Genetics* 8:1083–1095.

Post, E., J. Brodie, M. Hebblewhite, A. D. Anders, J. A. K. Maier, and C. C. Wilmers. 2009. Global population dynamics and hot spots of response to climate change. *BioScience* 59:489–497.

Richardson, D. M., J. J. Hellmann, J. S. McLachlan, D. F. Sax, M. W. Schwartz, P. Gonzalez, E. J. Brennan, A. Camacho, T. L. Root, O. E. Sala et al. 2009. Multidimensional evaluation of managed relocation. *Proceedings of the National Academy of Sciences of the United States of America* 106:9721–9724.

Root, T. 1988. Environmental factors associated with avian distributional boundaries. *Journal of Biogeography* 15:489–505.

Root, T. L., J. T. Price, K. R. Hall, S. H. Schneider, C. Rosenzweig, and J. A. Pounds. 2003. Fingerprints of global warming on wild animals and plants. *Nature* 421:57–60.

Rosenzweig, C., D. Karoly, M. Vicarelli, P. Neofotis, Q. Wu, G. Casassa, A. Menzel, T. L. Root, N. Estrella, B. Seguin et al. 2008. Attributing physical and biological impacts to anthropogenic climate change. *Nature* 453:353–357.

Schwartz, O. A. and K. B. Armitage. 2004. Weather influences on demography of the yellow-bellied marmot (*Marmota flaviventris*). *Journal of Zoology* 265:73–79.

Sinclair, W. J. 1904. The exploration of the Potter Creek Cave. *American Archaeology and Ethnology* 2:1–27.

Sinervo, B., F. Méndez-de-la-Cruz, D. B. Miles, B. Heulin, E. Bastiaans, M. Villagrán-Santa Cruz, R. Lara-Resendiz, N. Martinez-Mendez, M. L. Calderon-Espinosa, R. N. Meza-Lazaro et al. 2010. Erosion of lizard diversity by climate change and altered thermal niches. *Science* 328:894–899.

Smith, F. A. and J. L. Betancourt. 1998. Response of bushy-tailed woodrats (*Neotoma cinerea*) to late Quaternary climatic change in the Colorado Plateau. *Quaternary Research* 50:1–11.

Solomon, S., D. Qin, M. Manning, R.B. Alley, T. Berntsen, N.L. Bindoff, Z. Chen, A. Chidthaisong, J.M. Gregory, G.C. Hegerl et al. 2007. Technical summary. In S. Solomon, D. Qin, M. Manning, Z. Chen, M. Marquis, K.B. Averyt, M. Tignor, and H.L. Miller (eds), *Climate Change 2007: The Physical Science Basis. Contribution of Working Group I to the Fourth Assessment Report of the Intergovernmental Panel on Climate Change.* Cambridge University Press, New York, NY.

Springer, K., E. Scott, J. C. Sagebiel, and L. K. Murray. 2010. Late Pleistocene large mammal faunal dynamics from inland southern California: The Diamond Valley Lake local fauna. *Quaternary International* 217:256–265.

Stock, C. and J. M. Harris. 1992. *Rancho La Brea: A Record of Pleistocene Life in California.* Natural History Museum of Los Angeles County Science Series No. 37. The Natural History Museum of Los Angeles County, Los Angeles, CA. 113 pp.

Stralberg, D., D. Jongsomjit, C. Howell, M. Snyder, J. Alexander, J. Wiens, and T. L. Root. 2009. Re-shuffling of species with climate disruption: A no-analog future for California birds. *PLOS ONE* 4:e6825.

Thomas, C. D., A. Cameron, R. E. Green, M. Bakkenes, L. J. Beaumont, Y. C. Collingham, B. F. N. Erasmus, M. F. de Siqueira, A. Grainger, L. Hannah et al. 2004a. Extinction risk from climate change. *Nature* 427:145–148.

Thomas, J. A., M. G. Telfer, D. B. Roy, C. D. Preston, J. J. D. Greenwood, J. Asher, R. Fox, R. T. Clarke, and J. H. Lawton. 2004b. Comparative losses of British butterflies, birds, and plants and the global extinction crisis. *Science* 303:1879–1881.

Vermeij, G. J. 1991. When biotas meet: Understanding biotic interchange. *Science* 253:1099–1104.

Vrba, E. S. 1992. Mammals as a key to evolutionary theory. *Journal of Mammalogy* 73:1–28.

Willerslev, E., A. J. Hansen, J. Binladen, and T. B. Brand. 2003. Diverse plant and animal genetic records from Holocene and Pleistocene sediments. *Science* 300:791–795.

Willerslev, E., J. Davison, M. Moora, M. Zobel, E. Coissac, M. E. Edwards, E. D. Lorenzen, M. Vestergård, G. Gussarova, J. Haile et al. 2014. Fifty thousand years of Arctic vegetation and megafaunal diet. *Nature* 506:47–51.

Williams, J. and S. T. Jackson. 2007. Novel climates, no-analog communities, and ecological surprises. *Frontiers in Ecology and the Environment* 5:475–482.

Williams, J. W., S. T. Jackson, and J. E. Kutzbach. 2007. Projected distributions of novel and disappearing climates by 2100 AD. *Proceedings of the National Academy of Sciences of the United States of America* 104:5738–5742.

Zachos, J., M. Pagani, L. Sloan, E. Thomas, and K. Billups. 2001. Trends, rhythms, and aberrations in global climate 65 Ma to present. *Science* 292:686–693.

Zachos, J. C., G. R. Dickens, and R. E. Zeebe. 2008. An early Cenozoic perspective on greenhouse warming and carbon-cycle dynamics. *Nature* 451:279–283.

CHAPTER 13

Historical Data on Species Occurrence

BRIDGING THE PAST TO THE FUTURE

Morgan W. Tingley

Abstract. Understanding climate change impacts on the environment necessitates adapting monitoring strategies to evaluate long-term changes on the order of decades and centuries. Historical data, such as museum specimens, original field surveys, and other records, provide one potential source of information that can provide inference on changes over long periods of time. Historical data are often undervalued, however, as these data were typically collected for other purposes and are seen as difficult to integrate with current data collection efforts. Here, I describe sources of historical data and key considerations for using them, and give examples of how these diverse sources are informing our current understanding of how species respond to change.

Key Points

- Historical data are increasingly used to understand how climate change is currently affecting biological resources.
- Museum specimens, atlas data, old surveys and studies, and field notes are all rich sources of occurrence data that can be used with modern comparisons.
- Differences in methods and other data-specific traits can create challenges for making valid comparisons, yet strategies exist to overcome many issues.
- Integrating historical data into ongoing climate change monitoring can benefit resource managers.

UNDERSTANDING THE FUTURE BY LOOKING TO THE PAST

Climate change during the 21st century is expected to impact species in a multitude of ways, some of which we can predict, but many of which are unknown. Based on what scientists have already observed, phenological mismatches, range shifts, and both local and regional colonizations and extinctions are occurring currently, and are expected to increase both globally (Root et al. 2003, Chen et al. 2011, Anderegg and Root, this volume) and in California (Moritz et al. 2008, Forister et al.

2010, Tingley et al. 2012). Climate change holds consequences for managed systems, as the biological value of these systems may be affected by shuffling species assemblages and modified biological processes. In addition to working toward reducing the rate of climate change, our ability to protect our natural resources may be significantly aided by investment in inferential methods that provide understanding of how these resources are likely to change in response to ongoing climatic processes.

The premise behind this work is simple: We have the opportunity to improve our understanding of what could happen in the future by exploring what has already occurred. Understanding species occurrences in the recent past allows the development of a strong baseline to measure both current occurrence patterns and judge future changes. Since climate change impacts are generally long-term trends that can be obscured through short-term variation (stochasticity), broad temporal ranges in baseline data provide greater inference and perspective on changes. Particularly if past data are rich across a range of time periods, it is possible to separate short-term stochasticity and natural variation in species' occurrences (e.g., due to resource cycles) from long-term trends associated with shifting environmental gradients.

Work on the American pika (*Ochotona princeps*) provides a highly relevant example. Historical data (in this case, the collection of museum specimens) from various isolated mountaintop populations throughout the pika's range have provided a comparative baseline for current surveys at these same locations (Beever et al. 2003, Erb et al. 2011). Comparing modern pika presences to areas of known historical occurrence, researchers have been able to document local extinctions over the 20th century. Importantly for conservation, different key factors (summer temperatures, drought stress) explain these extirpations in different locations. This is just one example of the many studies that have used historical data to under-

stand climate change impacts in western North America (Table 13.1).

The greatest stumbling block to understanding changes in the recent past in a particular landscape is simply the availability of data. Although they are not always available, baseline data can come from many sources, and the original purpose of the data need not be the same as the purpose now. The specific purpose for collection can bias contemporary comparisons, however, through determining how data were collected (e.g., random sampling) or which aspects of data (e.g., habitat) were collected. Additionally, all occurrence datasets (i.e., data that pinpoint a species to a location and a time) can include false presences (i.e., misidentifications) and false absences (i.e., failure to detect a species where it occurs) that can bias comparisons. Differentiating false presences and absences from true presences and absences is an important analytical step. Thus, to unlock the value of historical data, we need to address two questions. First, what are potential sources of historical data? Second, how can historical data be compared in an unbiased way with recently collected data?

FINDING HISTORICAL BASELINE DATA

The first challenge in establishing a historical baseline for species occurrences in an area is simply uncovering historical data. "Historical" in this context does not mean preindustrial, nor does it necessarily mean a century ago, as rich data sources from those periods rarely exist. While truly historical examples do exist [such as Henry David Thoreau's notes on flowering times at Walden Pond from the mid-19th century (Primack et al. 2009)], historical is used here to denote a time period that is relevant to climate change. Ideally, "historical" data provide a point of reference with which a temporal comparison to the present will be meaningful. Depending on the rate of climate change in a region, the extent of time between past and present necessary for a meaningful temporal comparison can vary. Throughout

TABLE 13.1

Published examples, 1990–2014, from the western United States of historical data that have been compared to contemporary data in order to inform on the effects of different temporal drivers of change

INFORMATION BUILDS ON TABLE WITHIN TINGLEY AND BEISSINGER 2009

Inference type	Citation	State	No. of species[1]	No. of sites[2]	Drivers explored[3]	Historical source	Max time span[4]	Biotic system	Taxa
Extinction & Colonization	Bradford et al. 2005	Multiple	7	128	Land-use change, invasive species	Specimens, literature	106	Desert, wetland	Amphibians
	Davidson et al. 2001	CA	1	237	Land-use change, climate change, pollution	Specimens, literature	119	Grassland, wetland	Amphibians
	Davidson 2004	CA	5	1083	Land-use change, climate change, pollution	Specimens, original surveys	17	Mountain, wetland	Amphibians
	Larrucea and Brussard 2008	CA, NV	1	105	Land-use change, invasive species, fire	Specimens, literature, field notes	129	Sagebrush	Mammals
	Morelli et al. 2012	CA	1	74	Land-use change, climate change	Specimens, field notes	109	Mountains	Mammals
Range change	Brusca et al. 2013	AZ	27	30	Climate change	Original survey data	49	Mountain, desert	Plants
	Crimmins et al. 2011	CA	64	8747	Climate change	Original survey data	75	Mountain	Plants
	Kelly and Goulden 2008	CA	∞	20	Climate change, fire, pollution	Original survey data	30	Desert, mountain	Plants
	Moriarty et al. 2011	CA	1	80	Land-use change	Field notes, literature, original survey data	28	Forest, mountain	Mammals
	Moritz et al. 2008	CA	28	56	Climate change	Specimens, field notes	94	Mountain	Mammals
	Rowe et al. 2010	NV	25	15	Land-use change, climate change	Specimens, field notes	79	Mountain	Mammals
	Rubidge et al. 2011	CA	3	39	Land-use change, climate change	Specimens, field notes	82	Mountain	Mammals
	Tingley et al. 2012	CA	99	77	Climate change	Field notes	98	Mountain	Birds
	Winker et al. 2002	AK	∞	3	Climate change	Field notes, literature	85	Tundra	Birds
	Yang et al. 2011	CA	2	12	Climate change	Specimens, field notes	91	Mountain	Mammals
Abundance & Community	Bakker and Moore 2007	AZ	∞	5	Land-use change	Original survey data	92	Forest	Plants

(Continued)

TABLE 13.1
(continued)

Inference type	Citation	State	No. of species[1]	No. of sites[2]	Drivers explored[3]	Historical source	Max time span[4]	Biotic system	Taxa
–	Forister et al. 2010	CA	159	10	Land-use change, climate change	Original survey data	35	Mountain, grassland	Butterflies
–	Franklin et al. 2004	CA	∞	649	Fire	Original survey data	71	Chaparral	Trees
–	Harrison et al. 2010	OR	181	185	Land-use change, climate change	Original survey data	60	Mountain, forest	Plants
–	Kopp and Cleland 2013	CA	∞	62	Climate change	Original survey data	49	Mountain	Plants
–	Lutz et al. 2009	CA	14	655	Climate change, fire	Original survey data	67	Forest	Plants
–	McLaughlin and Zavaleta 2013	CA	1	10	Land-use change	Original survey data	32	Forest	Trees
–	Rickart et al. 2008	NV	48	48	Land-use change, invasive species	Specimens, literature	74	Sagebrush, mountain	Mammals
–	Rowe 2007	UT	27	3	Land-use change, climate change	Field notes, specimens	53	Mountain	Mammals
–	Rowe et al. 2011	NV	22	9	Land-use change, climate change	Field notes, specimens	81	Mountain	Mammals
–	Rowlands and Brian 2001	AZ	∞	5	Land-use change	Original survey data	38	Sagebrush, forest	Plants
–	Shultz et al. 2012	CA	72	1	Land-use change	Field notes, original survey data	94	Urban	Birds
–	Thorne et al. 2008	CA	∞	3422	Land-use change, climate change	Original survey data	62	Mountain, forest	Plants
–	Tingley and Beissinger 2013	CA	025	77	Climate change	Field notes	98	Mountain	Birds
–	West and Yorks 2006	ID	∞	13	Land-use change, climate change, fire	Original survey data	32	Sagebrush	Plants

1 The number of species chosen for temporal comparison within the study. Infinity (∞) implies that all species (detected and undetected) within the community were studied.
2 The number of sites implies the number of sampling plots—whether lakes, forests, quadrats, or point counts—that are matched across temporal resurveys. The scale of individual sites depends on the resolution of historical data. Original survey data often have smaller discrete sampling units, allowing more detailed and more numerous resurvey samples.
3 The specific temporal drivers of species change that were explored or tested within the study by comparing historical data to contemporary data. Here, the term "land-use change" has multiple meanings, including: changes in land tenure, landscape, or habitat conversion, and changes in management (e.g., grazing).
4 The maximum time span of comparison from the earliest historical datum to the most recent contemporary datum.

much of North America, anthropogenic climatic warming can be detected as early as 1900, but the recent trend of accelerating warming is not obvious until 1960, with the greatest changes since the late 1970s (Mastrandrea and Anderegg, this volume). In the present context, historical occurrence data provide perspective for modern distributions, so the time frame that defines "historical" may vary by the movement and dispersal abilities of focal taxa. For example, data from 1980 would provide very little utility if one is comparing distributions of long-lived canopy trees, yet one might potentially detect strong changes if comparing flying insects.

The most useful historical data contain three components: A taxonomic identification, a geographic specification, and a date. These three components define a "collecting event" for all occurrence data. Identification challenges vary by data source and taxa; although misidentifications of species exist within natural history collections (Miller et al. 2007, Graham et al. 2008), physical specimens (and sometimes photographs) provide a means to reverify unusual occurrence records. With survey data, either trust has to be placed in the skills of the original observer or unusual records need to be sufficiently vetted using expert knowledge. The second component, geographic location, may not be precisely specified, and thus occurrence locations frequently have an associated uncertainty or error. Generally, the risks of misleading conclusions due to geographical uncertainty are greater in areas of increased habitat or topographical variability (Rowe and Lidgard 2009). For example, if it is unknown whether a sighting was at 500 or 1500 m elevation, then it may not hold much value for measuring climate-induced elevational range shifts. Similar to location uncertainty, uncertainty in the date of an occurrence record may affect the utility of that record in a comparative analysis, such as a phenological analysis of seasonal change.

Occurrence data with these three components can come from a wide variety of sources.

For many long-established managed areas (e.g., national parks and forests in the United States), routine species monitoring likely extends prior to 1980. These original surveys have the potential to provide high-resolution information that can be resurveyed at a fine scale, for example, the revisiting of historical American Marten census plots at Sagehen Experimental Forest in northeastern California (Moriarty et al. 2011). While unit-specific original surveys are a great source when available, occurrence data can also be held by outside sources as part of data collected for a different purpose or part of a larger survey effort. A wide variety of sources provide this historical occurrence baseline, including biotic atlases, field notes, published literature, and museum specimens.

Combining herbaria, invertebrate, and vertebrate collections, natural history museum specimens are an incredibly rich source of occurrence data, on the order of billions of independent specimens collected (Graham et al. 2004). While museum-based specimen collecting continues today throughout the world, specimen collecting is closely associated with geographic exploration and biodiversity cataloguing, and thus particularly in North America, most specimens are historical in nature, although this differs by taxa. In California, bird specimens were predominantly collected in the first half of the 20th century, while amphibian and reptile specimen collecting peaked in California around 1970 (Figure 13.1).

Historically, specimen collectors were motivated by a desire to create collections that were physical embodiments of nature's diversity (Box 13.1). By the early 20th century, Darwin's lessons on the sources of natural variation were well known, and western naturalists sought to maximize the taxonomic and phenotypic diversity of collections through broadening their geographic coverage. As a result, this period was a golden age in many zoological fields for describing geographical variation in taxonomy—and these discoveries were backed up with physical specimens and often further described in field notes. Collection efforts in

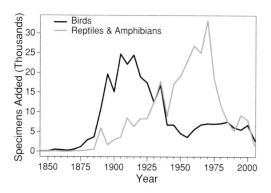

FIGURE 13.1: Temporal variation in the number of avian and herpetological specimens collected in California since 1850 and currently held in North American natural history museums. Bird-specimen collection was strongest during the first half of the 20th century as the natural history of California was being "explored." Collection fell off after the Second World War as collection efforts, when made, were focused internationally. Reptile and amphibian collections grew steadily through the 20th century with a drastic rise after 1950. This correlates with a reevaluation of morphology and taxonomy that resulted in the discovery of numerous cryptic species. Source: Data downloaded from VertNet.

the second half of the 20th century shifted toward documenting morphological variation within species, resulting in large specimen "series" of the same species. The cumulative result of 20th century scientific efforts is an extensive geographic coverage of museum specimens, including specimens from both public and private lands (Figure 13.2). Online databases combining specimen records from tens to hundreds of museums around the world now exist for most taxa, with prominent examples including the Global Biodiversity Information Facility (GBIF: http://www.gbif.org/), Map of Life (http://www.mol.org), and VertNet (http://vertnet.org/). VertNet works as a data portal, patching together taxa-specific occurrence data for birds, mammals, reptiles, amphibians, and fish. These databases have dramatically increased access to museum-vouchered occurrence data.

Field notes are an additional notable component to documenting historical species occurrences. While specimen records provide quick access to species occurrences, they often provide very little associated information, called "metadata," which provides context and background. In contrast, field notes made during specimen collecting expeditions or during other natural history and ecological survey work often provide a wide variety of useful metadata in addition to the essential collecting event for an occurrence. These metadata can provide more exact temporal and geographic precision (e.g., "A quarter till 8 we stopped along the west side of Long Meadow, 500 ft north of the stream outlet"), as well as other useful details, such as weather, survey effort, collecting method, ecological context, and behavior. While field notes are often the only source of occurrence data for observations, the metadata gathered from field notes can also be applied to associated specimen records, thus greatly increasing their analytical value (e.g., Moritz et al. 2008). While not all field notes are properly archived and made accessible, several field note repositories have recently begun the process of digitization and providing online access; prime examples include the Museum of Vertebrate Zoology at UC Berkeley (http://bscit.berkeley.edu/mvz/volumes.html) and the Smithsonian National Museum of Natural History (http://www.mnh.si.edu/rc/fieldbooks/index.html). Even in electronic form, field notes often require extensive research in order to pull out occurrence records and associated metadata. The reward can be proportional to the effort, however; field notes have been used successfully to understand long-term changes in occurrences and populations of a diverse group of species, from birds (Colwell et al. 2002, Martin et al. 2004) to fish (Labay et al. 2011) to invertebrates (Daniels 1998, Blalock-Herod et al. 2002) to plants (Brusca et al. 2013).

Historical occurrence data may not be the only historical data that will be important when analyzing long-term changes at one location, as we are often interested in relating occurrence changes of organisms to ongoing processes such as climate, land-cover, and land-use change. Historical climate layers are generally derived from long-term continuous weather

The flora and fauna of California and the western United States had already been well documented before Joseph Grinnell was named the founding director of the Museum of Vertebrate Zoology at the University of California, Berkeley, in 1908, but Grinnell ultimately changed the nature of the collection of natural history information. Prior to Grinnell, there was little effort invested in planning how specimen collections should be built. Generally, diversity was maximized, but additional information was rarely collected. Grinnell saw the value in metadata, as transcribed via field notes:

> Our field-records will be perhaps the most valuable of all our results. .... any and all (as many as you have time to record) items are liable to be just what will provide the information wanted. You can't tell in advance which observations will prove valuable. Do record them all!
>
> (Field notes of J. Grinnell, 1908)

Grinnell had the foresight to directly link the systematic collection of specimens and their metadata to the assemblage of information on occurrence and ecology:

> It will be observed, then, that our efforts are not merely to accumulate as great a mass of animal remains as possible. On the contrary, we are expending even more time than would be required for the collection of the specimens alone, in rendering what we do obtain as permanently valuable as we know how, to the ecologist as well as to the systematist. It is quite probable that the facts of distribution, life history, and economic status may finally prove to be of more far-reaching value, than whatever information is obtainable exclusively from the specimens themselves.
>
> (Grinnell 1910)

Grinnell expounded on his new method of collecting field notes when he could, trying to convince others to adopt the method:

> These notes were written 'on the spot' from time to time during the three or four hours of observation. They show the nature of a certain type of field observations, how these may be recorded in a running narrative style, and there is perhaps some information presented of general interest to the student of living birds.
>
> (Grinnell 1912)

Grinnell ended up succeeding in his efforts. His systematic and devoted collection of field notes became known as the "Grinnell Method" and greatly changed the means by which natural history information has been collected.

Literature Cited

Grinnell, J. 1910. The uses and methods of a research museum. *Popular Science Monthly* 77:163–169.

Grinnell, J. 1912. An afternoon's field notes. *Condor* 14(3):104–107.

---

monitoring stations. In the United States, data from these weather monitoring stations are aggregated and curated by the United States Historical Climatology Network (USHCN: http://cdiac.ornl.gov/epubs/ndp/ushcn/ushcn.html). They provide access to individual daily and monthly weather records for over 1200 unique stations with some records beginning as far back as 1895, although data quality from individual stations may deserve close attention. If continuous climate coverages are needed, there are several options. The Climatic Research Unit (CRU; http://www.cru.uea.ac.uk/) provides monthly and yearly climatic surfaces for the entire world (at large 5° × 5° grid cells) going back to 1850. Within the continental United States, the PRISM Climate Group (http://www.prism.oregonstate.edu/) produces monthly grids going back to 1890 at a much finer cell size (800 × 800 m²). Compared to climate data, historical land-use and land-cover data can be harder to find as they require the prior existence of detailed—often, hand-drawn—maps. The USGS provides access to resources that provide land-cover data from multiple time periods (http://landcover.usgs

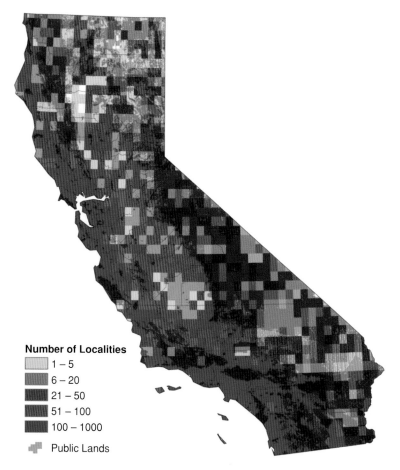

Number of Localities
- 1 – 5
- 6 – 20
- 21 – 50
- 51 – 100
- 100 – 1000
- Public Lands

FIGURE 13.2: Spatial variation in the availability of historical (pre-1980) avian and herpetological specimen-based occurrence data in California with respect to land ownership. Image shows California divided into 12 × 12 arcmin grid cells, colored according to the total number of unique geographic sites with bird-, reptile-, or amphibian-specimen data. Each unique site represents a place visited by museum collectors where information on one or more species is contained within the specimen record, and also likely within field notes. Grid cells with no specimen locations are not colored. The overall coverage shows the wide availability of species information from the specimen record, including areas currently predominated by both public and private land ownership. Note that this map is restricted to showing digitized and georeferenced records of specimens held in North American museums and is thus not a complete map of the true existing coverage of museum data for these taxa. Source: Map pools data for specimen records downloaded from VertNet.

.gov / landcoverdata.php), but specific data availability may differ by state and locality. In California, for instance, the Wieslander Vegetation Type Mapping Project (VTM: http://vtm.berkeley.edu /) is a highly informative dataset providing information on changes in forest composition and land cover in many parts of the state since the 1930s.

HISTORICAL RESURVEYS IN THE WESTERN UNITED STATES

The availability of historical data provides an opportunity to conduct a resurvey, where historical data are compared with newly collected data. Researchers are increasingly gaining inference on long-term changes in species distributions and abundances through resurveying locations with historical occurrence data (Shaffer et al. 1998, Tingley and Beissinger 2009). A review of published resurvey studies can help illustrate opportunities for learning from the past at both local and regional scales. In the western United States alone, there have been at least 30 published studies since 2000 involving resurveying for taxa using historical data (Table 13.1). These studies illustrate the diversity of potential inference available from historical data and can be lumped into three groups: Extinction and colonization, range change, and abundance and community composition.

Studies of extinction and colonization tend to focus on small numbers of species of conservation concern (fewer than 10 species), and use historical occurrence data to pinpoint where species used to be, and compare that to current, higher-resolution data. This type of inference has played a key role in documenting amphibian declines over the 20th century. For example, Davidson et al. (2001) explored extirpations of the California red-legged frog (*Rana aurora draytonii*) by determining the current population status of 237 sites of historic occurrence. Looking at the pattern of decline helped Davidson et al. to evaluate different causal hypotheses, including climate change, habitat loss, pesticides, and UV-B radiation.

To study climate change impacts in particular, researchers have been interested in using historical data to help describe changes in range (Table 13.1). This type of inference builds on extinction and colonization studies by using historical data to delineate species' distributions across gradients, whether latitudinal, elevational, climatic, or other. In order to make robust inferences on range change, these studies often take advantage of historical surveys that were repeated in the same place (see "strategies for using historical data"). Additionally, while inference on range change can be made from specimen occurrence data, field notes are often necessary in order to learn where species did not occur historically. For example, in order to understand altitudinal shifts in small mammals of the Ruby Mountains in Nevada, Rowe et al. (2010) supplemented specimen data with field notes to determine non-detections.

Studies focused on changes in species abundance and community composition (Table 13.1) require even more detailed historical data. To meet these information needs, abundance and community studies predominantly use original survey data. For example, Lutz et al. (2009) used data from historical vegetation plots in Yosemite National Park to examine declines in large-diameter trees. As they were unable to relocate the original plots from the 1930s, Lutz et al. compared the historical surveys to two modern surveys and attempted to control statistically for sampling bias. In this example, information content and data quality of historical data made up for highly uncertain geographical location.

Comparing resurvey studies in the western United States since 2000, several trends are evident. First, there is a negative correlation between the time span of inference and the average information quality of historical data. Extinction and colonization studies have the largest average time span of comparison (Table 13.1), but these studies also generally make the simplest types of comparisons. To make unbiased inference on range change or changes in abundance or community structure, one needs more and more detailed data. While one can go farther back in time with the specimen record, greater detail is often contained in more recent survey work. This trade-off in information content does not necessarily mean a trade-off in inference on drivers of change. On average, extinction and colonization studies examine just as many drivers (e.g., land-use change or climate) to explain temporal differences as other studies. In fact, range change studies generally examine fewer drivers despite focusing on more species and requiring broader sampling. This limited focus is probably due to the recent focus on climate change as a driver of range shifts, whereas other drivers have not historically been associated with these shifts.

Historical data are exciting because of their untapped potential, but by nature, historical data have extra uncertainties that can bias comparisons to contemporary data. Consequently, comparing historical data to modern data in a manner that limits the potential for bias is an important concern. The best strategies for addressing bias depend on the form of the assembled historical dataset.

The most basic form of historical data is a semi-random collection of species and their collecting events, with only broad spatial or temporal resolution. These "presence-only" data lend themselves to the development of baseline species lists, but must be used with the explicit understanding that the lists are incomplete because an unknown number of additional species may have been present at the site but not sampled. Consequently, presence-only data are best used for inference on whether a baseline species has been extirpated in an area. While false absences are unknowable solely from presence-only data, there are some ways to get around this limitation. For example, other studies that document how detectability differs by species and habitats may, if applied judiciously, provide transferable inference on the severity of false absence biases with a given dataset. Building distributional models off presence-only data may also provide a means to assess whether species occur in an area (e.g., Comte and Grenouillet 2013).

Information content can be added through field notes and other sources of metadata to increase the utility of presence-only data. A valuable piece of information is whether occurrences were collected as parts of unique-survey events where all species detected were recorded. For example, a researcher may have gone hunting one morning and returned with specimens of six different species. It is possible that she detected other species while out but did not collect them. Field notes could fill in these data gaps, informing whether any additional species

were detected, and if so, the identity of those uncollected species. This would then allow the bundling of all species identifications into a single collecting event where it is known both what was detected and what was not detected. At this point, the data have gained information content and become "presence and non-detection" data (Tingley and Beissinger 2009). Through knowing both what was detected and what was not detected, estimating colonization becomes possible.

Presence and non-detection data, however, suffer from the hazards of false absences. The problem is that the presence of something is ultimately knowable (if it is seen, it is present), yet the absence of something is rarely conclusive (if it is not seen, it may still be present). Since estimating colonization or extinction relies on a change in status from absent to present, or vice versa, false absences can lead to systematic bias by artificially inflating both colonization and extinction estimates. For presence and non-detection data, this bias is likely to be high for rare or cryptic species but may be negligible for always abundant or very visible species. On the other hand, ubiquitous species may be left out of nonsystematic survey records, as many observers are more focused on finding rare or target species. Comparing contemporary data to individual historical surveys may result in biased conclusions due to false absences.

Additional information can be added to occurrence data if discrete surveys are temporally replicated within a relatively short period of time at a location. For example, if a researcher went out collecting three days in a row at the same location and kept notes of what she found each day, the repeat surveys can be used to estimate detectability. Detectability, or the probability of detecting a species given its presence at a site, provides a means to reevaluate apparently vacant sites and ask whether a species was truly absent from the site or simply present and non-detected. Only through estimating the probability of true absence of a species from sites can occurrence data provide unbiased inference on both colonization and extinction of species over time.

Quantitative ecologists have arrived at a variety of statistical ways to estimate the true probability of presence or absence of a species, but one way, known as occupancy modeling (MacKenzie et al. 2006), has gained acceptance in recent years. Mackenzie et al. (2006) provide a full treatise on how to build and use occupancy models, while Tingley and Beissinger (2009) provide specific advice for using occupancy models with historical data. The great strength of occupancy modeling lies in both its power to provide unbiased estimates of an ecological state of great interest (i.e., the true absence of a species) and its flexibility to adapt to different data formats and questions. While they can be complex, occupancy models constitute an analytical tool capable of addressing the biggest challenge of using historical data—the task of differentiating between true occurrence changes and false changes that appear due to different methods, purposes, and changing detection probabilities over time.

INCORPORATING HISTORICAL DATA INTO CONTEMPORARY MONITORING

Historical occurrence data can play only a minimal role in management and conservation unless we are able to compare it to contemporary data. A challenge for land managers and others engaged in monitoring is in adapting existing data collection schemes so that data can be used to look both backward and forward. Implementing changes to any monitoring scheme will necessarily be site and program specific and may include numerous other considerations; but integrating historical data into a program for the purpose of monitoring contemporary change can be a gradual, multistep process that need not initially affect current monitoring. As a resource to managers, the following steps are presented as general guidelines.

STEP 1. BEGIN BY DOCUMENTING PAST OCCURRENCE Data on past occurrences have multiple uses, and provide a general baseline for assessing future occurrence changes. Many natural areas have compiled or partially compiled documentation of species occurrence over time. Pulling these information sources together opens the door for further exploration of historical data.

STEP 2. CLASSIFY DATA STRUCTURE What is the format of the historical occurrence data? What kinds of metadata are available? The answers to these questions will help you to classify data by information content, such as "presence-only" or "presence and estimable absence."

STEP 3. COMPARE DATA STRUCTURES How does the historical data structure compare with data being collected now? Assess the opportunities for determining colonization, extinction, range shift, community composition, and abundance.

STEP 4. DEFINE AND—WHERE POSSIBLE—FILL DATA GAPS Based on how historical and contemporary data differ, identify areas where inference can be improved through the filling of data gaps. For example, what data would be needed to upgrade from presence-only data to presence and non-detection data? Both historical and contemporary data can be enriched.

STEP 5. CONDUCT COMPARISON When both data sets have been organized and evaluated, a comparison between historical and contemporary data can be conducted. Conducting a comparison may necessitate normalization of data by survey effort or other factors, or incorporation of these factors into analytical methods. The results of this comparison can be used to prioritize future monitoring as well as to generate new hypotheses that can be evaluated using historical data.

STEP 6. REPEAT AND REFINE During the process of the first five steps, additional questions may develop, new data sources may be identified, or sampling frames may need to be expanded. All of these are valid reasons for repeating the process and incorporating additional data so as to refine the comparative process.

STEP 7. EXPAND THROUGH COLLABORATION Historical–contemporary comparisons at single locations may be more informative if they are expanded to a larger spatial perspective. To this end, single management units may ultimately benefit from working with other managed systems or collaborating with academic researchers to expand their comparison outward.

STEP 8. ARCHIVE AND SHARE COMPARISON If collaboration (step 7) is not feasible or logical, local studies of temporal change can be extremely valuable when incorporated as part of larger regional or continental meta-analyses. Single-unit studies can greatly benefit the broader understanding of long-term temporal processes just by archiving data in publicly accessible formats (see Box 13.2) and distributing or publishing findings.

OPPORTUNITIES TO LOOK TO THE FUTURE

Joseph Grinnell, lauded Western naturalist and advocate for extensive sampling and note-taking, commented in *Popular Science Monthly* in 1910:

> At this point I wish to emphasize what I believe will ultimately prove to be the greatest purpose of our museum. This value will not, however, be realized until the lapse of many years, possibly a century, assuming that our material is safely preserved. And this is that the student of the future will have access to the original record of faunal conditions in California and the west, wherever we now work. He will know the proportional constituency of our faunae by species, the relative numbers of each species and the extent of the ranges of species as they exist to-day.

Grinnell was tackling a problem that many managers face: How do we know what information will be the most useful in the future? Grinnell's solution was to collect as much information as possible because one cannot predict the future. While Grinnell's broad intensity and effort paid off (e.g., Moritz et al. 2008), the variety, detail, and types of data that can be collected today are vastly greater than in Grinnell's day, making his strategy impractical. Resource managers and conservation professionals today face a greater number of choices in monitoring schemes, yet as data are used in ever more ways, the trade-offs for choices are becoming less clear. This chapter has focused on using historical data, but today's data will all be "historical" soon. Thus, while we do not know exactly what data will be most useful in the future, we are at a good point to look to the past and see what lessons may be used to inform future decisions.

One lesson from the past is the importance of repeatability. While arduously collecting extensive species data can have valuable applications, contemporary data in the future may be valued by how easily comparative data can be collected. For example, Shavit and Griesemer (2009) explore the ambiguities in defining a resurveyable "locality" from Joseph Grinnell's specimens and field notes, noting that even specimens with assigned latitudes and longitudes may not be specific enough to confidently revisit the exact same spot. While dealing with uncertainty and ambiguity is part of why we use specialized tools for dealing with historical data, to the extent that we can make this process easier for ourselves or others in the future may be worth the effort. Repeatability, subsequently, comes from collecting metadata at multiple scales and resolutions. It is true that there have been advances in taking precise locations over the last several decades, but a physical location is better known if it is additionally described in archived field notes. Photographs of sites, taken at systematic directions or angles, can also help define locations (and their temporal condition) in ways that coordinates cannot. Considering repeatability can also help define what data to collect. While a highly specialized tool may collect useful data today, it is worth considering the potential lifespan of that tool and how easily the same or comparable data can be collected in the future.

Looking forward, the ability of scientists to understand how species and systems are responding to climate change may depend on the availability of data that are broadly distributed through both time and geographic space. This broad type of data cannot be collected by scientists alone, but relies on the efforts of hundreds or thousands of "citizen scientists." The datasets and websites below represent both sources of information and places to share observations collected in many current-monitoring programs.

Phenological Data

in the United States, the aggregation of phenological data is broadly overseen by the US National Phenological Network (USNPN: http://www .usanpn.org/). Within the USNPN, there are regional projects with more narrow spatial focuses that aim to build local capacity for phenological monitoring. For example, in 2010 the California Phenology Project (http://www.usanpn.org/cpp/) was launched to focus on building phenological monitoring in the state.

Occurrence Data

the systematic collection of species observations from scientists and citizen scientists is rapidly growing in capacity. Many states have online portals for the collection of information on rare or threatened species observations (e.g., California Natural Diversity Database: http://www.dfg.ca. gov/biogeodata/cnddb/). Taking advantage of the broad appeal of bird-watching, Cornell University and the National Audubon Society launched eBird (http://ebird.org) in 2002, which accepts data in a variety of formats. Since then, eBird has expanded to accept bird sightings from anywhere in the world and scientists are actively using data to monitor bird populations and phenology in real time. As of 2014, eBird only accepts bird records, but the idea is starting to expand, as exemplified by the North American "Butterflies I've Seen" database (http://www.nababis.org/) and the all-taxa iNaturalist project (http://www.inaturalist .org/). iNaturalist aims to collect observational data with the same level of information content as museum specimen data, including novel metadata such as photo vouchers and geographic uncertainty, with data quality assessment built-in.

A second lesson from the past is the importance of sampling gradients of variation. If the point of temporal analysis is to look for change, then sampling along gradients provides more opportunities for change as well as a fuller picture of change. The vast majority of published studies of climate change impacts have focused on one of three gradients: Shifts in phenological events (i.e., a seasonal gradient), shifts in elevational range, or shifts in latitudinal range. For those collecting data, the lesson here is that sampling only one relatively homogenous resource may not provide as much insight into the future as sampling a gradient of resources. Broadening a sampling scheme to include variation might require sampling resources that are not as ecologically intact, such as buffers or areas with greater human impacts. The value added

through the transformation of inference on a single resource to inference on a gradient of a resource may strongly contribute to understanding the dynamic impacts of future environmental change (Hargrove and Rotenberry 2011).

Finally, the clearest lesson from the past is the importance of information access. Data may have no value in the future if they cannot be found, accessed, and understood. Ironically, while technology can make archiving easier, it can also make access harder. Data from the past that were stored on paper and securely archived are much easier to access today than data from the 1970s and 1980s that are archived on now-archaic data storage devices. The internet and the development of global data depositories appear to provide a potential solution as online databases may be transitioned into new forms as technology changes (see Box 13.2). Countries in Asia and Europe have a head start on the United States as observational and phenological databases are often organized nationally and controlled by the government (Primack and Miller-Rushing 2012). Nevertheless, the past warrants caution. Multiple forms of data storage are recommended (both online and offline), and care should be made to consider storing data in ways that will be most accessible and understandable to future researchers. Even with archiving, metadata are almost as important as the occurrence data itself.

Altogether, the past has lots to offer: It can inform us about the present and it may help prepare us for the future. But dealing with the past can be difficult and the rewards of digging up historical records and struggling to use them are not always immediately obvious. Hopefully this chapter can work as both a tour guide and self-help manual for the exploration of historical data and the facilitation of discovery. Historical data may only be one tool in the process to understand long-term environmental change, but it is a tool that has much to offer.

ACKNOWLEDGMENTS

Kim Hall, Lori Hargrove, Mark Herzog, Chrissy Howell, and Terry Root provided helpful comments and thoughtful advice on this chapter. Many of the ideas presented here were developed in conversations with Steve Beissinger, Craig Moritz, Jim Patton, and members of the Grinnell Resurvey Project at the University of California, Berkeley.

LITERATURE CITED

Bakker, J. D. and M. M. Moore. 2007. Controls on vegetation structure in southwestern ponderosa pine forests, 1941 and 2004. *Ecology* 88:2305–2319.

Beever, E. A., P. F. Brussard, and J. Berger. 2003. Patterns of apparent extirpation among isolated populations of pikas (*Ochotona princeps*) in the Great Basin. *Journal of Mammalogy* 84:37–54.

Blalock-Herod, H. N., J. J. Herod, and J. D. Williams. 2002. Evaluation of conservation status, distribution, and reproductive characteristics of an endemic Gulf Coast freshwater mussel, *Lampsilis australis* (Bivalvia: Unionidae). *Biodiversity and Conservation* 11:1877–1887.

Bradford, D. F., J. R. Jaeger, and S. A. Shanahan. 2005. Distributional changes and populations status of amphibians in the eastern Mojave Desert. *Western North American Naturalist* 65:462–472.

Brusca, R. C., J. F. Wiens, W. M. Meyer, J. Eble, K. Franklin, J. T. Overpeck, and W. Moore. 2013. Dramatic response to climate change in the Southwest: Robert Whittaker's 1963 Arizona Mountain plant transect revisited. *Ecology and Evolution* 3:3307–3319.

Chen, I.-C., J. K. Hill, R. Ohlemüller, D. B. Roy, and C. D. Thomas. 2011. Rapid range shifts of species associated with high levels of climate warming. *Science* 333:1024–1026.

Colwell, M. A., T. Danufsky, R. L. Mathis, and S. W. Harris. 2002. Historical changes in the abundance and distribution of the American Avocet at the northern limit of its winter range. *Western Birds* 32:1–15.

Comte, L. and G. Grenouillet. 2013. Species distribution modelling and imperfect detection: Comparing occupancy versus consensus methods. *Diversity and Distributions* 19:996–1007.

Crimmins, S. M., S. Z. Dobrowski, J. A. Greenberg, J. T. Abatzoglou, and A. R. Mynsberge. 2011. Changes in climatic water balance drive downhill shifts in plant species' optimum elevations. *Science* 331:324–327.

Daniels, R. A. 1998. Changes in the distribution of stream-dwelling crayfishes in the Schoharie Creek system, eastern New York State. *Northeastern Naturalist* 5:231–248.

Davidson, C. 2004. Declining downwind: Amphibian population declines in California and historical pesticide use. *Ecological Applications* 14:1892–1902.

Davidson, C., H. Shaffer, and M. Jennings. 2001. Declines of the California red-legged frog: Climate, UV-B, habitat, and pesticides hypotheses. *Ecological Applications* 11:464–479.

Erb, L. P., C. Ray, and R. Guralnick. 2011. On the generality of a climate-mediated shift in the distribution of the American pika (*Ochotona princeps*). *Ecology* 92:1730–1735.

Forister, M. L., A. C. McCall, N. J. Sanders, J. A. Fordyce, J. H. Thorne, J. O'Brien, D. P. Waetjen, and A. M. Shapiro. 2010. Compounded effects of climate change and habitat alteration shift patterns of butterfly diversity. *Proceedings of the National Academy of Sciences of the United States of America* 107:2088–2092.

Franklin, J., C. L. Coulter, and S. J. Rey. 2004. Change over 70 years in a southern California chaparral community related to fire history. *Journal of Vegetation Science* 15:701–710.

Graham, C. H., J. Elith, R. J. Hijmans, A. Guisan, A. T. Peterson, B. A. Loiselle, and The Nceas Predicting Species Distributions Working Group. 2008. The influence of spatial errors in species occurrence data used in distribution models. *Journal of Applied Ecology* 45:239–247.

Graham, C. H., S. Ferrier, F. Huettman, C. Moritz, and A. T. Peterson. 2004. New developments in museum-based informatics and applications in biodiversity analysis. *Trends in Ecology & Evolution* 19:497–503.

Hargrove, L. and J. T. Rotenberry. 2011. Spatial structure and dynamics of breeding bird populations at a distribution margin, southern California. *Journal of Biogeography* 38:1708–1716.

Harrison, S. H., E. I. Damschen, and J. B. Grace. 2010. Ecological contingency in the effects of climatic warming on forest herb communities. *Proceedings of the National Academy of Sciences of the United States of America* 107:19362–19367.

Kelly, A. E. and M. L. Goulden. 2008. Rapid shifts in plant distribution with recent climate change. *Proceedings of the National Academy of Sciences of the United States of America* 105:11823–11826.

Kopp, C. W. and E. E. Cleland. 2013. Shifts in plant species elevational range limits and abundances observed over nearly five decades in a western North America mountain range. *Journal of Vegetation Science* 25:135–146.

Labay, B., A. E. Cohen, B. Sissel, D. A. Hendrickson, F. D. Martin, and S. Sarkar. 2011. Assessing historical fish community composition using surveys, historical collection data, and species distribution models. *PLOS ONE* 6:e25145.

Larrucea, E. and P. Brussard. 2008. Shift in location of pygmy rabbit (*Brachylagus idahoensis*) habitat in response to changing environments. *Journal of Arid Environments* 72:1636–1643.

Lutz, J. A., J. W. van Wagtendonk, and J. F. Franklin. 2009. Twentieth-century decline of large-diameter trees in Yosemite National Park, California, USA. *Forest Ecology and Management* 257:2296–2307.

MacKenzie, D. I., J. D. Nichols, J. A. Royle, K. H. Pollock, L. L. Bailey, and J. E. Hines. 2006. *Occupancy Estimation and Modeling*. Academic Press, Burlington, MA.

Martin, K., G. A. Brown, and J. R. Young. 2004. The historic and current distribution of the Vancouver Island White-tailed Ptarmigan (*Lagopus leucurus saxatllis*). *Journal of Field Ornithology* 75:239–256.

Mclaughlin, B. C. and E. S. Zavaleta. 2013. Regional and temporal patterns of natural recruitment in a California endemic oak and a possible "research reserve effect." *Diversity and Distributions* 19:1440–1449.

Miller, B. P., N. J. Enright, and B. B. Lamont. 2007. Record error and range contraction, real and imagined, in the restricted shrub *Banksia hookeriana* in south-western Australia. *Diversity and Distributions* 13:406–417.

Morelli, T. L., A. B. Smith, C. R. Kastely, I. Mastrosrio, C. Moritz, and S. R. Beissinger. 2012. Anthropogenic refugia ameliorate the severe climate-related decline of a montane mammal along its trailing edge. *Proceedings of the Royal Society B: Biological Sciences* 279:4279–4286.

Moriarty, K. M., W. J. Zielinski, and E. D. Forsman. 2011. Decline in American Marten occupancy rates at Sagehen Experimental Forest, California. *Journal of Wildlife Management* 75:1774–1787.

Moritz, C., J. L. Patton, C. J. Conroy, J. L. Parra, G. C. White, and S. R. Beissinger. 2008. Impact of a century of climate change on small-mammal communities in Yosemite National Park, USA. *Science* 322:261–264.

Primack, R. B. and A. J. Miller-Rushing. 2012. Uncovering, collecting, and analyzing records to investigate the ecological impacts of climate change: A template from Thoreau's Concord. *Bioscience* 62:170–181.

Primack, R. B., A. J. Miller-Rushing, and K. Dharaneeswaran. 2009. Changes in the flora of Thoreau's Concord. *Biological Conservation* 142(3):500–508.

Rickart, E. A., S. L. Robson, and L. R. Heaney. 2008. Mammals of Great Basin National Park, Nevada: Comparative field surveys and assessment of faunal change. *Monographs of the Western North American Naturalist* 4:77–114.

Root, T. L., J. T. Price, K. R. Hall, S. H. Schneider, C. Rosenzweig, and J. A. Pounds. 2003. Fingerprints of global warming on wild animals and plants. *Nature* 421:57–60.

Rowe, R. 2007. Legacies of land use and recent climatic change: The small mammal fauna in the mountains of Utah. *American Naturalist* 170:242–257.

Rowe, R. J., J. A. Finarelli, and E. A. Rickart. 2010. Range dynamics of small mammals along an elevational gradient over an 80-year interval. *Global Change Biology* 16:2930–2943.

Rowe, R. J. and S. Lidgard. 2009. Elevational gradients and species richness: Do methods change pattern perception? *Global Ecology and Biogeography* 18:163–177.

Rowe, R. J., R. C. Terry, and E. A. Rickart. 2011. Environmental change and declining resource availability for small-mammal communities in the Great Basin. *Ecology* 92:1366–1375.

Rowlands, P. G. and N. J. Brian. 2001. Fishtail mesa: A vegetation resurvey of a relict area in Grand Canyon National Park, Arizona. *Western North American Naturalist* 61:159–181.

Rubidge, E. M., W. B. Monahan, J. L. Parra, S. E. Cameron, and J. S. Brashares. 2011. The role of climate, habitat, and species co-occurrence as drivers of change in small mammal distributions over the past century. *Global Change Biology* 17:696–708.

Shaffer, H. B., R. N. Fisher, and C. Davidson. 1998. The role of natural history collections in documenting species declines. *Trends in Ecology & Evolution* 13:27–30.

Shavit, A. and J. Griesemer. 2009. There and back again, or the problem of locality in biodiversity surveys. *Philosophy of Science* 76:273–294.

Shultz, A. J., M. W. Tingley, and R. C. K. Bowie. 2012. A century of avian community turnover in an urban green space in northern California. *The Condor* 114:258–267.

Thorne, J. H., B. J. Morgan, and J. A. Kennedy. 2008. Vegetation change over sixty years in the central Sierra Nevada, California, USA. *Madroño* 55:225–239.

Tingley, M. W. and S. R. Beissinger. 2009. Detecting range shifts from historical species occurrences: New perspectives on old data. *Trends in Ecology & Evolution* 24:625–633.

Tingley, M. W. and S. R. Beissinger. 2013. Cryptic loss of montane avian richness and high community turnover over 100 years. *Ecology* 94:598–609.

Tingley, M. W., M. S. Koo, C. Moritz, A. C. Rush, and S. R. Beissinger. 2012. The push and pull of climate change causes heterogeneous shifts in avian elevational ranges. *Global Change Biology* 18:3279–3290.

West, N. E. and T. P. Yorks. 2006. Long-term interactions of climate, productivity, species richness, and growth form in relictual sagebrush steppe plant communities. *Western North American Naturalist* 66:502–526.

Winker, K., D. D. Gibson, A. L. Sowls, B. E. Lawhead, P. D. Martin, E. P. Hoberg, and D. Causey. 2002. The birds of St. Matthew Island, Bering Sea. *Wilson Bulletin* 114:491–509.

Yang, D.-S., C. J. Conroy, and C. Moritz. 2011. Contrasting responses of *Peromyscus* mice of Yosemite National Park to recent climate change. *Global Change Biology* 17:2559–2566.

GLOSSARY

ADAPTATION Evolution due to natural selection that results in individuals being more suited to their environment. We restrict our use of this term to mean genetically-based evolutionary adaptation, as opposed to non-evolutionary adaptations of organisms (e.g., adaptive changes in animal behavior without a genetic basis) or the management actions (e.g., policy adaptations) of humans. Evolutionary adaptation (like all mechanisms of evolution) can only be described in terms of populations of a species. In other words, populations may adapt as the overall genetic makeup changes over successive generations, whereas individuals cannot alter their genetic makeup and therefore cannot genetically adapt.

ADAPTIVE POTENTIAL A measure of a population's ability to evolve in response to selection for a given trait. Adaptive potential can be influenced by many factors: levels of genetic variation, population size, immigration and emigration, and life history characteristics. For example, a population with very low levels of genetic variation, little migration, and a long generation time is likely to have low adaptive potential.

ADVECTION The transfer of matter in a fluid.

ALEVIN Newly hatched juvenile salmonid with yolk sac that lives in streams.

ANADROMOUS FISHES Fishes that spawn and mature in freshwater but spend the majority of their lives in the ocean.

ANCIENT DNA DNA that is intact in the remains of ancient organisms.

ANTHROPOGENIC Resulting from human activities.

ATPase An enzyme that releases energy necessary for chemical processes in cells.

BASIN The land area where water drains to a river.

cDNA DNA artificially synthesized in the laboratory using RNA as a template. cDNA is more stable than RNA, so scientists usually store RNA information as cDNA for long-term purposes.

cDNA MICROARRAY This is a molecular tool used to study the expression of up to hundreds of thousands of genes simultaneously. Physically, the microarray is a glass slide that contains copies of many cDNA fragments arranged in a grid.

CHECK DAM A small dam, which may or may not be permanent, built across a rather small drainage ditch.

CLIMATE ENVELOPE Used in predictive model that relies on statistical correlations between existing species distributions and environmental variables to define a species' tolerance to the environmental variables. "Envelopes" of tolerance are then drawn around existing ranges. By using predicted future values of environmental factors such as temperature, rainfall, and salinity, new range boundaries can be predicted.

CONSPECIFICS Individuals belonging to the same species.

DEGREE DAYS A measurement of the length of a particular growth phase of a plant. It is calculated by multiplying the average temperature by the length of time involved.

DIEL VERTICAL MIGRATION A daily migration of zooplankton and small fishes to deep water during the day to avoid visual predators in well-lit surface waters, and to water near the surface at night to feed when less exposed to predation.

DOMESTICATED FISHES Fishes adapted to hatchery conditions and often maladapted to natural stream conditions.

DOWNWELLING The sinking of surface waters to deep water.

ECOTYPE A population adapted to a particular environment resulting from local natural selection pressures.

EQUATORWARD Moving toward the equator.

ESCAPEMENT The number of adult salmon returning to a river to spawn.

EUTROPHICATION The stimulation of aquatic plant growth due to increases in nutrients in the water.

EVOLUTION A change in the genetic makeup ("gene frequencies" or "genepool") of a population or a species over time. Generally, *microevolution* refers to any change in genetic makeup over small timescales within populations, whereas *macroevolution* refers to longer-term evolution affecting speciation.

FAUNA All of the animals in a given area.

FISHERIES The fishing industry.

FITNESS The survival and reproductive capacity of an organism in a given environment.

FRY Juvenile salmonids that emerge from stream.

GENE CLONES Copies of gene fragments stored in bacterial cells. Bacterial cells are able to maintain and easily copy gene fragments with very low cost.

GENE EXPRESSION The process by which information from a gene is used in the synthesis of a functional gene product, such as an mRNA or protein molecule. It is usually measured by quantifying the protein or the mRNA.

GENE FLOW The movement of genes among populations. It includes the physical dispersal of individuals (e.g., seeds) or gametes (e.g., pollen).

It is often referred to as "migration" in evolutionary literature, but is not to be confused with the regular movements (often seasonal) of organisms within their home range.

GENETIC LOAD The mismatch between a population's mean value for a given phenotypic trait and the optimal value for that trait in a given environment. If the genetic load is excessive, it could cause demographic bottlenecks and even extinction before the population can adapt.

GENETIC VARIATION The diversity of genotypes within a population. It is essential for evolution. "Adaptive genetic variation" refers to variation associated with the ability of a particular trait to adapt.

GENOTYPE The specific genetic makeup of an individual, usually in reference to a particular gene or set of genes influencing a particular trait. Individuals may share the same genotype for some traits, but have different genotypes for other traits.

HOLOCENE The last 11,700 years.

ITEROPAROUS Species with individuals that reproduce multiple times in their lifetimes.

LIFE HISTORY The overall "strategy" by which a species completes its life cycle. Differences in species' life histories are often in response to trade-offs, such as having many small offspring requiring little parental care or few large offspring needing a lot of parental care. A species' life history can substantially affect how populations respond to evolutionary processes.

LOCAL ADAPTATION Different populations of a species adapted to local conditions. It highlights evolutionary processes that operate within a species.

NATURAL SELECTION The differential survival and reproduction of individuals ("the survival of the fittest") based on traits that have a genetic basis. In order for natural selection to operate on a given trait, there must be genetic variability for this trait in the population—e.g., if all individuals have the exact same eye color, then it is impossible for that eye color to cause differences in survival and reproduction. Natural selection is the driving force behind adaptation in natural systems.

OUTSLOPING Downslope grading of road surfaces.

PALEOCLIMATE Climates of prehistoric time, usually reconstructed from different types of proxy data.

PHENOTYPE The physical, physiological, and behavioral trait(s) of an individual. An individual's phenotype, although often heavily influenced by genetic makeup (genotype), is also influenced by the environment in which it develops. Thus, a key distinction between an individual's genotype and phenotype is that natural selection can only act upon its phenotype.

PHENOTYPIC PLASTICITY The ability of organisms with the same genotype to display different phenotypes depending on environmental conditions. For example, genetically identical plants may have different growth rates or growth forms depending on their growing conditions. For example, taller plants on the foothills of a mountain and shorter plants of the same species growing higher up on the mountain.

PHOTIC ZONE The depth of water to which light penetrates that allows photosynthesis to occur.

PHOTOPERIOD Day length.

PLEIOTROPIC Having more than one effect. Usually pertains to a gene that has multiple phenotypic expressions.

PLEISTOCENE The period of time between 2.588 million years ago and 11,700 years ago, encompassing most of the glacial-interglacial cycles.

POLEWARD Moving toward the poles.

POPULATION A group of interbreeding individuals of the same species. Populations can have broad or narrow geographic ranges, and their boundaries can often be difficult to define. Species are normally composed of many populations. In general, the threat of extinction increases as the number of populations drop.

PRIMARY PRODUCTIVITY The rate at which new organic matter is created via photosynthesis.

QUATERNARY A geologic period of time spanning the past 2.588 million years, which encompasses both the Pleistocene and the Holocene.

REDDS Nest built by fishes in streams.

RESILIENCE The ability of an organism, population, or ecosystem to resist damage and rebound quickly from environmental perturbations.

RESISTANCE The ability of an organism, population, or ecosystem to remain essentially unchanged when subjected to adverse conditions.

RKM River kilometer.

SALMONIDS The group of fishes belonging to the family Salmonidae, including salmon and trout.

SEMELPAROUS Species with individuals that reproduce only once in their lifetimes.

SMOLTIFICATION (SMOLT) The physiological process that facilitates the excretion of excess salts.

SOIL COLLAR A cylindrical object inserted into the ground for the purpose of making soil respiration measurements within an enclosed chamber, using a device (i.e., the LI-COR 6400) for measuring the flow of CO_2 (or water) from the soil. Insertion depth is generally 3–5 cm, and the inside diameter of the collar corresponds to the outside diameter of the measurement chamber. The collar also defines a spatial area for repeated sampling over the measurement period.

STREAM INCISION Lowering of streambed elevation relative to the floodplain elevation.

STRESSOR The use of land (e.g., logging, mining) or a resource (e.g., fisheries) that degrades the habitat quality or causes decreasing species abundance.

TRAIT Any specific component of the phenotype (e.g., flower or fur color, height, temperature tolerance) that usually has a strong genetic basis.

UPWELLING The rising of nutrient-rich water from deep in the ocean to the surface waters.

INDEX

CONTRIBUTOR BIOS

WILLIAM R. L. ANDEREGG
Princeton University
Dr. William Anderegg is NOAA Climate and Global Change Postdoctoral Research Fellow at Princeton University. He received his PhD from the Department of Biology at Stanford University.

DEBORAH ASELTINE-NEILSON
CA Department of Fish and Wildlife
Debbie Aseltine-Neilson, Senior Environmental Scientist Specialist with the California Department of Fish and Wildlife, has more than thirty years of experience working in the marine environment.

JESSICA L. BLOIS
University of California, Merced
Dr. Jessica Blois is a paleoecologist interested in the effects of past climate change on biodiversity. She is currently Assistant Professor at University of California, Merced.

BRENDAN COLLORAN
Archimedes Incorporated
San Francisco, CA
Brendan Colloran holds an MA in Mathematics from SFSU and an MPhil in Economics from the University of Cambridge. He currently works as a data scientist in San Francisco.

ANDREA CRAIG
The Nature Conservancy
Los Molinos, CA
Andrea Craig is Preserve Manager with The Nature Conservancy in northeastern California. Her stewardship work is focused on rangelands and conservation easements.

JEFFREY G. DORMAN
University of California, Berkeley
Dr. Jeff Dorman is a biological oceanographer interested in how ocean physics drives biological productivity. He uses models to simulate the population biology of important zooplankton species in the coastal ocean.

MARK FISHER
University of California Natural Reserve System
Mark Fisher is Assistant Director and Staff Biologist at UCNRS P.L. Boyd Deep Canyon Desert Research Center, where he has been on staff since 1983.

SASHA GENNET
The Nature Conservancy
San Francisco, CA
Dr. Sasha Gennet is Senior Scientist with The Nature Conservancy in California. Her work focuses on developing science-based solutions to increase biodiversity and ecosystem services in working landscapes.

ALDEN B. GRIFFITH
Wellesley College
Dr. Alden Griffith is Assistant Professor of Environmental Studies at Wellesley College. His research interests address plant population ecology and include invasive species, plant-plant facilitation, and the effects of climate change.

ELIZABETH A. HADLY
Stanford University
Professor Elizabeth A. Hadly holds the Paul S. & Billie Achilles Chair of Environmental Biology at Stanford University. She investigates past, present, and future vertebrate diversity.

KIMBERLY R. HALL
Michigan State University
Dr. Kimberly R. Hall is a scientist with The Nature Conservancy and an adjunct faculty member at Michigan State University, and her work focuses on updating conservation strategies to address the risks of climate change.

LORI HARGROVE
University of California, Riverside
Dr. Lori Hargrove is a postdoctoral researcher with the San Diego Natural History Museum. She is currently leading the project "Centennial Resurvey of the San Jacinto Mountains" to document one hundred years of change.

MARK P. HERZOG
US Geological Survey
Dr. Mark P. Herzog is Wildlife Biologist at US Geological Survey. His research is on the demography and ecology of waterfowl and waterbirds in the western United States.

GRETCHEN E. HOFMANN
University of California, Santa Barbara
Dr. Gretchen Hofmann is a marine biologist whose research focuses on the responses of marine species to future ocean change such as ocean acidification and ocean warming.

DAN HOWARD
Cordell Bank National Marine Sanctuary
Dan Howard is Superintendent for NOAA Cordell Bank National Marine Sanctuary, a federally managed marine protected area west of Point Reyes in northern California.

CHRISTINE A. HOWELL
US Forest Service
Dr. Christine A. Howell is Regional Wildlife Ecologist for the Pacific Southwest Region of the Forest Service. She works on wildlife management and climate change on eighteen national forests in California.

ROB KLINGER
US Geological Survey
Yosemite Field Station
Rob Klinger is a population and community ecologist investigating interactions between climate shifts, mammal distribution and abundance, and vegetation transitions in the alpine zone of the Sierra Nevada.

LAURA KOTEEN
University of California, Berkeley
Dr. Laurie Koteen is currently Ecological Science Director for ANRFlux, a program that measures greenhouse gas exchange and climate-ecosystem interactions at the University of California's research centers across California.

GRETCHEN LEBUHN
San Francisco State University
Dr. Gretchen LeBuhn is Professor of Biology at San Francisco State University and Director of the Great Sunflower Project. Her work focuses on plants and pollinators.

TINA MARK
US Forest Service
Tahoe National Forest
Tina Mark is Forest Biologist for the USDA Forest Service, Tahoe National Forest. She is Program Manager for Wildlife, Aquatic Species, and Rare Plants.

MICHAEL MASTRANDREA
Stanford University
Dr. Michael D. Mastrandrea is Codirector, Science for the Intergovernmental Panel on Climate Change Working Group II TSU. His research centers on vulnerability and impacts assessment to inform risk management of climate change.

ALLAN MUTH
University of California Natural Reserve System
Dr. Allan Muth is Director of the Philip L. Boyd Deep Canyon Desert Research Center, a unit of the University of California Natural Reserve System.

CHRISTOPHER J. OSOVITZ
University of South Florida
Dr. Christopher Osovitz is Senior Instructor at the University of South Florida. He teaches courses in marine biology and animal physiology.

JULIE PERROCHET
US Forest Service
Klamath National Forest
Julie Perrochet worked for the USDA Forest Service. For the last nineteen years she worked on the Klamath National Forest, focusing on salmon population trends, aquatic habitat restoration, and endangered species.

ANNE POOPATANAPONG
US Forest Service
Idyllwild Ranger Station
Anne Poopatananpong is currently Regional Wildlife Biologist for the Forest Service in Portland, Oregon. She specializes in the management of habitats for threatened and endangered species across the national forests.

REBECCA M. QUIÑONES
University of California, Davis
Dr. Rebecca Quiñones's research focuses on evaluating the impacts of changing climatic conditions, land use, and resource management on native fishes in California.

JENNY RECHEL
US Forest Service
Pacific Southwest Research Station
Jennifer Rechel has a PhD in Geography. Her research focus is on changes in bird populations and habitat use over time in disturbed ecosystems and mapping biodiversity.

MARK REYNOLDS
The Nature Conservancy
San Francisco, CA
Dr. Mark D. Reynolds is Senior Scientist with The Nature Conservancy's California Chapter. His work is currently focused on conservation of migratory birds.

TERRY L. ROOT
Stanford University
Dr. Terry L. Root is Senior Fellow and University Faculty at Stanford University. Her work focuses on how plants and animals are affected by rapid climate change.

JOHN T. ROTENBERRY
University of California, Riverside
Dr. John T. Rotenberry is Emeritus Professor of Biology at the University of California, Riverside. His research has focused on avian communities in semiarid habitats, with emphasis on a landscape ecological perspective.

JASON P. SEXTON
University of California, Merced
Dr. Jason Sexton is Assistant Professor at University of California, Merced. His research focuses on plant adaptation, species range limits, biological invasions, and conservation science.

MARK STROMBERG
Hastings Natural History Reservation
Dr. Mark R. Stromberg was Resident Director of UC Berkeley's Hastings Natural History Reservation, part of the UC Natural Reserve System, from 1988–2011.

K. BLAKE SUTTLE
Imperial College London
Dr. Kenwyn ("Blake") Suttle is a community ecologist interested in questions of ecological prediction and public understanding of climate change.

MORGAN W. TINGLEY
Princeton University
Dr. Morgan Tingley is Assistant Professor in the Department of Ecology and Evolutionary Biology at the University of Connecticut. His research centers on how distributions of birds respond to large-scale environmental change.

ERIKA S. ZAVALETA
University of California, Santa Cruz
Dr. Erika Zavaleta is Professor and Pepper-Giberson Chair of Environmental Studies at the University of California, Santa Cruz, where she works to enhance the stewardship, understanding, and appreciation of wild ecosystems.